NATURAL HISTORY OF THE AMERICAN LOBSTER

This is a volume in the Arno Press collection

HISTORY OF ECOLOGY

See last pages of this volume for a complete list of titles.

NATURAL HISTORY OF THE AMERICAN LOBSTER

Francis Hobart Herrick

ARNO PRESS
A New York Times Company
New York / 1977

Editorial Supervision: LUCILLE MAIORCA

Reprint Edition 1977 by Arno Press Inc.

Reprinted from a copy in
The Pennsylvania State Library

HISTORY OF ECOLOGY
ISBN for complete set: 0-405-10369-7
See last pages of this volume for titles.

Manufactured in the United States of America

Publisher's Note: Plates XXVIII, XXXI and XXXII have been reproduced in black and white in this edition.

Library of Congress Cataloging in Publication Data

Herrick, Francis Hobart, 1858-1940.
Natural history of the American lobster.

(History of ecology)
Reprint of the 1911 ed. published by Govt. Print. Off., Washington, which was issued as Document 747 of U.S. Bureau of Fisheries.
From Bulletin of the Bureau of Fisheries, v. 29, 1909.
Bibliography: p.
Includes index.
1. American lobster. 2. Crustacea. 3. Lobster fisheries--United States. 4. Crayfish. I. Title. II. Series. III. Series: United States. Bureau of Fisheries. Document ; 747.
QL444.M33H47 1977 595'.3841 77-74228
ISBN 0-405-10398-0

INTRODUCTION.

The present work when originally undertaken in 1903 was designed to form the zoological part of a history of the lobster in both America and Europe, but subsequent events led to a modification of this plan, and when it was decided to issue this section separately, its character and scope were somewhat changed.

Dr. Hugh M. Smith, of the United States Bureau of Fisheries, had planned to deal with the lobster fishery and the economic questions which this great industry has raised, in a comprehensive manner, and hope is entertained that this design may still be carried out.

Though essentially a distinct work, this is in a measure both a revision and an extension of my earlier report upon The American Lobster, published by the United States Commission of Fish and Fisheries in its bulletin for 1895. But little from the latter, however, has been incorporated directly, and this only when newer or better research has failed to give us more light upon the subject. Six drawings of the young lobsters, three of which are in colors, have been reproduced, after slight revisions, from my former report; all of the rest are new and deal chiefly with the anatomy of the body and appendages, especially with torsion, reflex amputation, and the developmental history of the toothed and cracker claws, the sexual organs, and the germ cells. I have depended mainly upon the store of materials collected in former years, but have received accessions from the United States Bureau of Fisheries, for which as well as for many courtesies, now extending over a long period, I wish to offer my sincere thanks. The Bureau has generously given me the privilege of a free lance, and all critical sections of this paper should be read in the light of individual opinion only, directed, it is true, in a friendly spirit, and as we believe from the standpoint of science.

Our knowledge of the lobster has increased to such an extent during the past fifteen years that in all probability there is no marine invertebrate in the world which is now better known. This result is due to the suggestive ideas or elaborate researches of a large body of naturalists in both America and Europe, and to their labors the reader will find abundant reference in the pages which follow. As a result of this advance in the biological field, a signal success has been achieved in the artificial propagation or culture of the lobster, and particularly in rearing the delicate young to the bottom-seeking stage, a success from which this fishery should not be slow to profit, and which it owes to experiments begun under the auspices of the United States Fish Commission at Woods Hole, Mass., and afterwards carried to a high degree of perfection by the Commission of Inland Fisheries of Rhode Island, under the direction of Prof. Albert D. Mead, at Wickford. Through the aid of such a practical method there is ground for hope, not only of restoring our depleted fisheries on the Atlantic coast, but of establishing new ones on the Pacific, as well as in other parts of the world.

While many dark puzzles have been solved, and many questions, raised fifteen years or more ago, can now be answered with assurance, no enterprising or resourceful worker need be told that the field is still fertile for fuller or more exact researches in many directions. We hope that some of these subjects will be suggested by the imperfections of the present work when attention is not called to them directly.

F. H. HERRICK.

CLEVELAND, OHIO.

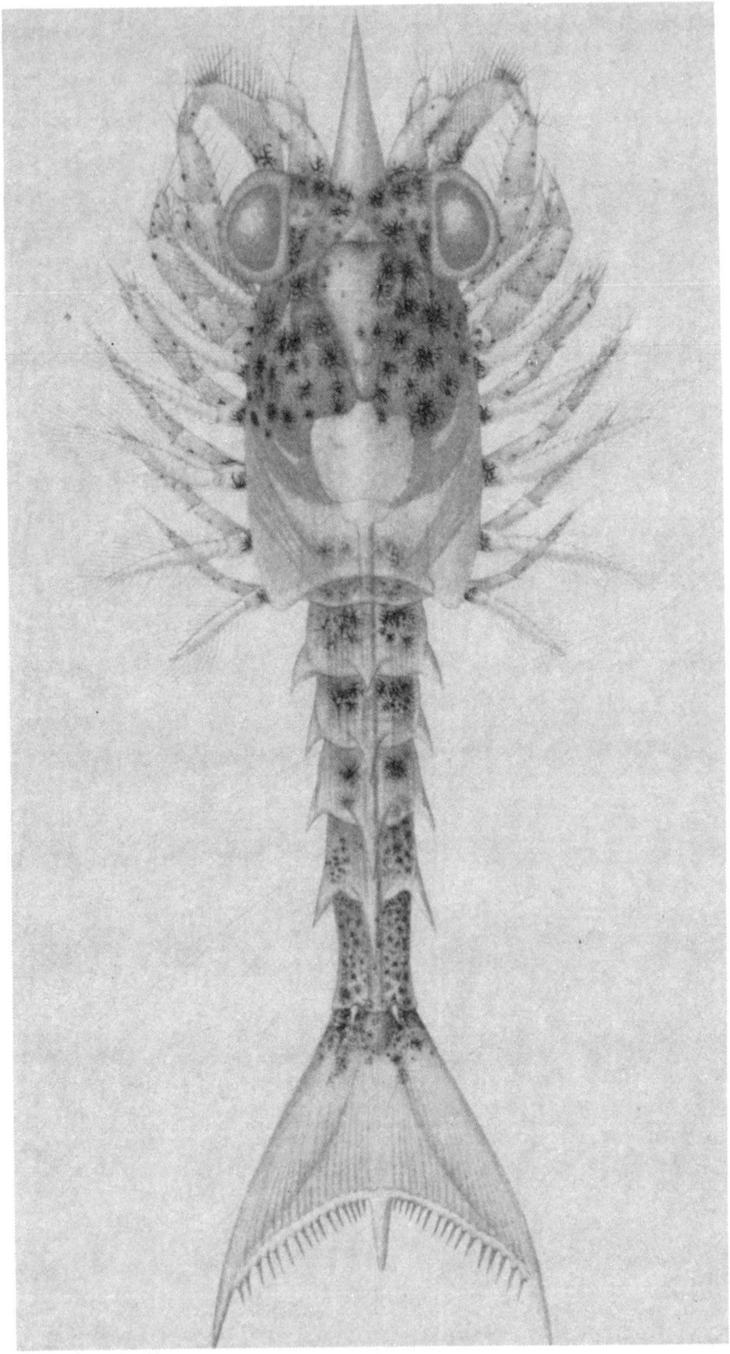

A Hoen & Co Baltimore

FIRST LARVAL OR SURFACE-SWIMMING STAGE OF THE LOBSTER
LENGTH 7.8 MM.

Chapter I.—THE LOBSTERS AND ALLIED CRUSTACEA; THEIR ZOOLOGICAL RELATIONS, HABITS, DEVELOPMENT, AND USE AS FOOD.

NATURAL HISTORY OF THE CRUSTACEA.

Nature works according to definite principles, and with a degree of uniformity which for most of our purposes is practically absolute. Accordingly we find that whenever an animal or plant has been successfully domesticated or whenever the young of any form have been successfully reared by the artificial impregnation and subsequent care of the eggs, as in the case of the oyster or the whitefish, this has been accomplished by acting, whether intelligently or not, in accordance with the principles of nature. The mollusk or the vertebrate is made to yield to experiments which a knowledge of its habits and structure would suggest. In the lobster we have to deal with another and distinct type, for although this animal swims in the sea, it is not a fish, but an arthropod, and a knowledge of the ways of fishes and mollusks will help but little in the study of its habits or in the propagation of its race.

The following paragraphs on the general characteristics of the arthropods will be of little or no use to professional zoologists, but may help to set our subject in a clearer light for other readers.

Of the eight or more animal types recognized by naturalists the arthropods are distinguished for their complicated structure and wonderful diversity of form, for the wide range and specialization of their instincts, their almost unparalleled fertility and corresponding activity. In the latter respect, at least, some of the insects are not surpassed by birds, the most active vertebrates.

The body of the arthropod is composed of a series of successive segments, the somites or metameres, which in conformity to vertebrate anatomy are divided into three groups, pertaining to the head, thorax, and abdomen. (Pl. XXXIII, and table 4.) Theoretically, each somite at one time possessed a pair of jointed limbs, and many of the segments still retain them. In the living adult state, the body is normally maintained in a definite upright position, which is often one of unstable equilibrium, whether the animal is in motion or at rest. These characteristics are shared in some degree by the annelid worms, their nearest allies, as well as by the vertebrates. The arthropod possesses in addition a dorsal brain, united by a ring-commissure about the esophagus, to a ventral chain or "ladder" of paired ganglia, a character also shared by the higher worms; the heart is dorsal and overlies the food canal; the cuticle, which encases the body and lines every inward fold, is secreted by the outer layer of the skin, the epidermis or hypodermis, and is chitinous—that is, contains chitin, a complex nitrogenous substance, by some chemists regarded as analogous to cellulose and lignin, which occur typically in plants and form the basis of all their woody tissues. This cuticle of the Crustacea is often reinforced by thick deposits of lime and other minerals, thus forming a hard external skeleton, to which every peripheral muscle is directly or indirectly attached, and by which every soft and delicate organ in the entire body is protected. No other animals possess all the several characteristics just enumerated. Since the arthropods embrace the insects, with their hundreds of thousands of species, it is not surprising that according to some estimates they include three-fourths of all the known species of living animals.

Of the five commonly recognized classes of arthropods the Crustacea are the lowest and most primitive. They fall into two principal subclasses: (*a*) The Entomostraca, embracing all the simpler, more primitive and generally smaller forms, such as water fleas, copepods, and barnacles, and (*b*) the Malacostraca, to which pertain the larger and the most highly organized of living Crustacea, such as lobsters, shrimps and crabs. The ancient name of the class served the older zoologists to distinguish those animals which possessed a "crust," or a shell flexible at certain joints, from the Testacea, or animals like the oyster and clam in which the shelly covering was a hard and unyielding "test."

Eight orders of Malacostraca [a] have been recognized, of which the more important, in view of their size, numbers, economic and general zoological interest, are the Amphipoda and Isopoda, which embrace the beach fleas on the one hand and terrestrial wood lice on the other; the primitive Stomatopoda, of which the edible mantis or "praying" shrimp are well known representatives, the small Schizopoda, or cleft-feet, and the ten-footed and stalk-eyed Decapoda, which mainly interest us.

In both the isopods and amphipods the eggs are carried in a brood chamber on the underside of the thorax, formed by membranous plate-like outgrowths from the thoracic legs in the female; the schizopods also carry their eggs in a similar way.

The breeding habits of the stomatopods are highly peculiar; although celebrated for their widely dispersed pelagic larvæ, and although it was understood that they dwelt in mud burrows under water, and did not carry their eggs attached to the body as in decapods, little was known of their early life history until the studies of Professor Brooks upon *Gonodactylus chiragra* of the Bahama Islands appeared in 1893, when he gave the first full account of their habits, and the first record of the rearing of a young stomatopod from the egg. Fortunately this animal does not deposit its ova deep in the mud, but in a burrow, apparently of its own making, in the soft coral rock; they are glued together by a viscous cement and molded to fit the convex form of the mother's

[a] In the classification briefly outlined in this chapter we shall follow mainly the excellent work on Crustacea by Geoffrey Smith, in vol. IV of the Cambridge Natural History. London, 1909.

body. With its egg cluster on its back *Gonodactylus* stands or sits on guard at the mouth of the burrow, awaiting its prey, and meantime keeping its eggs aerated by the fanning movements of the swimmerets. Says Professor Brooks:

When the burrow is broken open she quickly rolls the eggs into a ball, folds them under her body in a big armful, between the large joints of her raptorial claws, and endeavors to escape with them to a place of safety. The promptness with which this action is performed would seem to indicate that it is an instinct which has been acquired to meet some danger which frequently presents itself.

The decapods have the general characteristics given for the lobster in chapter VI. All glue their eggs to their swimmerets and carry them thus attached, protecting and aerating them for a period of weeks or months with unerring instinct until they hatch. After pairing, the sexes frequently separate, as is possibly the case with lobsters (see p. 302), or they remain together, swimming side by side, and receiving mutual aid as in *Stenopus*, for as long at least as the period of fosterage lasts. The young, upon hatching, usually either swarm together for a time, or are immediately dispersed, as in the lobster. A long and perilous metamorphosis awaits the young of most of the decapods, during which they are pelagic or free surface swimmers, but every degree of abbreviation of this development exists, and in the crayfishes and certain other species, both fluviatile and marine, the young resemble the parent at birth, and a complex family life, which will receive attention later, may be developed.

The decapods are divisible into three intergrading suborders: (1) The Macrura, or long-tailed Crustacea like the shrimp and true lobsters; (2) the Anomura or hermit lobsters and hermit crabs, and (3) the Brachyura or true crabs, the most highly specialized of the entire class, in which the tail is not only very short but is even rudimentary in the male.

To follow out the Macrura only and in brief, they embrace numerous families possessing both zoological interest and economic value, of which the most important are (1) the Nephropsidæ (Astacidæ of many authors) or true lobsters; (2) the fresh-water crayfishes of the world, or Astacidæ of North America and Europe, and the Parastacidæ of the Southern Hemisphere; (3) the other decapods known collectively as prawns or shrimps, including the Peneidæ, Alpheidæ, Pandalidæ, Crangonidæ, and Palæmonidæ; (4) the Palinuridæ, variously known as spiny, thorny, or rock lobsters, and (5) the Scyllaridæ, which are sometimes classed with the Galatheidæ, and are known as warty lobsters. Representatives of some of these families will now be briefly considered, before dealing more fully with the special subjects of this work embraced in the family of Nephropsidæ.

The crayfish (of the family Astacidæ) has become a favorite subject in zoology, and very few invertebrates have received the degree of attention which naturalists have paid to every phase of its history. It is well known that the common crayfish, *Astacus fluviatilis*, has been used for centuries as food all over the continent of Europe, while in France the farming of crayfish in order to increase the natural supply of this crustacean has been successfully practiced for some time. For many years also crayfish have found their way to the markets of American cities which possess large populations of foreign birth, as New York, New Orleans, Chicago, Milwaukee, and San Francisco; but many persons would probably be surprised to learn the present status of the crayfish industry in this country, where vast numbers are not only eaten but used to supply classes in zoology or some phase of nature study in nearly every State of the Union.

Professor Andrews,[a] from whose paper the following statistics are taken, thinks that the demand for the fluviatile crayfish is likely to grow steadily, and may help to counterbalance the waning supplies of marine food, especially in the form of lobsters and crabs.

The crayfish of the eastern central regions belong to the genus *Cambarus*, the Potomac supplying *C. affinis;* Chicago, *C. virilis;* New Orleans, *C. blandingii;* and Montreal, *C. bartoni*. A considerable fishery for the large and handsome American species of *Astacus*, a counterpart of the European form, has been developed on the Pacific coast. This centers in Portland, Oreg., where, in 1899, the product reached 117,696 pounds, valued at $19,556.

Andrews has shown that the common *Cambarus affinis* not only breeds annually, but that its young reared from spring eggs may in turn lay eggs the spring following, when under a year old, while at the age of 3¼ years they attain the average market size of 4 inches. It is further suggested that the large 6-inch Oregon *Astacus*, which is more lobster-like in appearance, could doubtless be successfully introduced into Eastern waters, and, with a growing demand, profitably reared, since there is no reason to suppose that climatic changes would offer any obstacle to its development.

The prawns and shrimps distributed among the various families enumerated are undoubtedly the most active and most graceful, as well as the most plentiful of all the decapod Crustacea. Many species are highly valued as food, and are netted and sent to market in vast numbers over a large part of the world. The most important shrimp fisheries of the United States center in the Coast States of the Gulf of Mexico and Pacific Ocean.

Among the best-known species in North America are the edible shrimp of the South (*Penæus setiferus* and *P. brasiliensis*), the still more abundant common shrimp (*Crangon vulgaris*), found on both coasts and closely related to the common European shrimp The California shrimp (*Crangon franciscorum*), the largest and most important of the edible species on the western coast, attains a length of 3½ inches. It not only supplies abundantly the local markets, but occupies an important place in the export trade of San Francisco, being boiled, dried, and shipped to China in large quantities.

Prawns are extremely abundant in the East Indies from Japan to Australia, and, commercially considered, are the most important crustacea of the Orient. Thirteen species of the genus *Penæus* alone are taken in Japanese waters. "They are highly prized and extensively used as food and bait, and dried prawns annually exported to China amount to about 900,000 kilograms in weight and to about 200,000 yen ($131,000) in value. The dried prawns belong almost exclusively to the genus *Penæus*."[b]

Closely allied to prawns, though placed in a distinct family, are the Alpheidæ, of which over 100 species of snapping shrimps belonging to the genus *Alpheus* and *Synalpheus* alone have been described. They are essentially tropical, and abound in

[a] Andrews, E. A.: The future of the crayfish industry. Science, n. s., vol. XXIII, 1906, p. 983–986. New York.

[b] Kishinouye, K.: Japanese species of the genus Penæus. Journal of the Fisheries Bureau, Tokyo, vol. VIII, 1900, no. 1, p. 1–29.

the coral seas of both hemispheres. The Alpheidæ have no commercial value, but are of great biological interest, on account of their wide variation in form, coloring, and development, as well as for their remarkable instincts and habits.

The large and handsome spiny or thorny lobsters (family Palinuridæ) are represented chiefly by the single genus *Palinurus*. The langouste of the French, which has been celebrated from antiquity, is noted for its great size, brilliant coloring, and formidable appearance, though claws are lacking, as well as for its small and numerous eggs and grotesque transparent larvæ. Its flesh, which is mainly confined to the thorax and tail, is considered by many quite as delicate as that of the true lobsters. From 13 to 16 species have been described from the temperate and tropical seas of the world. According to Spence Bate,[a] this genus is represented in the South Indian Ocean by *Palinurus edwardsii*, the range of which extends from the Cape of Good Hope to New Zealand, by *Palinurus trigonus* and allied forms in Japan, by *Palinurus frontalis* on the coast of South America, and by *Palinurus longimanus* and related species in the West Indies. The common spiny or rock lobster (*Palinurus vulgaris*) of southern and western Europe is an important article of marine food, particularly in France and on the coasts of the Mediterranean Sea and its islands. It is commonly seen in the markets and restaurants of London, where it commands a good price.

According to Ritchie,[b] *Palinurus vulgaris* occurs on all the shores of the British Isles except a part of the east coast to the north of Flamborough Head. It is most abundant in the southwest, and scarcer northward, but is frequently debarred from entering traps on account of its stout, unyielding antennæ. *Palinurus* in the adult state is unknown in the North Atlantic Ocean north of the Bermuda Islands, but its pelagic larvæ are undoubtedly borne far to the northward by the Gulf stream. It is represented on the western coast of North America by *Palinurus interruptus*.

The carapace of the langouste is not "buttoned" to the tail so effectively as in the common lobster; all the thoracic legs end in long dactyls with indurated tips, which are studded with dense bunches of stiff setæ. The first two pairs of legs are greatly elongated, and the tactile setæ of their dactyls, which resemble bottle brushes, exhibit an extraordinary development.

The largest of the scaly or warty lobsters is represented by *Scyllarus*, which occurs both in the Mediterranean and the North Atlantic Ocean, and is said to attain a length of 18 inches and to excel all other lobsters in the quality of its flesh. Their quadrangular, flattened shell and small, slender legs give them a singular appearance, but specially remarkable are the short, scale-like antennæ, which are possibly used as shovels or scoops in burrowing. Their small and widely separated eyes are completely embedded in the carapace, which is studded all over with wart-like tubercles, thus giving it a granulated and leathery texture, while on the inside it has the appearance of a fine sieve of uniform pattern. Each hole gives passage to a bundle of tactile setæ, which spread in the upper layers of the shell and issue through minute pores

[a] Bate, Spence: Report on the Crustacea Macrura; Scientific results of the voyage H. M. S. Challenger; Zoology, vol. XXIV. London, 1888.

[b] Ritchie, James: Distribution of Palinurus in British waters. Proceedings of Royal Physical Society of Edinburgh, vol. XXIII, p. 68–71. Edinburgh, 1910.

upon the tubercles or around their margins. The last pair of thoracic legs, in the females only, bear claws, which led to the fanciful notion that they were used by the mother in rupturing the eggs and liberating the young. The eggs are very small, and, as in *Palinurus*, the young issue in the peculiar transparent larva known as phyllosoma.

The whole front of *Palinurus guttatus* is armed with stout spines culminating in a pair of rostral horns, which in large specimens rise vertically to the height of an inch or more in parallel planes, thus shielding the eyes and presenting one of the most effective types of protective armature to be seen in an adult crustacean. The antennules are extremely long and slender, while the antennæ have very stout basal stalks, and long stiff flagella, encircled at intervals with sharp teeth, like the war mace of a South Sea Islander.

The second segment of the antenna bears a notable structure, usually described as a stridulating organ. The inner surface of this division is free, and carries a pad and flap which, with the movements of the antenna, plays backward and forward over a smooth ridge or track on the somite. The sound, which it is said may be heard in or out of water and may be produced artificially after death, is evidently caused by friction of the hard chitinous surface of the pad on the track over which it slides. (See p. 240.)

The California spiny lobster, according to Rathbun, may attain a length of 14 inches, and an average weight of 3⅓ pounds, the greatest weight recorded being 11½ pounds. The usual length of *Palinurus vulgaris*, as given by Bell (*20*)[a] in 1853, was about a foot, but 18 inches was sometimes reached. His description was from a male of the latter size, which weighed 5 pounds. "I can not but think," said Bell, "that Dr. Milne Edwards is greatly mistaken in attributing to individuals of that size a weight of from 12 to 15 pounds." The Californian langouste is most abundant on the southern part of the coast. It is often trapped in great numbers, but even twenty years ago we are told by Rathbun that the species was in danger of extermination from overfishing.

Artificial propagation of the Japanese spiny lobster, *Palinurus japonicus* Gray, was undertaken by the fisheries institute, near Tokyo, previous to 1899, and a report of progress was published in that year. Great difficulty was experienced in handling the larvæ, on account of their minute size and long metamorphosis. The spawning and hatching periods of this lobster, as I am informed by Tasute Hattori, who conducted the experiments, extend from late April to late September. The larvæ were easily hatched, but gradually died off after the fifteenth or sixteenth day. No success had been attained in 1901, since which time no further information has been received.

The Nephropsidæ, the best known of the Crustacea, on account of their high commercial value as food, are represented by three species, the Norwegian lobster, *Nephrops norvegicus* Linnæus, the common lobster of Europe, *Homarus gammarus* Linnæus, and the common lobster of America, *Homarus americanus* Milne Edwards.

The technical names for the lobsters adopted in a former work (*149*) are here retained, pending a decision upon the question by the International Committee on Nomenclature of the International Zoological Congress, which met in Boston in 1907.

[a] Italic figures in parentheses refer to works enumerated in the bibliography at the end of this paper.

The question of the validity of Latreille's types in his "Considérations Générales . . ." of 1810, has been raised by Stebbing, who would restore the terminology of Leach, designating *Astacus Potamobius* and *Homarus Astacus*.[a]

Aside from the merits of this controversy, it may be well to point out again that Latreille and others who have followed him were wrong in asserting that Aristotle makes no mention of the river crayfish (*149*). On the contrary, the Father of Zoology uses the term ἀστακός to designate both crayfish and lobster, and so far as antiquity is concerned neither has the claim of priority.[b]

The Norwegian lobster is common not only to Norway but to the coasts of Scotland and Ireland. While essentially a northern form, it is found as far south as the Mediterranean but in much less abundance. It attains a length of from 7 to 8 inches, and in life is of a delicate flesh tint, boldly marked with light brown in symmetrical pattern over the abdomen and tail fan. Its slender form suggests the shrimp type, and its large kidney-shaped eyes remind one of *Penæus*, and of the adolescent lobster (*Homarus*) when from 1 to 3 inches long. The claws of the first pair of thoracic legs are slender, of nearly equal size and keeled above, below, and at the sides, each keel having a single, or at the sides a double row of spines. Bell, writing at the middle of the last century, said of this species that it was frequently on sale in the Edinburgh markets, and was occasionally seen in London.

The European lobster is found on the shores of the British Islands, and on the western coast of Europe from Norway to the Mediterranean. The southwestern coast of Norway appears to be the central point of its distribution and still supports the largest of the European fisheries, but the species is found northward as far at least as Tromsö, or to about 69°–70° north latitude. (See *306*.) It is very rare, if present at all, in Iceland. It does not appear to enter the Baltic, and is not common in the Mediterranean, being limited in its eastern range by the Adriatic Sea. In Great Britain it is chiefly confined to certain districts on the west and north coasts.

Of the three kinds of lobsters already described for the Atlantic and its tributaries, the Norwegian and common lobsters are typical northerly forms, while the langouste or *Palinurus* abounds only in the south. The best fishing grounds for the common lobster in the Scottish seas are said to be the Orkney and Outer Hebrides islands.

The common lobster of Europe resembles the American lobster so closely in every structural detail that the two might at first sight be considered as geographical varieties of the same stock rather than as distinct species. It has been pointed out that the under side of the beak or rostrum is smooth in the *Homarus gammarus*, while in the American form it is armed with a spine, a rather trivial distinction in view of the variable character

---

[a] This commission reported to the Congress, which met at Graz, August, 1910, in favor of accepting Latreille's type designations. The term *Astacus* should therefore be restricted to the crayfishes, and the names stand as designated in the text. See opinions rendered by the International Commission on Zoological Nomenclature. Publication No. 1938, Smithsonian Institution, Washington, 1910.

[b] Those interested in discussions of this character are referred to no. 225 and no. 260 of the bibliography at the end of this work, and also to the following: Rathbun, Mary J., List of the decapod Crustacea of Jamaica, Annals of the Institute of Jamaica, vol. 1, no. 1, 46 p. Jamaica, 1897; Faxon, Walter, Observations on the Astacidæ in the U. S. National Museum and in the Museum of Comparative Zoology, with descriptions of new species, Proceedings U. S. National Museum, vol. xx, p. 643–694, Washington, 1898; Stebbing, Thomas R. R., The late lamented Latreille, Natural Science, vol. xii, p. 239–244. London, 1899.

of such structures. In fact, either one, two, or three spines of inconstant size may be present in the American lobster, though this is a condition which in some cases might be attributable to an injury and its imperfect repair. In the slight differences observed in the development of the American form, however, there are more valid reasons for maintaining the specific names.

It has been the accepted belief that the American lobster attains a greater size than its European counterpart, but it is possible that in early days the maximum size was essentially the same. The fishing of lobsters in Europe is of great antiquity, and the average size of the adults taken has been reduced in consequence, while the industry in this country has been mainly developed during the last hundred years. The same gradual falling off in size, due to the same cause, has nevertheless been experienced on the New England coast and in the maritime provinces. It seems certain, however, that the American lobster has larger claws, and, length for length, it will weigh more than the European form. (See chapter III, p. 195.)

The slight differences in the development of the two forms, already referred to, are seen in the young at the moment of hatching. The abridgement of the larval period has been carried a step farther in the common lobster of Europe, so that its young issue from their eggs in a stage nearly comparable to the second larva of the American lobster.

DEVELOPMENT OF THE CRUSTACEA.

All the decapod Crustacea are developed from eggs which in the Macrura are fertilized outside of the body and are generally carried until hatched on the under side of the tail or abdomen of the female, where they are glued to certain hairs of the swimmerets. The sperm cells are vesiculate and often "rayed." The eggs vary in number from less than a dozen, as found in small species of *Synalpheus* with abbreviated development, to several millions, as in *Callinectes* and *Palinurus*, and from nearly $\frac{1}{6}$ inch, in certain deep sea shrimp, to less than $\frac{1}{100}$ inch in diameter.

The time of fertilization, so far as known, always coincides with that of oviposition and attachment. By means of a liquid cement the eggs are fixed, in a way to be later discussed, often to one another and always to the swimmerets under the abdomen. In life the swimmerets beat rhythmically backward and forward, whether the animal is in motion or at rest, and the attached eggs are thus constantly cleaned and aerated under natural conditions.

The ova are delicate and soon die if cut loose and left to themselves. In order to rear them successfully under such conditions, artificial aeration of some kind must be resorted to and conditions devised to prevent the accumulation of sediment or parasitic growths over the surfaces of the eggs. The best "brooder" of any decapod's eggs is undoubtedly the mother, whether lobster, shrimp, or crab.

The period of fosterage varies from a few days or weeks in some of the smaller tropical decapod crustacea to nearly a year in the lobsters. There is a similar variation in the frequency of spawning; certain Alpheidæ of the Bahama Islands apparently have a succession of broods the year round, while others may lay their eggs twice or once only each year. In the American lobster the breeding period is biennial, but it is possible that successive annual broods are occasionally produced, as has been known to occur in *Homarus gammarus* on the English coast, and after transplantation to New Zealand.

In many of the prawns the eggs all hatch in the course of a few hours, and at night or very early in the morning, as I have observed in *Pontonia*, *Stenopus*, and *Synalpheus*. The adult *Pontonia* lives in the mantle chamber of *Pinna*, a large bivalve mollusk. For a day or two its young move about in a dense cluster like a swarm of gnats.

The young in most crustacea are hatched in an immature state, and in most species they cut loose from the parent at once, proceed to the surface, and as pelagic larvæ lead an independent existence for days or weeks. Though as adults they may be sedentary and chained to the bottom, as larvæ they are usually most active, and it is during this period of free swimming that they undergo their metamorphosis, or series of changes by which most of their proper adult characters are acquired.

So remarkable are some of these larval changes, and so great is the difference of degree in which they are expressed, even in forms so near akin as lobster, crayfish, and prawn, that the fact when first affirmed was denied as incredible. The credit for the discovery of the metamorphosis in Crustacea, which has proved to be a most fruitful generalization in zoology, belongs primarily to a Dutch naturalist, who has not always received his just dues, and secondarily to an Irish zoologist, for the old observations of Martin Slabber,[a] made June 24–28, 1768, and published with excellent drawings in 1778, were not followed up and fully understood until J. Vaughan Thompson confirmed and completed them by studies began in 1822, continued for many years, and published at various times from 1828 to 1843. The sea-waterflea or *Taurus* of Stier, which Slabber figured and distinctly described as passing by metamorphosis to a different and higher form, was afterwards regarded as representing an independent genus of animals and renamed *Zoé* or *Zoëa* by Bosc[b] in 1802.

Bell, who has given a very fair account of this subject in the introduction to his work already referred to, thought that the zoëa which Slabber had under observation was the larva of the common ditch prawn *Palæmon varians*, later described by Du Cane.

Very shortly Thompson obtained in abundance larvæ resembling the *Zoëa taurus* of Bosc by rearing the eggs of the common English crab, *Cancer pagurus*. Again in 1835, by extending his studies to the common green crab, *Carcinus mœnas*, he showed that it not only was hatched as a zoëa, but passed from this larval state into a megalopa before acquiring the true crab-like form and characteristics, proving that this mythical genus which had been proposed by Leach was, like the zoëa, only a passing phase in the metamorphosis of the crab. Then it was shown that in the course of its development from the egg the crab passed through two consecutive stages which were so unlike each other and so unlike the adult form that former naturalists had placed them not only in different genera but in different families.

[a] Slabber, Martinus. Natuurkundige Verlustigingen behelzende Microscopise Waarneemingen van In-en Uitlandse Water-en Land-Dieren. Waarneeming van een Zee-Watervloo, genaamd Taurus of Stier, v. stukje, 5 plaat, p. I–XII, 1–166, pl. 1–18. Haarlem, 1769–1778.

[b] Bosc et Desmarest, Manuel de l'histoire naturelle des Crustacés, t. II, p. 237. Paris, 1830.

Few general laws are without exceptions, and the fact that metamorphosis, which is even more common in crustacea than in insects, is sometimes scamped or wanting altogether, led at once to confused and contradictory ideas. The abbreviated larval history of the crayfish which had been worked out with great care by Rathke in 1829 and that of the European lobster first announced by Thompson (*262*) in 1831, and confirmed by Brightwell in 1835, as well as that of the West Indian shore crab, *Gegarcinus ruricola*, determined at the same time by Westwood, led to temporary difficulties, which were eventually cleared away when the development of many kinds of both macruran and brachyuran crustacea had been studied with sufficient care.

It thus appears that the term "zoëa" was first applied to the larva of a prawn and crab, in which the swimming appendages are three pairs of claw feet or maxillipeds, the thoracic legs being rudimentary buds when represented at all. The abdomen is segmented, but bears no appendages and ends in a forked telson. There is a long depressed rostrum and a very long and sharp dorsal spine which springs from the middle of the carapace, both of which seem to be admirably adapted for protection. Though many variations occur in the larvæ of closely related genera and it is difficult to make general terms fit the varying degrees of modification which larvæ have undergone, it seems best to preserve the historical usage of the word zoëa as far as possible. For this reason we speak of the young lobster when hatched with its thoracic appendages well formed and using both its great maxillipeds and following thoracic legs for swimming simply as a larva rather than as a zoëa, however modified.

Most true crabs and prawns hatch as zoëas from minute eggs, and are commonly translucent and flecked with brilliant red and yellow pigment cells. They molt frequently during the first few weeks of life, passing in the case of crabs through a megalops stage, and then gradually assuming the structure and habits of the adult animal.

Entomostraca generally, and exceptionally certain of the Malacostraca, such as the decapod *Penæus* and the schizopod *Euphasia*, hatch from eggs still more minute and in a much simpler larval form called the nauplius. It is unsegmented, possesses but three pairs of appendages, representing the antennulæ, antennæ, and mandibles of the adult, and has a single median "nauplius" or "Cyclopean" eye. Upon the theory of recapitulation, the nauplius has been regarded as the representative of a primitive or ancestral form, but it seems more probable that existing larvæ of this type have become modified to meet the present conditions of their environment.

In every metamorphosis individuality is preserved from egg to adult, and development proceeds according to this simple formula: Egg = embryo = larva 1, 2, 3+ = adolescent stage 1, 2, 3+ = adult stage 1, 2, 3+ $\begin{cases}\text{eggs, or}\\ \text{sperm.}\end{cases}$

A long metamorphosis which entails a long pelagic life near the surface means greater risk and greater destruction than one of short duration. Consequently it is not surprising to find a general tendency to shorten this larval period, reducing the metamorphosis by shifting it to the egg, or, more exactly, by lengthening the period of egg development. In this case the supply of food yolk is increased to support a longer life within the egg membranes, and the larvæ or young issue in a more advanced state, and as a rule have a shorter pelagic period. The size of the individual egg is increased, but the number of eggs is diminished. The alternative lies between two extremes as follows:

a { Eggs small, but many of them.
Long metamorphosis.
Less chance for individual survival, but more individual chances.

b { Eggs large, but few in number.
Metamorphosis shortened.
Greater chance for the individual, but fewer individuals to take it.

Between these two types of adjustment many compromises have been made. The principal larval stages or types in decapods which have received definite names, being the survivals in some cases of a period when crustacean larvæ were considered adult forms, are the following:

(1) Nauplius and metanauplius. The shrimp *Penæus* is hatched as a nauplius and passes through the metanauplius, first and second protozoëa, first and second zoëa, and mysis stages, before attaining the adult form. *Lucifer* hatches as a nauplius, molts into a metanauplius stage, with buds of three more appendages present; then passes successively through the protozoëa, zoëa, schizopod or mysis, and mastigopus stages, and finally to the adult.

(2) Protozoëa, zoëa, and metazoëa. The shrimps *Sergestes* and *Stenopus* hatch as protozoëas, and pass the successive stages as given for *Lucifer*.

In the protozoëas the antennæ are large and are often used in swimming; the carapace is formed, and the abdomen is unsegmented or but incompletely marked off into somites. The telson is forked and garnished with plumose setæ.

A protozoëan stage has been assigned to the lobster, but erroneously, as will be later explained.

The zoëa characteristic of the crabs has seven pairs of appendages and a segmented abdomen. The last two pairs—first and second maxillipeds (*Callinectes*)—are swimming feet, which in the adult are converted into mouth parts. Many shrimp are hatched as modified zoëas with three pairs of locomotor maxillipeds, and the abbreviation is carried a step farther in some species of *Synalpheus* (*S. minus*) where buds of three pairs of thoracic limbs appear behind the maxillipeds, and still farther in others (*S. brevicarpus*), where the first young to appear are in a "mysis" stage similar to the second larva of the lobster.

(3) Megalopa. The changes which follow in the early development in the crab zoëa lead first to the metazoëa, with rudimentary thoracic limbs and pleopods, and then by a sudden leap to the megalopa, a form comparable to the fourth stage of the lobster. The megalopa has large, free, stalked eyes, large claws, and functional walking legs. The swimming exopodites or outer branches of the maxillipeds have atrophied and disappeared, and like a lobster from the fourth stage onward, it has a segmented abdomen with functional swimmerets. It has also well-developed statocysts or balancing organs and no longer reels in its motion through the water by day, but maintains a definite, upright position. In the course of succeeding molts the abdomen becomes reduced and modified, while the animal acquires the peculiar structure and habits of

the adult crab. The development is abbreviated in the *Gegarcinus ruricola*, the gaily colored terrestrial crab of the West Indies, the large eggs and young of which were a puzzle to the early observers.

(4) Mysis or schizopod stage. The biramous condition of the thoracic legs characteristic of this stage is transitory in the larvæ of the higher Crustacea, but permanent in the lower order of schizopods. The oar-like exopods of the larval thoracic appendages persist in the lobster until the fourth molt, when they are suddenly reduced to rudiments, and after the fifth stage no vestige of them remains.

(5) Larval period reduced in various degrees, and metamorphosis in some cases practically absent. In addition to the crayfishes, lobsters, and other illustrations of abbreviated development already given, we may mention *Synalpheus longicarpus* of the West Indies as a striking example, in addition to certain fluviatile and many deep-sea forms.

Like other animals, the Crustacea tend to recapitulate in some degree the history of their ancestors in the course of their own development, and to become modified in structure and instincts to fit them for a temporary pelagic life which is totally unlike that assumed when adult. Their history is further complicated, as has just been seen, by the tendency to abridge the larval period or lengthen the time spent in the egg.

Shortening the path of development is not a peculiarity of arthropods, but is common with both vertebrates and invertebrates. It depends in a large degree upon the relative amount of food yolk and protoplasm of the egg cell, both of which are derived from the parent, and primarily upon the unknown variations and conditions which have led to this result. The size of the egg is proportional to the amount of yolk which it contains, not the size of the animal producing it. Thus the egg of a snapping shrimp 1 to 2 inches long may be many times larger than that of the lobster, while the egg of the latter is hundreds of times larger than that of the blue crab. When the amount of yolk is small, as in the egg of the starfish or spiny lobster, the young hatch in an immature condition; at the other extreme, when the egg is relatively large, as in the crayfish or domestic fowl, the whole period of early development is passed at the expense of the egg substance, and within its envelopes. The chick hatches in the form and with many of the instincts of an adult bird, ripe for the experience of bird life and capable of using it with profit.

The yolk retards the progress of development up to the time of hatching, but greatly shortens the adolescent period. The chick of the domestic fowl spends 21 days in the egg, but in the hands of the poultry breeder it may later attain the weight of 1½ pounds in 3 months, when it is ready for market.

On the other hand, the egg of the starfish or sea urchin, which is unencumbered by a great mass of yolk, and very small in consequence, measuring about $\frac{1}{200}$ inch in diameter, hatches at ordinary temperatures in 24 hours. It must, however, lead a long life as a larva, make its own living, run the gauntlet of enemies, and keep up the struggle for months. Thus the handicap at the start may count for little in the end. The advantage gained by the fowl in having a few very large eggs is offset by that of a vast number of almost microscopical ova in the echinoderm.

In the lobster the conditions of development are intermediate between such extremes, but in weighing them the structure and habits of the animals at every stage, the environment, and their adjustment to it must be considered. The whole period of development is long, followed by a long period of adolescence, but the relative duration of the swimming life, which is about 3 weeks, is shorter than in the starfish or in *Palinurus* (see p. 160). This is a fortunate circumstance in view of the possibilities of artificial propagation, as will be later seen.

While the abbreviation of the metamorphosis is attended by an accumulation of yolk in the egg, it is impossible to explain either how this has been effected or why in any case such a course should have been followed to secure greater harmony or fitness to the environment.

In fresh-water forms and in deep-sea species the shortening of the metamorphosis may be more uniform and the advantage derived more apparent. In all cases, however, it is a question of the survival of the young, but no one can say why in *Palinurus* the problem has been solved by increasing the number of individual chances and in the lobster by lengthening the period of fosterage and reducing that of the larva. In any case the tax on the parent, when no parental instinct is involved, is essentially the same, though the items are changed, since the total amount of food yolk manufactured in the ovaries of a crab, which lays millions of eggs, is probably not relatively greater than that produced in the organs of the lobster, whose eggs are counted only by tens of thousands. The greater the size of the egg, however, the longer is the tax issue upon the energy of the young deferred and the greater the reduction of its rate.

The adjustment represented by either extreme is certainly advantageous in the long run, but probably neither is the best under all circumstances.

FAMILY LIFE IN CRAYFISH.

The crayfishes, which are now all inhabitants of fresh water or burrowers in soil where moisture is available, are undoubtedly descended from marine lobster-like ancestors, and, as we have seen, for reasons not fully understood have undergone a still greater reduction in larval development. They have, further, acquired an interesting family life, which was noticed by Rösel von Rosenhof over one hundred and fifty years ago. An adequate account of this relation has finally been given by Andrews,[a] and in concluding this chapter we shall give a résumé of one phase of it, based upon his work.

Metamorphosis has been curtailed to such an extent in *Astacus* and *Cambarus* that they are hatched in a form which suggests the fourth stage of the lobster. In reality the young crayfish presents a curious compound of embryonic, larval, and adult characters. The peculiar family relation which serves to tide the young over a helpless period of infancy to complete independence endures, according to Andrews, for about a fortnight, or until after the second molt in *Astacus* and after the third in *Cambarus*.

---

[a] Andrews, E. A. The young of the crayfishes Astacus and Cambarus. Smithsonian Contributions to Knowledge, vol. XXXV, no. 17, 18, p. 1–80; pl. I–X. Washington, 1907.

It is dependent upon a complicated chain of events, which suggests the story of the old woman who went to market to buy a pig. Thus if the egg stalk in *Astacus* does not adhere to a "hair" of the parental swimmeret or to another egg; if the two egg shells are not themselves adherent; if a certain delicate thread, which is spun, as it were, from an embryonic cuticle shed at hatching time, does not itself stick on the one hand to the telson of the young and on the other to the inside of the inner egg shell, and thus tether the little one to its mother; if again, a little later, when its leading string has broken, this young one has not been enterprising enough to hook on to some part of the egg glue with its great forceps, the tips of which have been bent into fishhook form, it comes to certain grief. The result is fatal at whatever point the chain weakens and snaps.

A few hours after hatching, the helpless little crayfishes, still dangling from the "telson threads" which secure each to the parent, begin to flap their abdomens and to open and close their big, hooked claws. In this way they manage to seize the old stalk of the egg and, with hooks embedded in its tough chitinous "glue," they hold on, literally for dear life, often grasping the same stalk with both chelæ.

At the second molt this crayfish is for the first time free, and soon begins to descend the parental pleopod, climbs over its mother's body, and makes short excursions in the neighborhood, returning again and again to the alma mater and the family brood. Hitherto it has been sustained solely by the generous amount of yolk inherited from its egg state, but since the egg stalks and cases, as well as the cast-off skins, which were attached to the mother, disappear at this time, it is thought that they are eaten by the young and constitute the first direct food they receive before beginning to forage for themselves.

In *Astacus* the "telson thread," according to Andrews, represents an embryonic molt or cuticle, and the abdominal part is turned inside out at the time of hatching and drawn out into the thread, the cuticle sticking on certain of the median marginal spines of the telson. The newly hatched *Cambarus* is tethered to its mother in a somewhat similar way by means of the partially inverted and telescoped "lost larval cuticle," which is shed at hatching and is in this instance an "anal thread," since it sticks at two points only—on the side of the mother to the egg membranes, which are adherent to her, and on that of the young *Cambarus* to a portion of the intestine where its cuticular lining is at first set free. As a result of the tension this embryonic molt is stretched and crumpled, with a tendency to turn the abdominal part inside out. This telescoping and partial inversion of the discarded cuticle is checked only by the molted plate of the telson, with the resultant production of a narrow creased ribbon, the "anal thread," which is firmly fastened to the intestinal wall.

Chapter II.—THE AMERICAN LOBSTER: ITS ECONOMIC IMPORTANCE AND GENERAL HABITS.

White men caught lobsters in Massachusetts Bay for the first time early in the seventeenth century. The Pilgrims and Englishmen who began to flock into the bay colony about the year 1630 were well acquainted with the products of the sea in their old home, and the coast of New England supplied their tables with essentially the same kinds, only in far greater abundance. It is said, indeed, that the Pilgrims began at once to pay their debts, due in England, out of the products of their fisheries.

In the chronicles of those early days the lobster is honored with frequent mention, and the early colonists must have enjoyed to the full both the new and the familiar kinds of American fish, lobsters, crabs, and clams, so big, so palatable, so abundant, and so cheap everywhere along that coast. Indeed, one would think there was no need of starvation, with lobsters and the other forms of sea food to be had on every shore. To quote from Mrs. Earle (*80*), the minister, Higginson, writing of Salem lobsters, said that many weighed 25 pounds apiece, and that "the least boy in the plantation may catch and eat what he will of them." Again, in 1623, when the ship *Anne* brought over many of the families of the earlier Pilgrims, the only feast of welcome which the latter had to offer was "a lobster, or a piece of fish, without bread or anything else but a cup of spring water."

The Pilgrim lobsters "five or six feet long," ascribed to New York Bay, take us back one hundred years further, to the time of Olaus Magnus. In a tabulated list of some fourteen of the biggest lobsters ever captured on the Atlantic coast (no. 9, table 1, p. 195) for which authentic weights or measurements have been preserved, the giant among them all weighed 34 pounds, and measured exactly 23¾ inches from spine to tail. No doubt the Pilgrims would measure a lobster as some fishermen do now, with the big claws stretched to their fullest extent in front of the head. In this condition the actual length of the animal is about doubled, so that the length of the New Jersey record breaker, when distended in this way, would reach nearly 4 feet, and the Pilgrim 6-foot lobsters have probably been stretched nearly a yard. (Compare fig. 1.)

In an account of marketing in Boston in 1740, "oysters and lobsters" are mentioned, "in course the latter in large size at 3 half-pence each," and this abundance continued for over one hundred years.

To revert at once to modern times, many no doubt remember when lobsters were sold by the piece, and at a few pennies at that. Five years ago, with a market price of 25 cents per pound, a lobster weighing 3 pounds 9¼ ounces, at an inland market in New Hampshire, cost 90 cents. The clear meat of the claws and tail of this animal, which had a fairly hard shell, were found to constitute but 27 per cent of the whole. (See table 3, p. 214) This would bring the cost of such meat to 90 cents per pound.

Even when every edible part of this animal was saved, which is seldom or never done, the total waste was found to be 45 per cent, and the cost of all edible parts 45 cents per pound. At the present retail prices of from 30 to 35 cents per pound, these estimates would have to be considerably increased.

GEOGRAPHICAL RANGE OF THE AMERICAN LOBSTER.

The American lobster (*Homarus americanus*) is found only on the eastern coast of North America. Its geographical range covers about twenty degrees of north latitude, from the thirty-fifth to the fifty-second parallel, and embraces a strip of the North Atlantic Ocean 1,300 miles long and 30 to 50 miles wide, and according to one estimate 7,000 miles in length when measured along the curves of the shore. Its vertical distribution varies from 1 to over 100 fathoms. The most northern point at which its capture has been recorded is Henley Harbor, Labrador (*209*); the most southern point, the coast of North Carolina.[a] Since the fishery was begun on the southern New England coast and was gradually extended northward, it is not surprising to find the lobster at the present time not only more abundant but attaining the greatest average size in the northerly parts of its range—in eastern Maine and the Maritime Provinces. It should be noted, however, that three of the largest lobsters captured in recent years are from New Jersey. (See fig. 1 and table 1, p. 195.)

HISTORY AND IMPORTANCE OF THE LOBSTER FISHERIES IN BRIEF.

According to Dr. Richard Rathbun (*227*), who was the first to give us a history of the American lobster fisheries, this fishery as a separate industry began toward the close of the eighteenth or the beginning of the nineteenth century, and was first developed on the coast of Massachusetts and in the region of Cape Cod and Boston, some fishing being "done as early as 1810 among the Elizabeth Islands and on the coast of Connecticut." "Strangely enough, this industry was not extended to the coast of Maine, where it subsequently attained its greatest proportions, until about 1840."

The early white men learned many lessons in fishing from the Indians, and those living upon the coast in the course of time began to supply settlers more remote, until the Cape Cod region, having become famous, attracted fishermen with their smacks from Connecticut and from other states, and furnished most of the lobsters consumed both in Boston and New York for fifty years, or until the middle of the nineteenth century. In 1812, as Dr. Rathbun remarks, the citizens of Provincetown, realizing the danger of exhausting their fishing grounds, succeeded in having a protective law enacted through the state legislature, apparently the first but not the last of its kind, for legal restrictions, including this statute, have been in force ever since. But this measure was designed to protect the fishermen rather than the lobster, for it was merely declared

---

[a] So far as known, the lobster has been taken but four times on the North Carolina coast during the past forty years, namely: One lobster in 1870 at Beaufort; one dredged by the *Albatross* in 1884 off Cape Hatteras in 30 fathoms; one said to have measured 18 inches, caught in a gill net at Nags Head in 1903 and exhibited for some time as a curiosity at Elizabeth, Virginia; and another, as noted by J. N. Cobb, was caught by a fisherman at Oregon Inlet, presumably not far from the latter date. For the last two notices I am indebted to Dr. H. M. Smith of the U. S. Bureau of Fisheries.

illegal for anyone not a resident of the Commonwealth to take lobsters from Provincetown without a permit. The laws later enacted proved of little or no avail; by 1880 the period of prosperity had long passed, and few lobsters were then taken from the Cape. Only eight decrepit men were then engaged in the business, and were earning about $60 apiece. This great local fishery was thus rapidly exhausted by overfishing, and it has never recuperated.

The history at Cape Cod has been repeated on one and another section of the coast, from Delaware to Maine, and is already well advanced in the greatest lobster fishing grounds of the world, the ocean and gulf coasts of the British Maritime Provinces of Canada, especially of New Brunswick and Nova Scotia, and in Newfoundland.

Every local fishery has either passed, or is now passing, through the following stages:

1. Period of plenty: Lobsters large, abundant, cheap; traps and fishermen few.

2. Period of rapid extension: Beginning in Canada about 1870, and much earlier in the older fishing regions of New England; greater supplies each year to meet a growing demand; lobsters in fair size and of moderate price.

3. Period of real decline, though often interpreted as one of increase: Fluctuating yield, with tendency to decline, to prevent which we find a rapid extension of areas fished, multiplication of fishermen and traps and fishing gear or apparatus of all kinds; decrease in size of all lobsters caught, and consequently of those bearing eggs; steadily increasing prices.

4. General decrease all along the line, except in price to the consumer, and possibly in that paid the fisherman.

The official statistics for the State of Massachusetts and for Canada afford pertinent illustrations of the older and newer phases of this history. Thus, in Massachusetts in 1890, 373 fishermen, working 19,554 traps, caught 1,612,129 lobsters of legal size and 70,909 egg-bearing females, with an average catch per pot of 82. Fifteen years later it required 287 fishermen, using 13,829 traps, to produce about one-quarter of this number, or 426,471, and less than one-seventh the number of egg-lobsters, or 9,865; while the catch per trap had diminished by nearly two-thirds, and was only 31. No substantial increase followed until 1907, when the legal length was reduced to 9 inches, and this was undoubtedly due to the large number of small lobsters caught.

The total product of the lobster fisheries in the United States for 1892 was 23,724,525 pounds, about three-fifths of which were furnished by Maine, and valued at $1,062,392. It is significant to notice that thirteen years later, in 1905, the total yield, according to Dr. Smith (*325*), had fallen to 11,898,136 pounds, with a value of $1,364,721; in other words, during this comparatively short interval, the supply was practically cut in two, but the value greatly enhanced.

The lobster fisheries of Canada, which next to those of the codfish and salmon are most valuable to the Dominion, have yielded, from 1869 to 1906, inclusive, a period of thirty-seven years, a grand total of $83,291,553. In 1897 the produce of this fishery was 23,721,554 pounds, valued at $3,485,265. Ten years later, in 1906, the yield had dropped to 10,132,000 pounds, but, though less than one-half as great, it had nearly the same estimated value, namely, $3,422,927. Notwithstanding the increased cost to the consumer, even in Canada the total value of the fishery has begun to fall, the product for 1906 being less by half a million dollars than that of 1905.

The lobster grounds of the Atlantic coast were the finest the world has ever produced. In Canada alone 100,000,000 lobsters have been captured in a single year. If properly dealt with, it would seem as if this vast natural preserve should have yielded lobsters in abundance and in fair size for generations and even centuries to come. But instead, lean and still leaner years soon followed those of plenty, first in the older and more accessible regions of the fishery, until the decline, which has been watched for more than three decades, has extended to practically every part of this vast area.

The lobster fisheries of the old world, and especially the more important industries of Norway and Great Britain, when they came to be pursued with the system and energy characteristic of modern conditions, have experienced a similar decline, and upon the whole attempts have been made to meet it in a similar way and with the same result. The treatment has been of the symptomatic kind, and the real cause of the difficulty has not been reached. Sweden, indeed, is said to have felt the need of protective measures two hundred years ago, and to have framed the first laws regulating her lobster fishery in 1686. In 1865 the export of lobsters from Norway, to England chiefly, reached nearly 2,000,000 in numbers. Already as early as 1838 protective measures were being vigorously discussed, and it was proposed to establish a gauge limit of 8 inches; but this was rejected, and a close season (July 15 to September 30, and later extended from July to November) adopted instead. From 1883 to 1887 about 1,000,000 lobsters were captured on the Norwegian coast yearly, having a value of 640,000 francs ($128,000), a large part of the product being consumed in the interior and the rest exported alive. While this small fishery has maintained itself better than most others, it has suffered still greater reduction in recent years.

The product and value of the lobster fisheries of Norway from 1815 to 1907 are given by Boeck (*24*), and Appellöf (*305*), the latter from official returns. According to these data the best single year in its history was 1865, with a catch of 1,956,276 lobsters, and the best periods from 1821 to 1830, with numbers ranging from 784,511 (1823) to 1,609,051 (1825), and 1860 to 1886, with numbers varying from 987,370 (1877) to the greatest record as given above. Since 1886 the annual catch has not touched the million mark, and the numbers have varied from 549,446 (1892) to 992,761 (1907). It is further interesting to note the steady rise in value of the produce of this fishery. Thus the catch of 1883, namely, 1,255,790 lobsters, though greater than that of 1907 by 263,039, had only about one-half its value, or 423,083 crowns ($114,232), as compared with 835,002 crowns ($225,450). Expressed in another way the average price of lobsters had increased from 28.50 crowns per 100 in 1878 to 92.41 crowns in 1905, or over 300 per cent.

Herbst (*136*) writing about 1790, thus speaks of the importance of the lobster fishery Norway at that time:

In the Stavanger district this trade brings every year more than 10,000 Reichsthaler into the country. Yet many maintain that it is detrimental to Norway, since owing to the extensive fishing of lobsters, other fish have left the Norwegian coast . . . The inhabitants of Zirkson, Holland, were the first to undertake this trade, and through it they have become very rich. Up to the present time also the English have brought many lobsters from Hittland. From 30 to 40 lobster vessels come each year from Amsterdam and London to Norway, and each carries from 10,000 to 12,000 lobsters . . . When a load is safely landed it is very profitable, since a lobster which is bought in Norway for 2 Danish shillings is sold in England for a crown. This is the fixed price for a lobster, 8 inches or over in length, the legalized gauge. If a lobster lacks a claw, it is then sold for only a shilling . . . The females are considered the best eating.

The lobster fisheries of Denmark, Holland, Belgium, France, Portugal, and Spain are relatively of minor importance at the present time, and in most cases wholly insufficient to supply the home markets. Roché, in 1898 (*237*) placed the total annual value of the French fisheries of the lobster and langouste at 3,114,317 francs ($622,863), of which 1,425,572 francs ($285,114) was represented by the lobster (*Homarus gammarus*).

The yield of the lobster fisheries in the British Islands has in some years reached a total of 3,000,000 lobsters, and complaints of a diminishing supply have been loud and frequent. This would be a little over a third more than the returns of the Massachusetts fishery in 1888, with its higher gauge of 10½ inches at that time. Prince maintains that lobsters are so dear in England that only one person in 15 has one to eat in the course of the year. (See p. 368 footnote.)

Restrictive measures of some sort have been in force in England for a long period. Thus, R. Brookes in "The Art of Angling,"[a] under "necessary cautions," is careful to state that "Lobsters must not be sold under Eight Inches from the Peak of the Nose to the End of the Middle Fin of the Tail; the Forfeiture is One Shilling for each Lobster." He remarks that "Lobsters are taken in Pots as they are call'd, made of Wicker-Work," baited and set in 6 to 10 fathoms of water, or deeper, and adds: "Their Flesh is sweet, restorative and very innocent."

A review of the measures which have been taken to propagate the lobster and to check the decrease in its fishery in recent times is given in chapter XII.

THE CAPTURE, TRANSPORTATION, AND ACCLIMATIZATION OF THE LOBSTER.

The principle of the modern lobster trap is that of the old-fashined rat trap adapted for taking an aquatic animal with as keen a scent as the rodent, but with far duller wits. The device is undoubtedly of great antiquity, but as modified and applied for the lobster it is apparently not over 200 years old. It was introduced to this country from Europe, where, as Boeck (*24*) plausibly suggests, it was first applied in this way by the Dutch in 1713, and was adapted from the eelpot then in use.

Primitively lobsters were speared, gaffed, or hooked, and for a long time on the coast of Norway were taken with wooden tongs about 12 feet long and adapted for use in shallow water only; lobster tongs had not wholly disappeared at the middle of the nineteenth century. All animals taken by such means were injured more or less severely and were unfit for transportation. The gaffing of lobsters from small boats was a common practice in the early history of the American fishery, and a fisherman in Maine once

[a] 2d edition, London, 1740.

told me that in the period of plenty, from 1855 to 1860, he had taken 150 in this way in a single morning.

Then followed the hoop net or bag, sometimes called "plumpers" in England, or "Fallenkörbe" (basket traps) in Germany, which were in extensive use at the middle of the eighteenth and locally to the middle of the nineteenth century, or even later. This was a simple iron hoop with bag net attached and often with crossed and arched half hoops over its mouth. When baited and sunk it had to be watched and pulled at frequent intervals in order to secure the lobsters before they could crawl out. About the year 1858 a giant male lobster, said to have weighed from 25 to 30 pounds, was taken in one of these hoop nets in Golden Cove, Vinal Haven, Me.

Travis (*264*) describes the use of hoops at Scarborough, England, in 1768, but Pennant a few years later remarked that lobsters were sometimes—

taken by the hand, but in greater quantity in pots, a sort of trap formed of twigs and baited with garbage; they are formed like a wire mousetrap, so when the lobster enters there is no return. They are fastened to a cord sunk in the sea, and the place marked by a buoy.

This English lobster trap undoubtedly came, as Boeck suggests, from the Norwegian "Tejner," or baskets, which were the Dutch adaptation of the eelpot, the Scandinavian name being derived from "tün," the long tough roots of the juniper tree (*24*). After 1713 they were made of plaited willow twigs. Linnæus saw similar baskets in 1746 in use on the coast of Bohuslän. Herbst (*136*), writing in 1790, says that lobsters were then caught in "Tüner," "Teiner," or lobster baskets ("Hummertienen" or "Hummerkörbe") made of birch twigs.

The tines in later use among the fiords of the Norwegian coast were sometimes made of slats or rods nailed to small hoops, and at considerable intervals, which were filled in with interwoven cords of hemp. There were entrance funnels at either end, a door at the top, and a flat stone lashed to the bottom for weight, while in the center of the trap was suspended a peg for attaching the bait. (See *309*, p. 733.) When a lobster was taken from the tine, his claws were securely bound with pack thread, and thus held until he was delivered to the submerged box or car to await final transportation to market.

Essentially this old-style trap has been retained in Europe, where it is to be seen at the present day. Those examined at St. Andrews, Scotland, where they are called "lobster creels," in July, 1896, were small cylinders, made of a wooden frame covered with netting, and were anchored by means of a flat stone tied to the bottom. A fisherman with whom I conversed on the beach had 40 of these creels, and was going to haul them at 5 o'clock that evening, but with no expectation of taking any lobsters, for, as he expressed it, the sea was too calm; rough weather brought better luck. The "tiner" of the Helgoland fishermen, according to Ehrenbaum (*84*), are birdcagelike, cylindrical or four sided, with the bottom weighted with stones, covered with netting or wirework, and with funnel-shaped ends, like eelpots. Each is sunk to the bottom with attached cord which is floated with corks. In Norway hemispherical wicker traps, with funnel at the top, were occasionally used.

The American lobster trap of the present time is simply a larger and more efficient modification of the old wicker "basket," but made of laths with netted heads or ends in the form of a funnel with entrance ring. On the outer islands and coast of Maine the half-cylinder form is preferred. They are 2½ to 4 feet long, 2 feet wide, and 18 inches high, the smaller sizes being now commonly used. A trap of this type which I measured on Great Duck Island in 1902 was 3 feet 9 inches in length and 25 inches in both height and width. The frame was of scantling, from which were sprung three arches or "bows" of spruce, and to these were nailed laths at intervals of 2 inches, one side being provided with a hinged door. The "heads" are made of netted cotton, or, preferably, of manila cord, tarred and strung to a "funnel bow" or entrance ring of spruce, 6 inches in diameter, and often, as in this case, set obliquely to the long axis of the trap, the whole head being drawn inward to form an upwardly directed funnel. The lobster, in order to get to the bait, must therefore climb up the funnel and pass through the entrance ring; when once a prisoner it is liable to crawl over the ring rather than through it to liberty. The spindle for holding the bait is an iron spike securely attached to the center of the floor. Flat stones or bricks are used as weights, and the trap is secured to a 6-strand manila warp, which serves to lower and raise it, as well as to mark its position. This cord, the length of which is determined by the depth of the water, is fastened by one end to a corner of the frame or "sill" of the trap and by the other to a wooden float or buoy, which bears the owner's color or mark. Traps are commonly set on single warps, but in summer are sometimes strung to an anchored ground line or trawl, to the number of 8 to 25 or more units and at intervals of about 30 feet, according to the depth, so that when one trap is hauled to the boat the next in line will be at the bottom. In this case the position of the anchor at either end of the trawl is marked by a buoy. Trawls were sometimes set across currents so that fine particles coming from the bait would be widely diffused, but the practice has been mostly given up. Fishermen tend from 50 to 125 traps, according to conditions, and some have two sets, the winter relay being left on the beach to dry out in summer. The "counters," or lobsters of legal size, are temporarily stored in floating cars until gathered up by well boats, which carry them to the large markets or to the numerous pounds along the coast, where they are stocked for the winter and summer trade.

The traps are baited with small herring, halibut, hake, or codfish heads or with fresh or salted fish of any kind. The fishermen try to follow the movements of the lobsters and in summer fish closer to the shores, ordinarily in from 1 to 10 fathoms, but in winter they often go out 5 or 6 miles and set their traps in 20 to 50 fathoms of water. The traps are pulled as often as possible, once or twice daily in summer, but in winter weather a week or even a fortnight may elapse before the traps can be visited, and many are destroyed by storms.

The fish commission of Massachusetts, in recommending the adoption of a double legal gauge for lobsters of 9 to 11 inches, inclusive, proposed a standard trap which should have an entrance ring not to exceed 3½ inches, with slats not less than 1½ inches apart, to work automatically to the extent of not permitting lobsters above legal size to enter and of allowing the undersized to escape.

Lobsters destined for inland markets are successfully transported with or without plugging the claws, packed in wet seaweed, and with ice at the bottom. For a long time nearly the entire product of the Norwegian lobster fishery (see p. 172) has been sold in England, the animals, usually with claws bound with cord, being carefully packed in small fish boxes, in heather wet with sea water, and in summer with ice at the bottom; care is taken not only to shield them from the drip, for they can not stand fresh water, but also by means of paper linings to protect them from excessive cold; always with the precaution of leaving suitable openings at top and bottom to allow the air to enter and the water to pass out.

Early in the nineteenth century, according to Prince (*219*), several barrels of lobsters were sent from Nova Scotia, as a present to King George III of England. Again in 1862 several tubs of lobsters in sea water were forwarded from the coast of Maine to the Emperor Napoleon III of France. The longest sea journey yet made by the living lobster was accomplished some time previous to 1896, when the Otago Acclimatization Society of New Zealand succeeded in transplanting 9 lobsters from England, 3 only having died on a voyage of 54 days, covering a distance of 12,000 miles through the Tropics, where water not artificially cooled reaches a temperature of 84° F. The experiment was repeated in 1906, and up to May 30, 1909, four shipments had been made from Plymouth, England, to Portobello (Dunedin), for the fish hatchery and biological station there. The last of these proved most successful, 31 out of 34 lobsters being delivered alive. Each of the animals was given a separate compartment in the wooden shipping tank, and was supplied with clean, well-aerated and cooled water, and was fed during the voyage.

From 1874 to 1889 five attempts to acclimatize the American lobster on the Pacific coast were made by the United States Fish Commission, when 590 animals of both sexes, and some with external eggs, were successfully transported across the continent and distributed at different points from Monterey Bay to Puget Sound. Accounts of these early experiments have been given by Perrin (*319*), Rathbun (*228*), and Smith (*253, a*).

> No positive results having appeared [says Dr. Smith], the experiment was renewed in the fall of 1906, when a special carload of brood lobsters, numbering more than all the previous plants combined, was dispatched to Puget Sound, and in 1907 a still more extensive plant, aggregating about 1,000 adult lobsters, was made in the same water. Further consignments will be made until the lobster is removed from the list of failures and recorded as a great financial as well as a gastronomic success (*325*, p. 1406).

We believe that the Bureau has taken a most commendable step, and in the right direction, the initial attempt being to find a water where the Atlantic lobster will thrive. When this primary question has been settled, further importations to that point, supplemented in time by artificial propagation, promise well for the eventual establishment of new and remote fisheries which, for all that is now known to the contrary, may at some future day enjoy a greater prosperity even than those nearer home.

HABITS AND INSTINCTS OF THE ADULT LOBSTER.

At this point we shall examine certain facts in the general natural history of the lobster, leaving, however, such important subjects as reproduction, growth, and development for special consideration.

The sea bottom is the natural abode of the lobster, as it is of all the large and heavy Crustacea, the source of its food and the scene of all its activities, from the close of free pelagic life to old age. Its external world is the ocean floor, to which it reacts, and it knows no other. While its powers of locomotion are considerable, it never forsakes the water of its own accord or leaves the bottom, to which nature has consigned it by giving it a heavy body and a sedentary disposition. Lobsters wander close to the shore and out to depths of over a hundred fathoms, and the nature of the bottom, or more directly the supply of food, as well as the physiological condition of the animals, especially in respect to their molting periods, determine their abundance within these limits in any locality.

The supply of food, the temperature of the water, and in general the physical conditions of the environment vary greatly throughout the range of this animal, as one might infer from a study of the coast line. From Labrador to Maine the coast is very rugged, deeply indented with bays, and studded with islands, some of which present perpendicular walls to the sea. The coast of Maine, particularly in its eastern and middle sections, is essentially bold, rocky, and diversified to an extraordinary degree by deep channels, extensive bays, and inlets of all kinds, and these are studded with rock-ribbed, spruce-clad islands. The geological formation is pre-Cambrian, the rocks being mainly granites. From 10 to 30 miles from the shore we find large and important islands standing alone or closely related, as Monhegan Island and the Vinal Haven and Matinicus groups. All are essentially masses of granite, which in some cases have been cut by glacial forces into archipelagoes; they abound in basins and channels of various kinds, into which fresh sea water is driven with every tide, and thus form admirable breeding grounds for food fishes, the lobster, and a host of invertebrates. The Cape Cod region is distinguished for its extensive sand shoals, which resemble those of North Carolina. The northern part of the Massachusetts shore is rocky, while the southerly portions are very diversified, abounding in submerged ledges, sandy and weedy bottoms, a great variety of bays and channels, as in Vineyard Sound and neighboring waters. Here lobsters were once exceedingly abundant, until they were nearly exterminated by the fishermen.

Under the variety of conditions indicated we should expect not only to find lobsters larger and more abundant in some localities than elsewhere, a condition greatly influenced by the number and persistence of the fishermen, but also to meet with variations in the time of egg laying and hatching, of molting, and in the rate of growth.

This animal spends most of its time in the search for food and in reproducing its kind. Its instincts are constantly leading it to secure protection through concealment, and we find it burrowing in the mud or sand, or hiding under stones, whether to await its prey or to pass in greater security the crises of its successive molts.

In traveling over the bottom in search of prey the lobster walks nimbly upon the tips of its slender legs, which are provided with brushes of sensitive hairs. The large claws are directed forward, a position which offers the least resistance to the water, or when at rest are held somewhat obliquely, their tips touching the bottom, while the long sensitive "feelers," or antennæ, sweep back and forth continually to give warning of a foe or of objects which its other sense organs fail to detect. In exploring its feeding grounds the movement of the body is chiefly maintained by the swimmerets, or pleopods, which spring from beneath the tail in the form of a double bank of paddles on either side. The swimmeret consists of a short stalk and two flexible blades, which beat rythmically with a backward stroke, and thus impel the animal forward even without the aid of the ambulatory legs. Each blade is further garnished with a fringe of long and strong hairs or setæ, which add to its efficiency as a rowing organ, and certain of which in the female catch and hold the egg glue by which her progeny, in the form of thousands of eggs, are tethered to her body.

The most primitive sense of animals being that of touch, it is not surprising to find tactile organs widely distributed over the body of this crustacean. As will be seen later, they occur by thousands in the form of tufts and fringes of hair-like setæ on the legs and free margins of the shell, and in any part subject to frequent contact either with the body itself, with its food, or the ocean floor. It will also appear that instead of being incased in a solid, impenetrable armor, the crustacean can receive stimuli and impressions from without as readily as if it possessed a soft and delicate skin.

When an enemy appears, or the lobster is suddenly surprised and cornered, it will immediately strike an attitude of defense. Raising itself on the tips of its walking legs, it lifts its powerful claws over its head, after the manner of a boxer, and, striking the offending object, endeavors to crush and tear it to pieces.

When transferred from sea to land the lobster can only crawl in its vain attempts to walk, owing to the great weight of its body, which the slender legs are unable to sustain. If turned on its back its discomfort is immediately shown by attempts to right itself, which are usually successful. When taken directly from the water and left to its own devices on the beach, I have seen it strike out by the nearest path to the sea with as keen a sense of direction as a turtle shows on land. It should be stated, however, that this experiment was tried only within short distances from the water.

By far the most powerful organ of locomotion in the lobster is its "tail," called also the "abdomen" (terms borrowed from vertebrate anatomy), and the "pleon." By the rapid flexion of this muscular tail, aided by its terminal fan, the lobster shoots backward through the water with astonishing rapidity, going, according to one observer, 25 feet in less than a second. If tossed into the water, the animal quickly rights itself, and with one or two vigorous flexions of the tail makes quickly for the bottom as if sliding down an inclined plane.

On calm summer evenings toward sundown lobsters are often seen close to shore, lying on little patches of sand or in eel grass, awaiting their chance to seize a passing fish or crab. When alarmed, they assume the defensive attitude; but press them close, or try to pin them down with an oar, and they will dart backward toward deeper water; if still pursued they flee in other directions, zigzagging their way over the bottom until safety is found at still greater depths.

Lobsters kept in aquaria of sufficient size and provided with running water often thrive, and if they receive proper care will live for a long period. If the tank is provided with a pile of stones, the lobster will examine this carefully until the most attractive holes are discovered. When several individuals are placed in the same aquarium, each soon selects a hole or corner, for the possession of which it is always ready to fight. This is true of the "lobsterlings" as well as the adults, showing that the power of association or of the formation of habits, which is the mark of intelligence, is well developed. When the occupants of the same aquarium are of equal size and show no weakness, they usually live in peace; but should one become disabled, as by the loss of a claw, it is quickly attacked by the strong and forthwith destroyed.

As the lobster lies in its corner of the aquarium, usually with the tail folded, and always so if a female in "berry," it slowly sweeps the water with its long, sensitive antennæ, which are now held erect, now lowered, until they lie horizontal and extend directly forward in front of the body. The smaller antennæ are elevated, while the stouter outer branch of each beats with a rythmical up-and-down movement; this branch carries the delicate hairs or setæ, which are regarded as the organs of smell. One often sees the animal deliberately lower the whip-like branches of the first pair of antennæ and clean them by drawing them through the brushes of the large maxillipeds; the great claws when not extended and ready for immediate use are turned obliquely inward and downward, with their tips touching the bottom.

All animals that play the part of scavengers must have strong powers of scent or keen eyes to guide them to their prey, and lobsters are no exception to this rule. The turkey buzzard sees, but, according to Audubon and Bachman, can not scent its prey, while the lobster, though dull of sight, has a keen chemical or "olfactory" sense. This is illustrated by the way in which it can be enticed into the traps. It is asserted that when traps are set on a trawl placed across the tide, the catch is greater than when the trawl is set in the direction of the current, since in the former case the chemical substances, or fine particles coming from the bait, are more widely diffused. Lobsters are sometimes wary and shy of entering a trap, and have been seen to crawl about it several times and examine it cautiously on all sides before, too weak or too hungry to resist temptation, they finally enter. When the pots are hauled, lobsters sometimes escape by darting backward through the narrow opening of one of the funnels, but this seldom happens and may be set down to accident.

Sluggish as the lobster may appear when out of the water and partially exhausted, it is quite a different animal, as we have just seen, when free to move at will in its natural abode on the bottom of the sea. In the water it is agile, wary, pugnacious, capable of defending itself against enemies often larger and more powerful than itself, and on occasion of exhibiting a high degree of speed. It often captures its prey by stealth and with concealed weapons. Lying hidden in a bunch of seaweeds, in a rock crevice, or in its burrow in the mud, it waits until the victim is within reach of its claws. Though far less active and keen witted than many of the higher crabs, and sedentary in the sense of being restricted in its range, it is sluggish only at the period of the molt or in very cold weather. The sense of hearing is probably absent and that of sight far from acute, but this animal possesses a keen sense of touch and smell, possibly a sense of taste, and is quite sensitive to changes of temperature and light (see p. 184).

MIGRATORY INSTINCTS.

Adult lobsters never migrate up and down the coast at definite periods or in considerable numbers in any degree comparable to the semiannual movements of many fishes and birds; in April and May, however, they come in toward the shore, and again in fall retire to deeper water. Such migratory instincts as they possess are of a very diffuse type and are far from being generally displayed. The abundance of food and periodic necessity of molting and laying eggs, and the temperature of the water, may one and all enter with more or less force into bringing about local and restricted movements. When the question of food is paramount, lobsters will pass the winter in considerable numbers in the shallow waters of harbors, but usually only on a rocky bottom where food is to be found. The extent of their journeys is influenced by the slope of the bottom and the depth of water, as well as by the nature of the bottom itself, and varies in different sections of the coast as well as at the same point in different seasons.

Movements of tagged lobsters.—In order to test the extent and rapidity of the adult lobster's movements along the coast, as well as to and from deep water, some interesting experiments in tagging lobsters have been made by Bumpus (*43*) at Woods Hole, Mass., Mead and Williams (*195*) at Wickford, R. I., and by Meek (*316*) and Appellôf (*305*) in Europe.

In the summer of 1898 Bumpus tagged 479 lobsters from which eggs had been removed, and liberated them at various points about Woods Hole. Seventy-six of these were recaptured and the tags returned for identification. The valuable data thus obtained showed a great variation in the "migratory" impulse and remarkable rapidity of movement in individual cases. Some had not strayed far after gaining their freedom for from 3 to 4 weeks, being recaptured near the points where they had been set free, while others had moved at the daily rate of a mile for a period of 10 to 12 days. One of them which had been freed at Woods Hole on July 2 entered a trap at Cuttyhunk Island, 12 miles to the southwest, on July 13, having covered this distance in 11 days. It does not seem probable that such sporadic movements are determined by the search for more abundant food, or for more favorable conditions as regards the temperature and depth of the water or character of the bottom, but are to be set down to individual initiative and general restlessness of behavior. In this connection it would be interesting to learn whether the more sedentary or the more active individuals had showed any evidences of preparation for the molt, which is due in female lobsters shortly after the hatching of the eggs (middle May to middle July at Woods Hole).

Tagging experiments were undertaken by Mead in the summer of 1902 and 1903 at Wickford. Of the 16 released in the first season, the most enterprising traveler had covered 10 miles in less than 8 days. Out of 385 lobsters tagged and set free in 1903, 30 were later reported, most of them having taken a southerly or southwesterly course down the Narraganset Bay. Eight which had been free from 9 to 31 days had traveled only a mile when captured, June 11 to July 3; 6 had wandered from 10 to 12 miles in the course of 22 to 58 days, having been liberated June 24 to July 26. Further systematic experiments in this interesting subject have been carried on at the Wickford station, and are recorded by Barnes (*15* and *16*, *a*). One of the fastest travelers made 4 miles in a single day.

Movements off Cape Cod and at Woods Hole.—If there were any considerable coastwise migration, it is evident that regions once depleted could be restored under favoring conditions by accessions from neighboring parts. Apparently this does not occur, and, as Rathbun has observed, we may regard each geographical section of the coast as inhabited by a more or less distinct colony, which tends to hold its ground fairly constantly, so that if its numbers be once seriously depleted, recovery under nature must needs be a slow process at best. The history of the Provincetown region on Cape Cod, already referred to, seems to support this idea.

The region about Woods Hole, Mass., including the western end of Vineyard Sound, No Man's Land, and the Elizabeth Islands, was studied for a period of 5 years, from 1890 to 1894, with reference to the general natural history of the lobster, and the following conclusions were then reached regarding its migratory habits: The general movement of lobsters toward the shore in the spring is modified by reason of females with old eggs finding it advantageous to remain on rocky ledges until their young are hatched, while the males press onward to shallower water. After hatching is over, the females make their appearance in large numbers in the sound toward the last of June or 1st of July, and form a large part of what fishermen call "school lobsters" or "buckle shells." Their appearance is probably not as sudden as it often seems. Fishermen as a rule work only one set of traps, setting them now here, now there. In order to follow the movements of these animals systematically, it would be necessary to set traps simultaneously in different places and on different bottoms, and to record the catch for a considerable time.

Some females with old eggs come into the sound before the young are hatched, but the majority do not. It must also be borne in mind that many lobsters remain in the sound and harbors the year round, and that these observations refer only to the movements of the larger number. Toward the latter part of August the pendulum begins to swing the other way, and the lobsters move into deeper water or to a rocky bottom. This outbound movement is continued during the months of September and October, but, as already remarked, it is by no means general and may be more pronounced in cold than in mild seasons.

Aside from their in and off shore movements, the lobsters must be regarded as essentially sedentary or stationary animals. Yet their occasional sudden appearance in great numbers, and often at points where a previous scarcity had been noted, creeping toward the shores in veritable swarms of thousands of individuals, as already reported by Sars (*244*), Appellöf (*305*), and myself (*149*, p. 21), indicate that at certain times and under certain conditions not at present completely understood, movements of a somewhat different character may take place. The "traveling lobsters" of Sars probably belong to this category, and my former suggestion that they might represent "some large species of surface-feeding shrimp" (*149*, p. 19), may be an error, as Appellöf asserts. Sars's account, if correctly translated, is somewhat ambiguous; it is as follows: "The hard-shell and ponderous lobster must always make an extra exertion in moving about, and its movements can therefore not be of long duration. People certainly talk of the 'traveling lobsters' ('Faerd-hummer') which are said to come from the open sea in large schools, and some even say that they have seen such schools many miles from the coasts moving about rapidly near the surface of the sea. If this is really so, I consider it as absolutely certain that these schools come from no very great distance, possibly from some of the elevated bottoms off the coast." (No. *244*, p. 675.) We consider it highly probable that the "swarms" referred to represent only more concentrated movements of the usual inshore character, the animals coming from elevated areas not hitherto discovered and fished.

In general we conclude that since lobsters as a rule spawn in warming water the migratory impulse must be regarded as primarily correlated with the development of the reproductive organs, which periodically respond to a rising temperature. Incidentally the carriage of eggs, the abundance of food, and molting which occurs in the female shortly after the eggs are hatched, tend to disturb the regularity of these movements.

OPTIMUM TEMPERATURE.

While the question of food supply must be of paramount importance to all bottom-feeding animals like the lobster, the temperature of the water can hardly fail to exert some influence upon their movements. Whether there is a direct reflex response in the lobster to the warming waters of the shores in spring or not, it is a fact that it shows a marked tendency, as we have seen, to move shoreward at this time. Further, without any doubt, there is a certain optimum temperature, under the influence of which, when other conditions are favorable, growth is most rapid, and those dependent processes of reproduction and exuviation most accelerated. The data available, however, do not enable us to determine this point with much accuracy.

The physical conditions of Woods Hole region have been made the subject of special study by Sumner,[a] from whose account the following facts have been gathered. The temperature of sea water at Woods Hole for May ranges from 50° to 60° F. The warmest period extends from approximately July 12 to August 24 (which corresponds with the height of the spawning period of the lobster at this point), with a temperature of 70° to 71°. The September range of 69° to 65° is about the same as that for the first half of July. In the latter part of October the water cools to about the same temperature it had reached during the first half of May. The lowest daily temperature, of about 30°, is recorded for mid-February. The bottom temperature at the western end of Vineyard Sound, at the period of maximum summer heat, was found by Sumner to be 60.2°,

[a] Sumner, Francis B. An intensive study of the fauna and flora of a restricted area of sea bottom. Proceedings of the Fourth International Fishery Congress, Bulletin of the Bureau of Fisheries, vol. XXVIII, p. 1223–1264. Washington, 1910.

or about 10° lower than the average at Woods Hole at a corresponding period. A station in that part of the sound which showed in August a bottom temperature of 55° (60.3° at surface) gave in March 36.7° (at the surface 37.4°).

The temperature of the surface water of Winter Quarter Shoal, Virginia, ranges from 35° to 76° F.; at Five Fathom Bank, New Jersey, the range is 37° to 76°. Delaware Breakwater, which at one time was practically the southern limit of the lobster, is situated between the lightships anchored upon these two shoals. In the Gulf of Maine the mean annual range is approximately 32° to 62°, while at some points the maximum is only 54°. (*228.*)

The average temperature on the north shore of Prince Edward Island has been given as 56.56° in June, 63.40° in July, and 62.27° in August, the bottom temperature in 6 to 8 fathoms being estimated at 55°.

The temperature of the sea on the Labrador coast is said not to exceed 46.05° F. on the warmest summer days. The lobster thus seems to be debarred from this coast east of the straits of Belle Isle by the Arctic current and the lingering ice.

From the facts given above we may infer that the optimum temperature of the lobster lies between 50° and 60° F. When the temperature of the sea water marks from 50° to 55° in spring large numbers of these animals have already begun to creep nearer the shores into shallower and warmer places, and again in fall, when the temperature has fallen to this point, many have already been impelled to recede to greater depths. Many lobsters, however, remain in the relatively shallow water of harbors all winter, a fact already emphasized; so it is certain that temperature is not the only influence at work in directing these semiannual movements. The question of food or nature of the bottom may at times be of equal or of even greater importance.

The lobster, like many other marine invertebrates, is very sensitive to the extremes of heat and cold. If exposed to direct sunlight out of the water, or to the nipping air of a winter's day, it weakens or succumbs in a short time. On the other hand, if packed in seaweed with ice it will live for days or weeks, a fact daily illustrated in the transportation of this crustacean alive to inland markets far from the coast. (See p. 176.)

Lobsters which pass the winter in relatively shallow water often seek protection by burrowing in the mud, as usually happens when they are confined in pounds. In such cases a long period of severe cold may prove fatal. On March 10, 1882, a number of lobsters were taken through the ice by the scoop of a mud-digging machine off the coast of Prince Edward Island. They were said to be sluggish but not torpid.

INFLUENCE OF LIGHT AND NOCTURNAL HABITS.

The lobster is essentially an animal of the twilight, and in its semiadult and adult condition explores the bottom in quest of food mainly after sundown or at night, when it is generally far more active than by day. This may be proved by anyone who watches its behavior when confined in either lobster cars or pounds. These animals it is true on occasion move about by day, but at night they become exceptionally restless. It is probable that the eggs are laid and that pairing takes place as a rule under the cover of darkness, either at night or in early morning. Such indeed is known to be the common habit of crayfish, shrimp, and many other Crustacea.

While the lobster is very sensitive to light throughout every stage of its existence, its reactions to this stimulus are of a complex character, especially during its free swimming career, as will be seen in a later chapter. It will appear that the young shun or avoid light of a greater intensity or move toward or away from a source of light and in the direction of its incident rays as a result of the varying state of the animal itself and of its environment. There seems to be ever a struggle among competing impulses, now one set of reactions winning the day, now another. In general the young seem to seek the light, as their swimming habits might lead us to expect, and are usually captured in the day time, but they are sometimes caught at night.

After the discovery of the bottom has been made, through all their later adolescent and adult stages they practice concealment, and prefer the twilight of their rock caves or tangles of weed amid the sand. Yet, under exceptional conditions, the adult may expose itself to stronger light.

According to Forel, light can not penetrate the ocean below a depth of 400 meters of tolerably clear water, but even in fifty fathoms off the Atlantic coast the difference between day and night can not be very considerable. This is not the case in shallow bays or sounds with sandy bottoms, which lobsters frequent in summer, and where we may expect to find the greatest difference between their diurnal and nocturnal habits.

The large floating cars in which lobsters are generally stored in readiness for market are always kept closed. When they are particularly shallow and the lobsters are exposed to the glare of the sun they are sure to suffer, and sometimes die in consequence. The majority of lobsters probably spend the greater part of the year at depths where the effect of sunlight is but slight, and during the course of its evolution the eye of this animal has become sensitive to a minimum quantity of light. For this reason alone we should expect that adults would tend to avoid intense sunlight.

BURROWING HABITS.

The lobster not only digs up the sea bottom in its search for shellfish and covers itself with mud in cold weather, but burrows under some conditions as extensively as the muskrat. Impounded lobsters will sometimes burrow during both summer and winter, and this habit is no doubt freely practiced when they roam at will.

The burrowing habit was typically shown in one of the pounds at Southport, Me., where the lobster holes were driven horizontally into a mud bank for a distance of from 1 to 5 feet. When we did not see the feelers and claws of a lobster projecting from its hole, the occupant could usually be felt by inserting the end of an oar, and it would sometimes grip the blade and allow itself to be dragged out clear.

The holes had an opening of from 8 to 10 inches in diameter, which allowed of their being readily probed and measured with an oar blade. I did not observe that they ever had an upward or downward curve, but they sometimes swerved to the right or left, which might be due to the presence of some obstacle in the path. In some cases the burrows were under rocks, and the entrance was often much larger than that described, possibly owing to the union of the mouths of two originally distinct burrows. The pile of dirt and the broken clam shells which are sometimes seen near the hole of the lobster recall the excavations of the muskrat. It was exceptional to see a lobster with his tail projecting from the burrow, and when disturbed in this position they were quick in disappearing.

In digging, lobsters probably make use of their large claws and walking legs, and possibly the tail fan may be brought into service as a scoop or shovel, but we have no observation in support of the latter supposition. Yet, in some cases we have noticed the underside of the tail fan to be scratched and scarified, and the marginal fringe of hairs worn down in a way to suggest the probability of such use.

Mead (*193*) found that the young lobster sometimes burrows in its fourth or lobsterling stage, and this instinct is very pronounced in all its later phases. It removes bits of gravel presumably with its claws and deposits them short distances away, thus digging to a depth of 2 or 3 inches. Young lobsters, like the old ones, hide in their holes, and issue stealthily in search of prey. Indeed, it may be said that such commanding instincts of the adult as preying, concealment, and fear, are manifested suddenly and for the first time in the fourth stage.

The burrowing habits of certain species of crayfish are well known, while those of the stomatopods (see chap. 1) are equally characteristic. We meet with the same habit in many snapping shrimps, expressed in a greater or less degree in terrestrial crabs, and in a great number of the lower Crustacea.

FOOD AND PREYING HABITS.

The food of the adult lobster consists principally of fish, alive or dead, and of invertebrates which inhabit the bottom and come within its reach. It is not unusual to find bits of algæ or common eel grass in its stomach, and at times in such quantities as to suggest that it may not be an accidental occurrence. Vegetable matter, however, forms at most but a small and casual part of its diet. Fragments of dead shells, coarse sand, and gravel stones as large as duck shot are also swallowed. The former yield lime, which is in some measure absorbed; the latter are not needed in grinding the food as in the gizzard of the domestic fowl, since the lobster's stomach has, as is well known, a mill admirably adapted for this purpose, and their occurrence is probably accidental.

I have dissected soft lobsters, with fragile papery shells, from 3½ to 4½ inches long, in which the stomach was literally crammed with water-worn calcareous fragments of the dead shells of crustaceans and mollusks such as one can gather on the beach, besides other shells of mollusks which had undoubtedly been eaten alive. This suggested the possibility that the supply of lime for hardening the new shell might at times be obtained in this way (see *149;* p. 89–90) for it seemed hardly probable that they would be swallowed to be immediately regurgitated. The lobster undoubtedly regurgitates the insoluble and indigestible parts of its food, as is the known habit of crayfish. Some such outlet for waste matter is absolutely necessary in an animal where the fluid or finely divided and digestible parts of the food only can pass to the delicate intestine. The hard parts of fish, mollusks, and crustaceans, however, appear to be retained until they have given up a good deal of their lime, thus contributing to the calcareous supply of the exoskeleton.

An analysis of the stomach contents of lobsters captured at Woods Hole from December to June revealed the following organisms, which are named in the order of their relative abundance: Fish (procured independently of the traps); crustaceans, embracing chiefly isopods and decapods; mollusks, consisting largely of small univalves; algæ, echinoderms, and hydroids. The bones of the fish eaten belonged as a rule to small individuals or species. Among the crustacean remains parts of small mud-crabs, *Panopœus* (*P. sayi* and *P. depressus*, the common species in Vineyard Sound), were almost invariably recognized, and it was not unusual to find parts of the skeletons of small lobsters. The isopod *Civolana concharum* is frequently eaten by the lobster, and often in large numbers. It is a scavenger, and devours the bait used in the traps, a fact which explains its common occurrence in the stomachs of lobsters newly caught. In the case of a female, captured in January, the stomach was filled with fresh lobster eggs in an advanced stage of development. These eggs were not stolen from any lobsters in the trap, but under what circumstances they were obtained one can easily conjecture. The egg-lobster is undoubtedly a shining mark, not only for predaceous fishes but even for members of its own species. The larger mollusks are eaten by crushing the shells and picking out the soft parts, while many of the smaller kinds are swallowed entire, and presumably pulverized in the gastric mill. Echinoderms probably enter largely into the diet of the lobster wherever they abound. Parts of the common starfish (*Asterias forbesii*) and rarely a few spines of the sea urchin (*Arbacia punctulata*) were detected, but it might be that the latter were swallowed together with other calcareous fragments. Very little change in the food was noticed during the winter and spring months, and there was little evidence that the appetites of these animals sensibly abated during cold weather, yet it is probable that food if not less abundant is less necessary in winter.

That lobsters catch fish alive there is no doubt, but few observers have ever seen the feat performed. Fish that inhabit the bottom, like the flounder, would naturally fall an easy prey to the powerful claws of the lobster, which is said to catch the sculpin; and I have known a lobster when confined in an aquarium to seize and devour a sea robin (*Prionotus evolans*).

While lobsters are great scavengers, it is probable that they always prefer fresh food to stale. Some fishermen maintain that there is no better bait than fresh herring. Fresh codfish heads, flatfish, sculpins, sea robins, menhaden, and haddock are also used, as well as salted fish. The flesh of sharks was occasionally used by the Gay Head fishermen on account of its firmness and lasting qualities. Nothing could be more offensive to the human nostril than the netted balls of slack-salted, semi-decomposed herring, which are commonly used as bait on the coast and islands of Maine, but by the wonderful chemical processes which are continually going on in the laboratory of its body, the lobster is able to transmute such products of organic decay into the most delicate and palatable flesh.

Lobsters are very fond of clams, as they are of mollusks of all kinds, and when kept in pounds are constantly scoring and digging up the bottom in search for these shellfish. In a large lobster pound at the Vinal Haven Islands I have seen the muddy bottom scored in all directions, the work of lobsters in their search for clams. One was reminded of a pasture in which the soil had been rooted up by pigs. As a fisherman remarked, if you put lobsters in a pound and do not feed them they will soon turn over the bottom as effectively as it could be done with a plow. Some of the holes which the lobsters had made in digging clams were 2 feet in diameter and 6 inches or more in depth. Here they had dug up the eel grass, or loosened it so that it had floated to the surface, and cartloads had been cast ashore. We have already seen that the lobsters sometimes eat parts of this plant, but they had plainly rooted it up in this case with another object in view. The broken and often comminuted shells of the long-necked clam (*Mya arenaria*) could be seen strewn everywhere about their excavations.

The lobster probably attacks such large and powerful mollusks as the conchs, which live upon hard bottom in deep water, and devours their soft parts. An illustration of this was afforded in an aquarium at Woods Hole in the summer of 1892, when a conch (*Sycotypus canaliculatus*) was placed in the same tank with a female lobster which was nearly 10 inches long and which had been in captivity about eight weeks. The conch, which was of the average size, was not molested for several days, but at last, when hard pressed by hunger, the lobster attacked it, broke off its shell, piece by piece, and made quick work of the soft meat.

If a lobster that has fasted for a number of hours is fed with a little fresh meat, such as a piece of clam or fish, the process of feeding will be found to be one of no little interest. The lobster eagerly seizes a piece of food with the chelæ of the third and fourth pairs of walking legs, and passes it up to the third pair of maxillipeds, which are held close together, each being bent at the fifth joint and folded on itself. With the third maxillipeds thus pressing against the mouth, the food is kept in contact with the other mouth parts, all of which are in motion, and their action is thus brought to bear upon it. By means of the cutting spines of the appendages external to the mandibles—chiefly the maxillæ and second pairs of maxillipeds—the meat is as finely divided as in a sausage machine, and a stream of fine particles is passed on toward the mouth, to be finally subjected to the cutting and crushing action of the mandibles before entering it.

If one wishes to watch the movements of the complicated mouth parts more closely, one has only to take a lobster out of the water, place the animal upon its back, and when it has become sufficiently quiet stimulate the mandibles or the broad plates of the second pair of maxillipeds with the juice of a clam or the vapor of ammonia, which can be squirted with a pipette. Masticatory movements are immediately set up in the appendages, those belonging to the side stimulated usually working independently. The two small chelate legs are also drawn up rapidly to the mouth, as if to hand up pieces of food.

When stimulated in this way, the plates of the first pair of maxillæ come together over the lower posterior half of the mandibles. The movements of the masticatory parts of the second maxillæ are synchronous with the beating of the scaphognathite. These leaf-like plates project somewhat obliquely over the convex surfaces of the jaws, and are directed inward and slightly upward. The large plates of the first maxillipeds work up and down and at the same time inward toward the middle line, describing an ellipse. The second pair of maxillipeds move alternately or together, inward and outward, with slight up-and-down movement. The large maxillipeds move together, the toothed margins meeting like the jaws of a nutcracker, while the three terminal joints are bent inward and somewhat downward, as in the case of the second maxillipeds, so as to meet on the middle line below and hold the food up to the mouth. (For full analysis of the mouth parts, see ch. vi, p. 227.)

CANNIBALISM.

Lobsters are cannibals from birth, owing, primarily, to their strong instinct of pugnacity. The small, as well as the large, are ever ready to prey upon those still smaller or weaker than themselves. This is certainly true of all the lobsters which have been kept under observation in the restricted space of hatching jars or aquaria and where suitable food in suspension was either lacking or insufficient. In their natural environment in the sea, however, where the young are quickly and widely dispersed, opportunities for the display of this tendency could seldom arise. In the early stages, at least, it is questionable whether cannibalism would occur under any conditions, provided the larvæ were properly fed.

When crowded in cars or pounds, lobsters play the rôle of cannibals at a great rate. As Mr. F. W. Collins remarked to me in 1902, persons not understanding this will lose 20 per cent of their stock in a very few days. He usually counted on a loss in crowded cars of 5 per cent in the course of three days, the larger feasting on the smaller, even when the precautions of supplying them with food and separating the "soft shells" had been duly taken.

REVIEW OF THE INSTINCTS AND INTELLIGENCE OF THE ADULT LOBSTER.

The instincts of fear and of concealment by burrowing or hiding in seaweed or under stones; the restless activity of the lobster in exploring the bottom for food, feeling its course by whipping the water with its long antennæ, and testing all objects with both these and its sensitive feet, or smelling its way to food by beating its antennules, even seeming at times to stalk and approach its prey by stealth; storing up food or, at least, dragging dead prey into its burrows or sometimes burying it, to be afterwards exhumed, thus recalling a well-known trait of the dog; the fighting instinct so often displayed between members of its own race and not confined to captives, which brings into play all its caution and characteristic attitudes in attack and defense; its incessant activity at night whether in search of prey or not; its irregular migratory movements to secure, it may be, a rocky bottom where food and better places of concealment abound, deeper or warmer water, or, in a word, those conditions which for the time suit better certain individuals of one or the other sex for feeding, spawning, or shedding the shell—these may all be observed in either free or captive animals.

In every movement the lobster is guided chiefly by the chemical sense and that of touch, and, least of all, by its eyes. Thus vision, which is never keen, is probably almost nil in bright lights. This explains its nocturnal activity and its frequent retreat from light to shadow.

Of the habits of the European lobster, Williamson (*282*) remarks that it has the sense of light and shade, that it will test a strong shadow with its antennæ, and will even jump at it with outstretched and snapping claws. It is guided mainly by its antennæ, with them finding and exploring every cavern, and with them searching its depths before entering or inserting a claw. As I shall point out elsewhere, the wary lobster, "tiptoeing" over the bottom, feels its way at every step. If food is thrown to the captive, no appeal is made to its sense of sight. The bait remains unnoticed unless it happens to touch one of the antennæ or legs; but a lively whipping of the antennules seems to announce the awakening of the chemical sense. The lobster immediately takes notice and begins to explore the water with its long "feelers," at first without leaving its hole. The antennules begin to whip in the direction of the food and explorations become more active. The lobster cautiously leaves its hole, goes straight for the bait, feeling its way. The food is usually picked up and handed to the mouth parts by the second pair of legs.

Meanwhile, says Williamson, the expected feast has by association stimulated the maxillipeds, which are actively working as if they were already masticating the food. Once this is seized it is conveyed to the maxillipeds and the lobster retreats to its hole, there to enjoy its meal. Two lobsters were noticed to have stored up in one case some mussels and in the other a dead sand eel (*Ammodytes tobianus*) in the inner recesses of their caves.

In regard to the interesting question of storing food, we give the account of a lobster which was kept at the Rothsay aquarium in England (*302*):

A flounder was unintentionally left in one of the aquaria, in which three lobsters were living. The largest animal immediately appropriated the fish, which was then dead, and buried it beneath a heap of shingle, over which it mounted guard. Five times within 2 hours was the fish unearthed, and as often did the lobster shovel the gravel over it with his huge claws, each time ascending the pile and turning his bold defensive front to his companions.

To this catalogue of instincts we must add the parental instinct of the mother lobster in protecting her cargo of eggs during the long period of fosterage. The parental instincts of birds are, as a rule, far keener than in the invertebrates; but it should be added that in many of our commonest birds they endure for a time which is only an eighth or a tenth as long. Through her inbred caution the mother lobster saves not only herself but her progeny from many a strong and clever adversary. Barring the fisherman's trap, she will run the gauntlet of daily life, escape a thousand perils, and after 330 days or more of successful fosterage deliver her young to the teeming and merciless sea. She shows this parental instinct not only by keeping to cover but by folding her tail in emergencies, so that the inquisitive cunner and insidious eel and other troublesome neighbors can not pick off her eggs or pull them out of her brood pocket. Further, by the incessant beating of the egg-laden swimmerets, the lodgment of destructive parasites is discouraged. The lobster also instinctively cleans her antennæ by drawing their whips through the brushes of the great maxillipeds and applies the "broom," the tips of the last pair of slender legs, to the swimmerets and under side of the tail when ready to deposit a new batch of eggs.[a] Sexual union is largely, if not wholly, indiscriminate, and it is possible that the males "try" every lobster which they meet, or at least every female, whatever her condition (see p. 303).

Lobsters about to molt, and possibly after the shell is cast, often conceal themselves in sand or seaweed, and the soft lobster will instinctively eat its own cast or swallow a miscellaneous mass of calcareous fragments, presumably for the purpose of obtaining an immediate and abundant supply of lime for the hardening of its new shell (see p. 185).

Most important to the welfare of the lobster race no doubt is the instinct of fear upon which all their characteristic actions of burrowing, hiding, and what we have described as "stealth" and "caution" depend. Moreover, it is as important for the life of the young as of the adult, for this instinct manifests itself with comparative suddenness, as in birds, at the close of the larval swimming life, in the fourth-stage lobster, when, as if by magic, the lobsterling casts aside its larval habits, together with its characteristic larval organs, and appears in a new rôle, with new armor to suit the part which it is to play. It betrays fear and caution, and now goes to the bottom, digs burrows, and hides. The possession of the instinct of fear gives ground for the hope that the method of rearing the young to the fourth or fifth stage before liberation, which has met with complete success, may yet furnish a means of restocking our coastal waters, and of thus reviving the decayed lobster fisheries of the northern Atlantic States.

The intelligence of the lobster is shown in its power of associating things with actions or of forming habits in the technical sense; in other words, in a power, however limited, of profiting by experience. Thus the lobster habitually returns to its burrow or place of hiding, which it recognizes and claims as its own, being ready to fight for its possession. There can be little doubt that it finds its way back by the same process that the fox returns to its hole or the bird to its nest, through the power of association, though not necessarily through the mediation of the same sense.

But this rudimentary power of using experience as guide does not carry the lobster very far any more than it does many of the fishes and lower vertebrates generally. It does not enable it to escape from a trap or to avoid this engine of destruction in the future when once set free.

[a] It may be noted further that Coste, who made some remarkable statements about the European lobster which are not confirmed by later observers, says that "In order to favor incubation the brood lobsters can expose at will their eggs to the light or keep them in shadow, according as they bend or straighten their tails; when assuming the latter attitude they will now bring their eggs to rest, or now wash them by gently moving the swimmerets." (*55*, p. 204.)

COLOR IN THE ADULT LOBSTER.

The color of the adult lobster is due primarily to the presence of pigments, either in a state of solution in the blood or in the form of granules in the protoplasm of certain cells, particularly the chromatoblasts, which lie beneath the cuticular epithelium. The chromatoblasts are richly supplied with blood, which flows in a system of irregular sinuses through the spongy tissues underlying the epidermis.

In the adult lobster the hard shell is an opaque lifeless substance, and the pigments to which it owes its characteristic coloring are excreted by the chromatoblasts of the soft underlying skin. These are immediately exposed upon removing the shell. The delicate skin is seen to be flecked or mottled with scarlet, and with the aid of a simple magnifying glass it is readily perceived that its color is due to branching pigment cells, groups of which correspond to the blotches of color on the shell itself. The excreted pigments undergo physical and possibly chemical changes in the hard cuticular shell and may thus come to differ markedly in color from the parent chromatoblasts. Since the colors of the lobster reside in a lifeless body, the pigment layer of the shell, it is evident that no changes of a vital nature can take place after this is definitely formed.

The coloration of the lobster is fairly uniform in plan, but extremely variable in details, even more so than we find in the case of the color patterns of many insects. The brilliancy and purity of the shell pigments depend largely upon the age of the shell or upon its condition with respect to the molting period. These pigments are usually most brilliant just after the molt, when the cuticle is thin and translucent, and dullest before ecdysis begins, when the old shell still encumbers the body.

The pigment cells themselves, as we have seen, reside in the soft skin, and when the shell is once hardened the color of the animal is more or less fixed and permanent. It is certain, however, that under the action of light and possibly from other natural causes the shell pigments undergo molecular or chemical changes. Men who handle lobsters have frequently observed that when they are exposed in shallow cars to unusually intense light they become decidedly bluer in color.

According to MacMunn (*185*) the coloring of the skin of the lobster is due to the presence of chromogens, which may be converted on slight provocation, as by dehydration, oxidation, or some molecular change, into a red lipochrome resembling rhodophan. Everyone is familiar with the wonderful change in color which the living lobster undergoes when boiled, and according to the same writer the beautiful pigment of the larval lobster is converted by alcohol into a true lipochrome.

Alcohol quickly converts the chromogens in the lobster's shell into lipochromes and dissolves them at the same time. This is seen when a recently molted lobster with brilliant coloring is placed in alcohol for preservation. The soft shell is first reddened, and then in a short time completely bleached, while a hard lobster treated in the same way will retain much of its shell pigment for years, if not indefinitely.

Lipochromogens are found in a natural state in the gastric glands, blood, soft skin (as the blue prismatic cyano-crystals, which are reddened by alcohol or by boiling), and in the exoskeletons of crustaceans generally. MacMunn is of the opinion that they are "built up in the digestive gland and carried in the blood current to be deposited in other parts of the body." If this is true, it would not be remarkable if the color of the animal were affected by the nature of its food, yet this does not seem to be often the case.

Following the classification of Bateson (*19*) we distinguish between (*a*) variations in colors themselves, and (*b*) variations in color patterns. The variation in colors, which Bateson calls "substantive variations," may be the result of a physical or chemical change, and has no vital significance. The different colors themselves are further liable to different discontinuous variations, as when crustaceans occasionally lay bright, golden-yellow eggs, while the normal color is dark green.

The following substantive variations have been met with: (1) Blue lobsters, in which the prevailing color is blue; (2) red lobsters, which are pure red or reddish yellow; (3) cream-colored lobsters, characterized by the almost entire absence of color; and we should also add (4) black lobsters, to include possible cases of melanism, where the colors are extremely dark. A specimen of this kind was reported to me at Beal Island, near West Jonesport, Me., where a fisherman recently captured, in 3 fathoms of water among the eelgrass, a lobster about 6 or 7 inches long with moderately hard shell and almost jet black. He supposed at first that it was covered with coal tar. It did not appear to be preparing to molt. Malard speaks of meeting with cases of melanism in crabs, where in consequence of a lesion of the skin the animal becomes entirely black.

Changes in color pattern are more elusive. There are (1) the normal variety, in which the upper part of the body is mottled with green, blue, and cream color; (2) spotted or "calico" lobsters, the coloration of which is a bold pattern of green and light-yellowish or cream-colored spots; (3) pied or parti-colored varieties, in which the contrast of tints is abnormally pronounced. This may perhaps be better classed under substantive variation. The changes are due apparently to vital or physiological causes, which have at least no adaptational significance.[a]

There is no sexual color variation in the lobster, and such substantive variations as the eggs undergo are not of an adaptive character. The freshly laid egg is dark green, sometimes almost black, due to the presence of dissolved lipochromogens. Occasionally the ova are nearly pea-green, grayish-green, or greenish-straw color, but the golden-yellow variation, so striking in some of the snapping shrimps, has never been observed in the lobster.

If the eggs are treated with hot water, alcohol, or other killing reagents, the green lipochromogen is quickly converted into red lipochrome. When the water is heated gradually, the red color appears slowly, and it is interesting to observe that if these red eggs are now plunged into cold water the green color is restored. This change may be somewhat analogous to the breaking up and reconstruction of the blue compound of of starch and iodine upon the successive application of heat and cold, and to the variation in color which sometimes appears in the living animal at the time of the molt.

[a] For fuller account of red living lobsters and other color variations, with illustrations, see *149*.

Soon after the water has been brought to the boiling point the red color becomes permanent.

The colors of deep-sea animals that live in total obscurity can not be of any utility to the animal as a source of protection. The color may be very brilliant, red, scarlet, orange, rose color, purple, violet, and blue, which is less frequently reported, but they appear to be developed quite independently of the light. It has been shown by experiment with sensitive photographic plates that luminous rays do not penetrate ordinary sea water to a greater depth than 400 meters, as noted above. In depths of 50 fathoms or more there might be an appreciable amount of light on clear days, but even then, when the water was loaded with sediment and the bottom composed of dark materials, it seems hardly probable that colors would have any protective value whatever.

The normal colors of the lobster, which are spread like a mantle over the whole upper surface, tend undoubtedly to obliterate it and to screen its movements while crawling over a weedy or rocky bottom. The absence of all color or a more generous display of bright pigment would make it a more conspicuous object, especially upon sandy bottoms in shallow water, which it is usually careful to avoid in the daytime. The vivid red of the claws appears to be overlaid by a darker pigment in spots, particularly on the upper surface. The underside of the pleon, which rests upon the bottom when the tail is not folded, is very meagerly supplied with pigment, as is usually the case with marine animals which inhabit the bottom.

48299°—Bull. 29—11——13

Chapter III.—GIANT LOBSTERS.

Stories of gigantic lobsters made their appearance at a very early period, and one could probably gather as many exaggerated accounts of this animal now as in the days of Olaus Magnus. Time, however, has narrowed the bounds of credulity, even among the ignorant, and we no longer hear some of the interesting legends which the old writers have carefully handed down. Thus Olaus Magnus tells us in his description of northern lands and seas,[a] published in 1555, that between the Orkneys and Hebrides there lived lobsters so huge that they could catch a strong swimmer and squeeze him to death in their claws. His curious figures were copied by Gesner, who has many others equal to any which are described in the old mythologies.

Giants are met with in all the higher groups of animals. They interest us not only on account of their actual size, but also in showing to what degree individuals may surpass the mean average of the race. It may be a question whether lobsters weighing from 20 to 30 pounds or more are to be regarded as giants in the technical sense, or simply as sound and vigorous individuals on whose side fortune has always fought in the struggle for life. I am inclined to the latter view, and look upon the mammoth lobster simply as a favorite of nature, who is larger than his fellows because he is their senior; good luck never deserted him until he was stranded on the beach or became entangled in some fisherman's gear.

Gesner gives a poor likeness of a lobster, but an excellent drawing of the large crusher claw of one which he had preserved in his collection on account of its great size. The length of this claw was 8¾ inches, and its breadth at the junction of the dactyl about 4 inches, so that it was borne by a lobster which weighed not far from 8 pounds.

The European lobster of to-day seldom or never attains so great size as the American species, as already remarked, and its average weight is considerably less. Buckland gives an account of large lobsters from the British Islands, in which the greatest weight recorded was 14 pounds, and European lobsters of this size are undoubtedly now very rare. The Academy of Natural Sciences of Philadelphia possesses a skeleton of *Homarus gammarus*[b] which, judging from its measurements, must have weighed from 23 to 25

[a] Historia de Gentibus Septentrionalibus, Rome. 1555.

[b] It is possible that a mistake has been made in attributing the Philadelphia specimen to the European species The determination was made by Prof. John R. Ryder, who evidently relied upon the character of the rostrum (see p. 161) in basing his opinion. Regarding this specimen, Professor Ryder wrote under date of March 10, 1894, as follows: "It turns out to be European instead of American. I send the data obtainable. The catalogue does not give weight or locality. At one time there was a label stating the weight; now that has also disappeared." Again on March 15, he wrote: "There is no doubt of the large lobster being *H. vulgaris*. I found no spines on the under side of the rostrum of the large specimen; perfectly smooth, as was also another smaller specimen of the same species. I made a very careful examination to-day and can assure you that the facts are as I state." He further added that the large skeleton "is also perfectly symmetrical and must have been a beautiful specimen originally, as it now is."

pounds. (Table 1, no. 15.) There may also be seen in the museum of Bergen, Norway, a lobster which Prof. S. O. Sars in 1878 described as an "immense specimen," the living weight of which could not have been much over 12 pounds.

Though it has been an accepted belief that the American lobster attains a greater size than its European counterpart, it is possible, in view of comparison of no. 10 and no. 16 of table 1, that the maximum size of each species is nearly the same. The data are not at hand for determining the question with certainty. It seems certain, however, that American lobsters of average or medium size are considerably stockier and have larger claws than the European, and that length for length, such animals will weigh more. The lobster fishery of Europe, though pursued for ages by primitive methods, is still very much older than that of America, and it is probable that the larger lobsters have been more effectually weeded out there than here. At the time Sars's paper was written (*244*) it would not have occurred to one familiar with the American species to speak of a 10 or 12 pound lobster as in any way remarkable, yet at present few of this size find their way to our markets. In fact the same gradual falling off, due evidently to the same cause, has been experienced for many years in Maine and Canada.

TABLE 1.—RECORD OF GIANT LOBSTERS.

[No. 1–14 refer to *Homarus americanus*, No. 15–16 to *H. gammarus*.]

No.	Sex.	Place of capture.	Date.	Length.	Length of carapace.	Crushing claw.		Toothed claw.		Where preserved.	Living weight.
						Length.	Girth.	Length.	Girth.		
				Inches.	*Inches.*	*Inches.*	*Inches.*	*Inches.*	*Inches.*		*Pounds.*
1	......	Gloucester, Mass.	1840	....	....	13	17.50	....	....	Peabody Academy of Science, Salem, Mass.	[a] 28
2	♂	Salem, Mass.....	1850	21.75	9.94	12.50	15	12.37	....	.....do................	[a] 23–25
3	♂	Boothbay, Me...	1868	20.25	9.37	12.50	15.25	13.25	11.12	Land Office, Boothbay Harbor, Me.	[a] 24
4	......	Small Point, Me.	....	....	....	12.50	15.50	....	....	Portland, Me..........	[a] 24
5	♂	Delaware, Conn..	....	....	....	13.12	16.12	12.87	8.50	Smithsonian Institution, Washington.	[a] 25
6	♂	Belfast, Me......	1891	20+1	9+1	13.75	16.87	13.87	12.50	Adelbert College, Cleveland, Ohio.	[b] 28
7	♂	Lubec, Me.......	1892	20.62	9.25	12	15	11.50	11	Campobello Island, New Brunswick.	[a] 23
8	♂	Provincetown, Mass.	1894	20–21	....	12–13	....	....	....	Formerly at St. Nicholas Hotel, Boston, Mass.	[a] 23–25
9	♂	Atlantic Highlands, N. J.	1897	23.75	12.24	15	20.50	15.50	15.25	American Museum of Natural History, New York.	34
10	♂	Newport, R. I.[c]	1898	19.50	....	11.75	....	11.87	....	Rhode Island Commission Inland Fisheries, Providence, R. I.	19
11	♂	Atlantic Highlands, N. J.	1899	22.50	10.28	14.66	17.68	14.40	13.54	American Museum of Natural History, New York.	31
12	♂	Monhegan Island, Me.[d]	1899	23–24	....	....	....	....	....	....	[a] 32+
13	♂	....	....	19.80	....	12.33	15.60	12.40	11	American Museum of Natural History, New York.	[a] 24
14	♂	Near Bayonne, N. J.	1898	20.37	....	14	16	13.50	11.50	....	[a] 25–28
15	♂	Europe..........	....	19.40	9.29	13.10	16.80	12.40	10.15	Museum of University of Pennsylvania, Philadelphia.	[a] 23–25
16	♂	Coast of Norway.	1850(?)	18.73	8.58	10.23	10.62	10.03	8.07	Bergen Museum, Norway.	[a] 12+

[a] Living weight estimated. [b] Living weight estimated from weight when boiled. [c] After Hadley. [d] Body length estimated.

The large Belfast lobster (no. 6, table 1), which came into my possession in 1893, was captured in Penobscot Bay, near Belfast, Me., in 1891. (For full account with photographs see *149*.) Its total length, had the rostrum been perfect, would have been 21 inches. The body seems surprisingly short for so powerful an animal, and it is indeed in the large claws that the greater part of the weight and strength resides. This may possibly be explained by the fact that as age advances the increase in length at each molt becomes less, while there is a corresponding gain in the volume of the body and of the claws. Thus Ehrenbaum mentions a lobster 42.2 cm. long, which showed an increase in length of scarcely 1 mm. on molting. The length of the crushing claw of the Belfast giant is 13.75 inches, and its greatest girth 16.87 inches.

GREATEST SIZE ATTAINED BY THE LOBSTER.

It is difficult to obtain exact data regarding the true weights and measurements of all big animals, and the lobster seems to be particularly deceitful in this respect. Remembering the decision of the judge that "affidavits are not lobsters," I endeavored to take a conservative position on this subject, when writing in 1895 (see *149*, chapter v). Fortunately since that time two specimens of the mammoth class have been added to the collections of the American Museum of Natural History in New York. Through the kind offices of the museum I have been able to obtain data and to present a sketch of one of the biggest known lobsters in the world. The larger (no. 9, table 1), when received in the fresh state, weighed, according to Whitfield (*278*), "about 34 pounds;" the weight of the smaller (no. 11 of table) is given as "about 31 pounds." Both were taken alive by fishermen off the Atlantic Highlands in New Jersey in the spring of 1897. The larger animal was exhibited in one of the tanks at the Castle Garden Aquarium, but neither lived more than a few days in captivity. Both specimens have been remounted at the museum, the smaller to show the upper (fig. 1) and the larger the under side.

The most important measurements upon which we can rely for exact comparisons are: (1) The length of the carapace from the tip of rostrum to hinder border, (2) the length of each of the big claws, taken with callipers from the short spur near the proximal end of the larger division of the claw to its apex,[a] and (3) the greatest girth of the propodus, measured in a line at right angles to the last. These values should be fairly constant by whomsoever made, and in whatever form the skeleton is mounted.

Knowing the measurements in the American Museum specimens to be correct, and assuming that the weights as given by Whitfield are correct also, I have taken these data as a new basis for estimating the weights of other large lobsters recorded in table 1, and believe them to be a closer approximation to the facts in each case than I was able to make in 1895. The former estimates were founded on the measurements and supposed weight of the Belfast lobster (no. 6, table 1), the largest specimen known at the time. I was assured that this animal weighed 23 pounds after it had been boiled, and allowing a shrinkage of 40 per cent in the process, its living weight was estimated at 28 pounds. Notwithstanding the doubts cast upon this statement at the time, com-

[a] Or from the spur near the proximal articulation to apex of propodus, the last measurement being somewhat less. Where big claws are chopped off for preservation, the joint is apt to be defective.

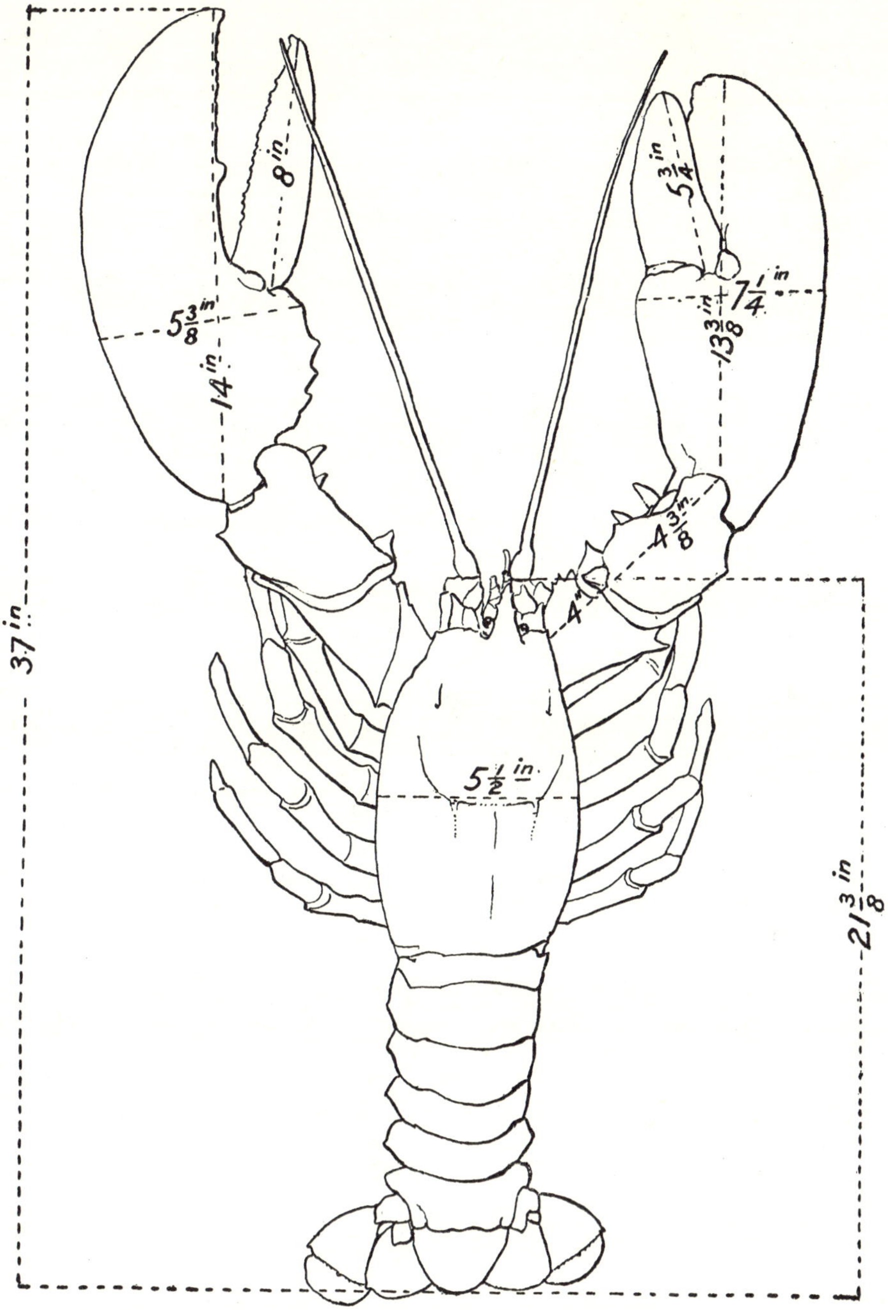

FIG. 1.—Giant lobster from New Jersey; living weight, 31 pounds. (See table 1, no. 11.) Outline after photograph, and reproduced through courtesy of the American Museum of Natural History, New York. General measurements of skeleton as mounted, indicated in inches. About one-fifth natural size.

parison with the measurements of lobsters 9 and 11 shows that it must have been substantially correct. It will be seen that this animal approaches closely the 31-pounder from New Jersey, the lengths for the carapace being 10 inches (allowing for 1 inch of the rostrum missing) and 10.28 inches, respectively, and the girth of the crusher claw 16.87, as opposed to 17.68 inches.

After taking account of the facts so far as ascertainable at present, my former statements regarding the weights of giant lobsters are revised to the following effect; the greatest known living weight of the American lobster is 34 pounds and that of the European lobster about 25 pounds. (See note, p. 194.) Altogether six or seven individuals of the American species weighing 25 pounds or more are known to have been caught on the Atlantic coast during the last 70 years.

The lobster (no. 12, table 1) which was seen by Cobb at Peak Island, Maine, in 1899, is said to have measured 44 inches with claws extended in front of the head. It was caught off Monhegan Island, Maine, and exhibited about the country by fishermen of that region. If this measure was correct, it would correspond to a body length of 23 to 24 inches and a corresponding weight of upward of 32 pounds, thus being one of the largest lobsters on record. The ratio of body length to the total length with extended claws varies greatly in small and large lobsters, being as high as 72 per cent in a female of 3 inches and 38 per cent in a male of 10.37 inches, while in the big Belfast lobster (no. 6) this ratio is somewhat under 55 per cent. On the other hand the ratio of carapace length to total body length for the average 10.5-inch lobster, as applied in the gauge law adopted in Maine, is 45 per cent (see chapter IV, p. 212).

In addition to the lobsters given in table 1, Cobb (*52*) has noticed a male said to have measured 25 inches and to have weighed 25 pounds. It was caught in a hake trawl off the Matinicus light, Maine, at a depth of 60 fathoms, in 1898. The given length in this case does not accord with the given weight, and is probably much too great. Another lobster is mentioned by Hadley (*126*) as having a length of 22.5 inches, but weighing only 19.5 pounds; the same kind of difficulty is presented here, the length calling for a much heavier individual. Waite (*274*) has also recorded the measurements of a large male lobster, which was captured at Block Island April 10, 1896, measured 21 inches and weighed when alive slightly over 22 pounds. The length and girth of the cracker and toothed claws were 13.25 and 16.5 inches, and 12.75 and 12.25 inches, respectively.

In June, 1898, Dr. H. M. Smith called my attention to a large lobster which had been recently captured in New Jersey and which was reported to have measured 23 inches in total length and to have weighed 36.5 pounds. Through the kindly aid of the late E. G. Blackford of the Fulton Market, New York, we were able to obtain a reliable account of this interesting specimen, together with the necessary measurements, which are given in table 1. This lobster was caught on June 21, 1898, by a fisherman in New York Bay, off East Forty-sixth street, near Bayonne, N. J., and was taken alive to the Bayswater Hotel, where it was on exhibition in a tank for several days. The man who was sent by Mr. Blackford to take the required measurements found that the animal was then dead and partly dried out, the owner claiming that it had shrunk 2.62 inches in consequence. It is hardly necessary to show that this was impossible, since the body of a lobster can be distended at only one point, namely, at the articulation of the carapace and the tail, and there only to the limit of the articular membrane, which is inelastic. Drying would tend rather to contract this membrane and to give more accurately the true length, but the difference would not in any case be very great. The measurements taken from the dead shell show that this animal probably did not weigh over 25 to 28 pounds. In his letter to Dr. Smith Mr. Blackford remarked that the owner asked the modest sum of $250 for the specimen. We do not know what finally became of it.

In August, 1891, according to Mr. F. W. Collins, a lobster of undetermined sex was caught at Blue Hill Falls, Maine, which weighed 18.5 pounds, and in November, 1892, a perfect female lobster weighing 18 pounds was taken at Green Island, Maine. This outer island has long been noted for its fine lobster fishing. Mr. Collins stated that in August, 1891, he had 50 lobsters at one time in his establishment which would weigh from 10 to 18.5 pounds. About half of these came from Castine and the remainder from Blue Hill Falls. All of them were "new shell lobsters," or those which had shed in the year, probably in July.

After the lobster has attained a length of 20 inches and a corresponding weight of 23 to 25 pounds or more, we may be certain that the stage periods, or intervals between each molt, are long, and probably several years apart, and that this interval is gradually increased with advancing years. The relative increase in length seems to slow up with increasing age, but volumetric increase still goes on, and the animal becomes stockier and its big claws more powerful. There is no fixed limit to age, growth, or molting power, but the practical limit is probably not far from that of the largest animal on record. Whether giant or pigmy, the fighting strength is apparently renewed at each molt, when a brand new suit of armor is acquired.

The shell of the crusher claw of the Salem lobster (for full-sized drawing, see *149*, pl. 15) weighed but a trifle over a pound, and the living weight of this animal is now estimated at about 28 pounds. The skeleton of the crusher of a 12 to 15 pound lobster with very dense shell weighed 8.25 ounces. The Salem lobster had probably molted within less than 3 months from the time it was caught. The Lubec lobster (no. 7) had a clean shell, which indicated that not over 6 months had intervened between the time of its capture and the last molt. It was light for its length and the most perfectly proportioned large specimen I have seen.

In general it is undoubtedly true that the older the adult lobster the longer its stage periods and the less the increase at each molt. Yet it is almost equally certain that both may vary greatly in the giant as in the pigmy. At present our data regarding the molting of large lobsters is insufficient to enable us to estimate their age. Giants weighing from 25 to 35 pounds have possibly weathered the storms of life for half a century or more.

Chapter IV.—MOLTING.

Molting is an incident and expression of growth. The crustacean does not "grow by molting," as is sometimes said, but it molts because it has grown. It has outgrown its inelastic shell, which is cast off in one piece, normally without a break in any of its hard parts. Other animals molt or shed a part of their cuticle and its products, but nowhere is the process so striking, so abrupt, or so critical as in the higher crustacea. In these animals the span of life from infancy to old age and death may be divided into a series of stages, varying in length, each stage-period of life culminating in a molt.

Any influence which retards growth or unduly taxes the vital energies prolongs this period, and conversely the more vigorous and the more rapid the growth the shorter the interval between molts. Shortly after molting the body increases in size, probably in part through the absorption of water, but this expansion should be distinguished from the change that has already taken place, which is due to cellular growth, and is the primary cause of the molt. Thus in molting the animal parts with its old shell or epidermic exoskeleton at one stroke, and presently attains to greater size.

Molting begins on the second day after hatching and lasts throughout life or at least as long as there is any growth. The first three molts are passed in from 12 to 15 days. From first to last the cuticle is cast as one piece (excepting only the gastroliths), the animal escaping through a rent of the membrane between the tail and back. In healthy young animals molting lasts but a few minutes, but at all times the process is critical and it is frequently fatal. It often leads to the distortion or the loss of limbs and to a variety of deformities such as duplications of a limb or of its parts.

It is difficult to avoid repetition in dealing with the molting process since it has modified the habits of the animal at so many points, but we shall now consider the subject in regard to the adult animal as a whole. In order to understand the process it will be necessary to examine the structure of the shell and of the soft skin, of which the former is a product.

THE SKIN AND SHELL.

The skin as a whole is composed of the soft dermis, the soft epidermis, and the shell or cuticle which the latter secretes. The epidermis is typically composed of a single stratum of chitin-producing cells, and often rests upon a thin basement membrane, which then forms a distinct boundary between the two layers and like the outer shell is a cuticular product. The dermis is composed of connective tissue cells, which are often attached to the basement membrane, blood vessels, nerve fibers, pigment cells, and glands, which are apparently of epidermic origin. Wherever muscles are attached to the shell, the epithelium is greatly modified or reduced (see ch. VI, p. 241). The shell in sectional view shows four layers, namely, (1) a thin outermost stratum, which is structureless, called the enamel layer; (2) an underlying and lamellated pigment layer, transversed by vertical canaliculi, abounding in pigment and impregnated with mineral salts; (3) the calcified layer proper, devoid of pigment, but otherwise like the last, and forming the greater part of the shell substance; and (4) a noncalcified inner stratum composed of very thin lamellæ.

The chitinogenous epithelium may be compared to the Malpighian layer of the epidermis of the vertebrate, while the layers of chitin represent its horny cuticle, though formed in a different manner. The vertical canaliculi of certain decapods, according to Vitzou (*272*), correspond to the boundaries of the epidermic cells, but this is not the case in the lobster, where they are close together and very numerous.

During the molting period the cells of the chitinogenous epithelium undergoes a great change, its cells being extended vertically into very long and slender rods (pl. XLVI, fig. 2). The epithelium developed over the surface of a budding limb is of a similar character. The chitinous layers of the new shell are formed by discontinuous thickenings of what, according to Vitzou, may be regarded as the upper wall of the epithelial cell. Thus are formed parallel lamellæ of varying density, which fuse with those of adjoining cells and make a continuous shelly crust.

At the time the shell is ready to be cast the tegumentary coverings consist of (1) the old shell, (2) the new shell, (3) an intermediate structureless membrane, besides the chitinogenous epithelium, and (4) the dermis. The new carapace, according to Vitzou, is composed of the enamel and pigment layers only. The calcified layer is not formed until after the molt.

Certain peculiar cells which have been referred to as connective tissue become very conspicuous at the molting period, particularly in the dermis, and experimental evidence seems to show that they secrete glycogen which is used in the production of the new shell, but no exact knowledge concerning these structures is available at present. The enamel layer is the first formed, and when once laid down can not be removed except by the shedding of the entire shell. However, it is worn away by abrasion, as seen in the old hard-shelled animals, and its function is purely protective.

The surface of the shell has a punctate appearance, due to hair-pores, which mark the points where hairs or setæ now pierce the shell or where they were present at an earlier stage of development. In the adult lobster the setæ of the carapace have disappeared more or less completely except upon its margins and in the orbital region.

The dense shell of this animal is in reality a veritable strainer, being perforated by hundreds of thousands of minute passages, which lead from the surface to the parts below it—to the tegumental glands on the one hand or to the sensory cells which lie at the roots of the hairs on the other.

PERIODS, CONDITIONS, AND SIGNIFICANCE OF MOLTING.

The hard-shell lobster is heaviest, has the firmest flesh, stands transportation best, and is therefore most valuable for the market. A large percentage of all lobsters taken during the fall and winter months are of this character, and nearly all lobsters caught in March, April, and May belong also to this class. Shedders and soft-shell lobsters are taken in greater or less abundance from June to October, varying somewhat with the season and surrounding conditions, such as the nature of the sea bottom and the temperature of the water. By far the greater number of lobsters cast their shells during the months of July, August, and September. The time of shedding, however, varies considerably on different parts of the coast, being from 4 to 6 weeks earlier in some seasons in western Maine than in the extreme eastern section. Shedders are not fit for the market, being lean and watery, and soft lobsters are in a similar condition and will not bear much handling or transportation. Until the shell becomes tolerably hard the soft lobster is easily wounded and killed. Lobsters with very soft shells and those that have been mutilated are often kept in the lobster preserves or pounds until the shell is hardened or the injury repaired.

Traps set by Mr. Vinal Edwards at fixed points on the rocky bottom in the harbor of Woods Hole, Mass., for a period of 7 months, from December 1, 1893, to June 30, 1894, were daily hauled and the conditions of the shell of each lobster noted. The significant data thus obtained were as follows:

TABLE 2.—DATA FOR LOBSTERS EXAMINED AT WOODS HOLE, MASS., WITH REFERENCE TO MOLTING CONDITION.

Number of lobsters caught.		Lobsters recently molted or preparing to molt.	Shell hard and dull.	Shell soft.
Males	1,313	77	33	44
Females	1,344	33	7	26
Total	2,657	110	40	70

Of the entire catch, 110 lobsters had either recently molted or were preparing to molt; 77 of these were males and 33 females. The total number of males was smaller, yet the number of soft shells among them was nearly twice as great as in the other sex. This fact implies that the males molt oftener than the females, which would be an *a priori* deduction from the greater size which the male attains, or that they molt more frequently during this period, assuming that the distribution of these animals was uniform for the time and place.

In the fullest sense the molting process consists of two distinct phenomena: (1) The formation of a new shell and (2) the rejection of the old. When once formed the shell admits of no increase in size, since it is a dead structure, excreted by the soft skin below it, and when it is outgrown it must be cast off and give way to a new and larger covering. The new shell is gradually secreted under the old one, and when the latter is discarded the new cuticle is soft and flexible, so that it is easily distended to meet the requirements of growth. The growth of the lobster, as of every arthropod, is thus measured by a series of stages characterized by the growth of a new shell under the old, by the shedding of the outgrown old shell, a sudden expansion in size, and the gradual hardening of the shell newly formed.

Not only is the external shell cast off in the molt, along with the linings of the masticatory stomach, the esophagus, and the intestine, but also the internal linkwork of hard tendons described in chapter VI. The sloughing of the latter is rendered possible, first by the presence of absorption areas and secondly from the fact that the inner skeleton is in origin an infolded part of the cuticle; in molting the lobster withdraws its soft body from the mold of its old and hardened skeleton. It is thus easy to see why the molting act is a continually recurring crisis in the life of the decapod crustacean, for it is both dangerous and expensive, not only calling for a considerable excess of energy, but demanding that a long series of preparatory changes, to be later considered, must be exactly executed. Since it is dependent upon the condition of the individual, which is subject to wide variation, the molt does not take place at any stated time, but may occur in any month of the year. In general, molting in either sex is rare in winter and spring and most frequent in summer. Warmer weather, a more active life, a greater abundance of food, and a more vigorous appetite, which are characteristic of the lobster or its environment during the warmest part of the year, are most favorable to the renewal of the shell. The lobster, though a carnivorous animal, feeds less in winter, when its habits are relatively sluggish. Broken limbs and injuries to the shell are then but slowly repaired, and there is less energy to be drawn upon in molting.

As a rule, the adult female that lays her eggs in August of any given year carries them for 10 or 11 months, until they hatch in the succeeding June. Since the spawning periods are 2 years apart, Hadley (*126*) infers that the molting periods can not oftener occur and that the rate of growth in the female is consequently diminished. In average cases this rule may hold, but exceptions occur. Thus, I have recorded two cases (*149*) where soft-shelled lobsters with eggs were taken in which the molt could not have preceded ovulation by more than two or three weeks; still further, in exceptional cases, a second molt may possibly take place in late autumn or in the early winter, following the hatching of a brood.

It is several weeks before the new envelope becomes as hard as the one rejected, so that the lobster is, for a large part of its life, either preparing for a molt or recovering from one. Therefore it is not remarkable that lobsters have acquired many popular names among fishermen, such as "hard shell" or "old shell" lobster, "shedder," black shell," or "crack shell" (lobster preparing to molt), "soft shell," "new shell," "shadow," "rubber shell," "paper shell," "buckle shell" lobsters, etc. (animals which have recently molted).

Shedders can be readily distinguished by the dark, dull colors of the old shell, hence the common name of "black lobster," and by the deep reddish tint of the membranes at the joints, where the flesh is seen through the old and new cuticulæ. The lobster is now naturally sluggish and takes but little food, but it can not be said that the shedder never breaks its fast. It is not a very unusual experience for the fisherman to take both the soft lobster and its cast from his traps. When in this condition lobsters commonly haunt shallow water, with a sandy, muddy, or weedy bottom, and at low tide have been taken out of bunches of eelgrass at a depth of a few inches only. They frequently dig a shallow hole in the mud under stones, where they can await the coming change with greater security from enemies. Fishermen have frequently seen a cast shell lying on the bottom and have found a soft lobster near by, protected by a rock or bunch of kelp.

Many of the prawns habitually molt in the early morning while it is yet dark, but lobsters which we have kept in aquaria have cast both by day and at night. Considering the nocturnal habits of the lobsters, we should expect to find the latter practice the commoner in a state of nature. In those captives which Brook (*37*) observed with great care, the shells were cast off in the night time and partially buried.

Anderton (*5*) found that the lobsters transported from England to New Zealand molted mostly at night, their cast shells being usually seen lying upside down on the bottom. The shedders retired to some secluded spot where the water was shallow, and appeared vicious upon the approach of intruders. On the 3d of September, says Anderton, "a male lobster was seen to be behaving in a very peculiar manner in the shallow end of the pond. It would walk alongside the concrete dividing wall for a distance of about 5 feet, halt, and then turning round would retrace its steps the same distance in the opposite direction. In this manner a rut several inches deep was formed in the gravel and at one end of this the lobster scooped out a hole about 4 inches deep and 12 inches in diameter." The water had to be temporarily withdrawn from the pond, but as soon as permitted to do so this lobster resumed its peculiar walk, and continued it through the night and the following day. Molting began at 4.30 p. m. of that day and lasted 35 minutes. The lobster at first lay on its side, with its large claws extended in a direct line with its body, and later turned on its back when the tail, the last part to be withdrawn, was released. The habit of scooping a hole in the gravel was noted on several occasions, when the soft lobster was found lying beside its "shadow." As noted in chapter IX, molting in the females was almost immediately followed by copulation, whenever a male was available, and the interval between this act and the laying of the eggs was in two cases observed—65 days. Molts in both sexes were recorded from November 18 to March 3, but rather more frequently in the warm months of November and December.

THE MOLTING ACT.

A male "shedder" was caught in the harbor of Woods Hole July 13 and placed in an aquarium. At exactly 2.48 p. m. this lobster began to molt and in 6 minutes was out of its shell.

When the lobster is approaching the critical point the carapace or shell of the back gapes away a quarter of an inch or more from the tail. Through the wide chink thus formed the flesh can be seen glistening through the old and new cuticle, giving it a decidedly pinkish tinge. Take the lobster up in the hand now and the tail drops down as in death, the strong muscles which bind the pleon to the carapace being completely relaxed. When this stage is reached the time of exuviation is at hand and the process becomes purely automatic, the animal having no control over its own movements.

The period of uneasiness, which foreshadowed the molt and was very marked, ended in this lobster by its rolling over on its side, briskly moving its legs, and bending its body in the shape of the letter V, the angle of the V corresponding to the gaping chink between the dorsal shield and tail. Presently the old cuticle, holding these parts together, began to stretch, the wall of the body pressing against it with considerable force, and the hinder end of the shield being slowly lifted up, while its anterior part remained attached to the rest of the skeleton. The slow but sure pressure of the parts within cause an increasing tension in the yielding cuticular membrane, which finally bursts, revealing the brilliant colors of the new shell. The legs and other appendages are occasionally moved, but no marked convulsive movements are to be seen. The carapace has now become raised to an elevation of perhaps 2 inches in its hinder part, in consequence of which, the anterior end being fixed, the rostrum is bent downward and the animal presents a very singular appearance.

When this stage has been reached the lobster becomes quiet for a few seconds and then resumes its task with renewed vigor. From this time on until free its muscles work intermittently. The doubled-up fore part of the body, with each effort of the animal, is more and more withdrawn from the old shell, and this implies the separation of the skin from the intricate linkwork of the internal skeleton, and particularly in its release, together with a part of the nerve cord, from the closed archway of this structure, as well as the freeing of the 28 separate appendages from their old cases and tendons, for the accomplishment of which special adjustments are made in advance. The cuticular sheath of every ectodermic structure is stripped off. The exoskeleton folded to fit so complicated a mold is virtually a continuous structure, and from the method of its regeneration the sloughing of one part necessitates the shedding of the whole.

The carapace is now elevated to such an extent from behind that the rostrum is directed obliquely downward and backward. The lobster is still lying in comparative quiet upon its side, but the muscles of all its appendages are undergoing violent contraction as the animal tugs and wrestles violently as if to free itself from ropes which bind it down firmly on every side. The carapace is unbroken, yet the two halves bend as upon a hinge along the median line, where the lime of the shell has been absorbed. Presently the pressed-down bases of the antennæ, the eyestalks, and the bent-down rostrum of the new shell can be clearly seen. No part of the covering of the large claws or of any of the legs have been split or cracked. The muscular masses of the powerful claws have been withdrawn through their narrow openings without a rent. Finally a few kicks free the entire forward half of the body, the antennæ, chelipeds, and varous other parts, which now lie above or to one side of the old covering. The tail has been gradually breaking away from its old case, and as soon as the forward part of the body is withdrawn the lobster gives one or two final switches and is free.

The newly molted lobster has a very sleek and fresh appearance, and its colors were never brighter or more attractive. Try to take it up in the hand, after some time has elapsed, and it feels as limp as wet paper; but immediately after casting the shell the muscles of the crustacean are hard and tense, probably from being in a state of cramp or tetanus. Every part of the old shell down to a microscopic hair has been reproduced in the new one, but in the latter the fringes of stiff setæ are as soft as silk, the stony ends of the claws, the rostrum, and every spine of the body so soft as easily to bend beneath the finger. Possibly the hardest parts of the newly molted lobster are the horny surfaces of the teeth of the stomach sac. The large claws are considerably distorted, as well as some of the other parts, being compressed and drawn out to an unnatural length. After getting clear of the old shell the animal is not inclined to activity. It soon orients itself, however, resting in the usual way, and is capable of moving about with some degree of agility by the flexure of the tail. Fishermen who have had lobsters shed in their cars or traps have often been surprised by the ease with which they sometimes slip through their fingers.

The length of the cast shell of this lobster was 11.25 inches, and shortly after the molt the animal measured 12 inches from tip to tip. On July 17, four days after molting, the length was a little short of 12.5 inches. The increase in length was thus very nearly 1.25 inches. Very soon after molting the lobster is ready to take food, the body plumps out to its natural shape, and no further increase in volume can take place until another molt.

The increase in length of body at each molt in lobsters between 5.5 and 11.5 inches is between 11 and 12 per cent. Increase in length diminishes beyond this period, yet the volumetric increase of the entire body, especially the big claws, may be as great or even greater. Beyond the twenty-second stage, according to Hadley, the male grows more rapidly than the female.

WITHDRAWAL OF THE BIG CLAWS.

The shell of the large claw is molted entire without a rupture in any part. This means that the great mass of muscles which fill its terminal joints must undergo distention and compression to an extraordinary degree, since it is all drawn through the constricted base of the limb as wire is pulled through the holes of a drawplate. What this implies will be best appreciated when it is realized that the cross sectional area of the biggest part of the cheliped is more than four times greater than that at its narrowest point, in the second joint.

The lobster is aided in accomplishing this feat by the elasticity of the muscles and other tissues and by the removal of blood from the fine meat of the claw (pl. XL, and fig. 3, pl. XLVI), as well as by the development of absorption areas in the shell of the third and fourth segments of the cheliped. (Pl. XXXVII, fig. 2, *abs. a.*) The muscles of the big claw, which are pulled out like a stick of candy, are at first quite tense. Very soon, however, they relax and, filling with blood and presumably taking up some water, they assume their natural form, with proportional increase in size. The absorption areas, from which mineral matter is removed preparatory to the molt, are easily distinguished in the hard-shell lobster, though less clearly defined. The shell of the basal joint becomes a slender ring, but does not break.

At the time of the casting of the shell the large claws must be practically free from blood, since, as Vitzou has pointed out, if the claw were to be increased in size it would be next to impossible for it to be withdrawn without rupture. The older naturalists used to explain the withdrawal of the large claws by a wasting of the tissues. The lobster was supposed to become sick and emaciated, which was, of course, an error. The most significant fact in this process is the displacement of the liquids which normally belong to these appendages. The terminal soft tissues of the claw are essentially a sponge work of involuntary muscle fibers, to which the returning blood stream has free access.

The changes in the armature of the lock forceps, which attend each molt in both young and adult, are discussed in chapter VII.

MOLTING OF THE "HAMMER" CLAW IN THE SNAPPING SHRIMP, ALPHEUS.

It would be erroneous to infer that all relatives of the lobster in molting withdraw the flesh of their big claws through the "drawplates" of the basal segments of the limb. This is not true of certain species of the snapping shrimp, in which the great "hammer" claws are proportionately larger than in the lobster.

On November 13, 1896, while at the zoological station at Naples, a large male of *Alpheus dentipes* molted in a small aquarium at 3 o'clock in the afternoon. Preparations for this act had been going on for several hours, and were probably begun in the early morning. In this case the muscular mass of the claw was withdrawn through a crack, which extended along the outer margin of the propodus. This cleft was continuous, with a similar fissure involving the proximal segments of the cheliped and extending through the basal ring. The great muscular mass of the hammer claw was thus withdrawn without distortion. This fissure was assumed to correspond to a linear absorption area, but I have not been able to repeat the observation.

CHANGES IN THE SKELETON PREPARATORY TO MOLTING.

At the time of the molt there is an intermediate membrane which makes its appearance between the new and old shells. It is noncellular, has a gelatinous appearance, is very transparent, and may be found adherent to the old shell after the molt is past. It bears the impress of a mosaic of cells, which can be none other than the cells of the chitinogenous epithelium. Vitzou is thus in error in supposing that this substance is a secretion of chitinogenous epithelium underlying the new carapace, which it traverses by endomosis. It must be either the first secreted product of the new shell or the innermost layer of the old shell modified by absorption, if not derived from tegumental glands.

In this cuticular membrane the parts which correspond to the cell boundaries of the chitinogenous epithelium have the form of elevated ridges on the under side, and in the center of each polygonal area there is a slight thickening. Réaumur [a] had in view a similar structure in the crayfish when he spoke of a glairy matter "as transparent as water, which separated the parts which the crayfish was soon to cast off from the rest

[a] Additions aux observations sur la mue des écrevisses, Mémoires de l'Académie Royale des Sciences, p. 263–274, 1 pl. Paris, 1719

of the body, and which allowed these to glide smoothly over one another." The old shell becomes brittle, owing to the absorption of organic matter previous to molting, and if the carapace is pressed between the fingers it will sometimes split down the back in the longitudinal median furrow, but in most cases the shell does not crack in this plane unless artificially compressed. In the course of the preparation for the molt the lime salts of the shell are absorbed along the middle line of the carapace, leaving a narrow, perfectly straight gutter extending from the spine or rostrum to the posterior margin of the shield. The chitinous portion of the cuticle still remains, forming an elastic hinge, on which the lateral halves of the carapace bend without breaking asunder. In the molted shell there is also a linear membranous area on either side of the rostrum. Absorption of the hard matter of the shell at these points tends to give greater latitude to the movements of the two halves of the carapace. If you examine a hard-shell lobster, you will find in place of the median furrow a blue line, drawn as if with pen and rule. Below this line the epidermic cells of the skin become so modified as to bring about the total absorption of the lime salts of the cuticle.

Other areas of absorption besides those of the great chelipeds, already described, include the wide lateral margins of the gill covers or branchiostegites, which in life are colored light blue, parts of the endophragmal skeleton, especially the roof of the passageway, in which are lodged the sternal blood sinus and part of the nerve cord, and the endotergites, three small teethlike projections from the under side of the carapace, on which the posterior gastric muscles are partly inserted. Rupture in the rostral regions is further provided against by the narrow absorption areas on each side of it, while the softening of the margins of the carapace makes the lifting of this from the body an easy matter during the molt. The softening of the endotergites and apodemes of the internal skeleton is also necessary to prevent injury to the soft tissues and to permit their release.

The lobster, as we have seen, leaves its old envelope by drawing the anterior part of its body backward and the abdomen forward through a rent in the soft membrane between carapace and tail. The cuticular lining of the masticatory stomach and esophagus comes out by way of the mouth, while whatever is molted from the intestine is withdrawn from the anus. The intestinal molt of the larva is apparently much more extensive than that of the adult. When the discarded carapace falls back into its natural position we might, as Réaumur says of the crayfish, mistake the empty shell for another animal.

THE GASTROLITHS, OR "STOMACH STONES."

The gastroliths of Crustacea are found only in the lobster and crayfish, and according to Patrick Browne, as noticed by Stebbing (*259*), in certain land crabs of the island of Jamaica. Having been first discovered in the common river crayfishes of Europe, they figured in the old pharmacopœias as *oculi seu lapides cancrorum*, and have excited the interest of naturalists from early times. Owing to their transitory character, they are not commonly seen in the lobster.

If the shell of the lobster which is nearly ready to molt is removed, there will be seen two glistening snow-white masses, one on either side of the stomach. A gastrolith taken from a lobster 11 inches in length was an inch long, three-quarters of an inch wide, and a quarter of an inch thick. Its outer convex side was applied to the sac in which it lay, while its concave side was separated from the cavity of the stomach by the old cuticular lining of this organ. When the stomach is raised the gastroliths almost break through its delicate outer wall by their own weight. They lie between the old cuticular lining of the stomach, which may be stripped off, and its delicate outer wall, next to the body cavity. The impression of the gastrolithic plate (pl. XXXIII) is seen on the new cuticular lining only. If the sacs in which they are formed are cut open, each mass separates into hundreds of small ossicles or columns, the majority of which are slender truncated prisms of irregular shapes and about one-fifth of an inch long. Each ossicle resembles a piece of milk-white glass, with transparent edges, and is faintly marked with transverse and longitudinal striations, like those seen in the cuticle.

The gastroliths, though a part of the cuticle, are not regularly cast off during the molt, but are retained in the stomach; when the old lining of this organ is withdrawn, they are soon set free, and breaking up into their constituent parts are speedily dissolved. Consequently it has been supposed that they served the function of providing a supply of lime for hardening the new shell. Messrs. Irvine and Woodward (*165*), however, have proved that the amount of calcareous matter obtained in this way is only about one one-hundred-and-eighty-sixth part of that of the entire skeleton, and therefore too insignificant to be of any practical value. Lime, moreover, is at hand in abundance in the form of the shells and skeletal fragments of mollusks and other animals, which lobsters make free use of at the time of the molt.

We have suggested that the gastrolithic plates or sacs in the walls of the stomach are organs for the excretion of lime, and that the gastroliths represent the lime removed from the absorption areas previous to the molt. Upon this theory their retention and absorption is an incident of no special importance (see *149*, p. 93).

The gastrolith of one of the common crayfishes (*Cambarus robustus*) when 4 inches long is about the size of a split pea, 7 millimeters in diameter by 5 millimeters thick. It shows no divisions into ossicles, but is a hard mass. The convex face is dull white and nearly smooth, while the flattened side presents a brown circular scar with a white center. In form and appearance it suggests a small mushroom with the stem cut off close to the cap. In sectional view it shows concentric striations.

Chemical analysis [a] has proved that lime salts as carbonates and phosphates form about half the constituents of the hard shell, there being from three to five times as much carbonate as phosphate. We also find that in the cast shell of the lobster the proportion of organic matter present is considerably less than under other conditions. An absorption of organic matter thus takes place during the period in which the new shell is formed, and this fact explains the fragility of the cast-off shell. Small quantities of alumina and silica are normally present in both the shell and gastroliths.

---

[a] See article by Prof. A. W. Smith, *252* of bibliography.

The composition of the gastroliths is similar to that of the shell, a conclusion which we should be led to draw from the fact that these bodies are specialized parts of the dead chitinous integument. The same substances are found in both, but in different proportions. The gastroliths are far richer in lime, chiefly in the form of carbonate ($CaCO_3$), than is the shell, and the amounts of magnesium carbonate ($MgCO_3$), alumina (Al_2O_3), ferric oxide (Fe_2O_3), and silica (SiO_2) are more or less reduced.

Lime estimated as carbonate ($CaCO_3$) constitutes about three-fourths of the gastrolith, but less than two-fifths of the carapace. Lime reckoned as phosphate ($Ca_3(PO_4)_2$) forms about 10 per cent of the gastrolith and but little less in the case of the shell; about 10 per cent of the gastrolith is water and organic matter, probably mainly chitin, and the rest is made up of various salts and oxides. In the only molted shell analyzed about 38 per cent was water and organic matter, while in two hard-shell lobsters this percentage was considerably greater, 42.21 in one case and 51.80 in the other.

Since the total quantity of lime contained in the gastroliths is but a small fraction of the amount necessary for building up the hard crust, the rapidity with which the new shell hardens depends in some measure upon the individual, and particularly upon the quality of its food. Lobsters when young and sometimes when adult not only eat their own cast after molting, but swallow fragments of shells and other calcareous materials, which are dissolved in the stomach and help to strengthen the new shell.

Williams (*279*), who has recently studied this subject, has added some important facts to our knowledge of the gastroliths. He found that while absent in the larvæ they made their appearance at the fourth stage, when the shell begins to receive deposits of lime, and at about the middle of this period. After the next molt the gastroliths were dissolved in the course of a few hours, either remaining in place or falling to the bottom of the stomach sac, to be later broken up. With their dissolution there was observed a gradual hardening of the gastric teeth, mandibles, and later of the chelipeds and other parts.

> As soon as the gastroliths are dissolved [says Williams], the lobster attacks his cast, beginning to eat the bristles and small parts and proceeding to devour more or less of the harder parts. The newly molted lobsters seldom seriously attack their sloughs within three or four hours, and generally eat the greater part of the cast within twelve or eighteen hours.

He therefore supports the older view that the gastroliths represent a store of lime and other minerals reserved from the old shell for the immediate hardening of the new, with the additional statement that this reserve is destined for particular parts—gastric teeth, mandibles, and chelipeds—so that the cast and other calcareous matter within easy reach may be quickly available.

Stebbing (*260*), who also has criticised the view that the gastroliths are primarily excreted products, does not believe that such nicely adjusted structures can serve as "mere off scourings of the body."

The difficulties in the way of supposing that these interesting bodies are necessary rather than incidental sources of lime to the newly molted lobster are by no means removed by the observations quoted above. To be of service at all the carbonates of the gastroliths must be dissolved, absorbed into the general circulation, and converted into phosphates. There is no reason to suppose that the gastric teeth or any other part can make exclusive use of this lime, or use it at all except through the roundabout course open to all lime-absorbing cells. Moreover, the total amount of mineral matter in the gastroliths is so small that when equally disseminated it is difficult to understand how it could be of vital importance.

It seems altogether more probable that the parts mentioned by Williams are hardest in the end because they have the hardest chitinous base in the beginning, and that all parts receive only their due proportion of lime.

Assuming the problem of the gastrolith to be similar in both lobster and crayfish, the spicular character of the former may have no special significance. In the crayfish these bodies, as we have already seen, are solid stones, which, according to Chantran,[a] are slowly ground down rather than dissolved, their complete dissolution taking upward of three days in an adult animal.

Turning to the other side of the question, the absorption of lime from definite areas of the shell is of the utmost importance. Deformity or death awaits every animal in which the absorption areas are not duly formed. The production of such areas involves the excretion of lime through the medium of the blood. Their actual development proceeds, in some measure at least, with the growth of the gastroliths.

Accordingly, while the question may still be regarded as somewhat involved, we still believe that the theory earlier given, that the gastroliths are primarily excreted products and represent mineral matter removed from the shell in preparation for molting, and that their use for hardening the new shell is purely incidental, is the only one which meets all the facts in the case with any degree of success.

If it could be experimentally shown that the gastrolith is essential to life after the molt, as we now know it to be for the safe passage of the molt itself, a theory early maintained but not satisfactorily proved, the present status of the question would be changed.

HARDENING OF THE NEW SHELL.

A lobster which molted while under observation was watched particularly with reference to the hardening of the shell. One hour after the molt the cuticle seemed to the touch of the finger to be perceptibly hardened, but this may have been due to the turgescence of the tissues. Eighteen hours after shedding the cuticle had a leathery consistency, and the tubercles and spines had hardened slightly. The shape of all the parts was perfectly normal. Four days after the molt, when the animal died, the cuticle was still coriaceous, and but slight increase in the stiffness of any parts had occurred.

Another animal which also molted in confinement was kept for a period of 25 days. The carapace at the end of this time was easily compressible between the thumb and finger. The large claws could be made to yield in the same way, but not without

---

[a] Comptes rendus de l'Académie des sciences, t. LXXVIII. Paris, 1874.

using considerable force. It was in the state which the fishermen designate as a "paper-shell" or "rubber-shell" lobster. If sent to market it would have been classed as a soft-shell lobster. It is possible, of course, that in this space of time an animal under natural conditions would have become harder. It is safe to conclude, however, that from 6 to 8 weeks are necessary, under ordinary conditions, to produce a shell which is as hard as that cast off, and if the lobsters were destined for the market they would probably be in a still better condition in 10 weeks or 3 months. Many lobsters with soft shells are caught and sent to market, but their flesh is then watery and of inferior quality. When cooked, the fine meat of the claws, which will serve as a good index of their condition, shrinks to an almost unrecognizable remnant. According to the opinion of a canner of lobsters in Maine, 7 pounds of soft-shelled lobsters in summer or fall will yield no more than 4 pounds in spring, when the flesh is more solid.

RELATION OF WEIGHT TO LENGTH IN THE ADULT.

The lobster's weight does not bear a constant ratio to its length, but is very variable owing chiefly to the loss of limbs, and particularly of the great claw-bearing legs. These alone represent from one-fourth to one-half of the weight of the animal, and probably in all giants of the 20 to 30 pound class, which are invariably males, the weight of the great chelipeds is fully two-thirds that of the entire body. The lost limbs are promptly regenerated, as we have seen, but never completely without the intervention of one or more molts, so that a lobster with an undersized claw is a common occurrence.

The length of lobsters is commonly measured from apex of rostral spine to the end of the telson, not including its terminal fringe of hairs. More exact comparisons can be made from measurements of the nondistensible carapace or back shell alone. This method of measuring the lobster was adopted by the legislature of Maine in 1907, and should be generally followed. The Maine laws require the marketable lobster to measure 4.75 inches from the beak to hinder margin of the carapace, which is equivalent to a 10½-inch animal under the old standard, the ratio of carapace length to full body length being approximately 45 per cent for animals of average size. When the rostrum is defective the total body length can be taken. Under such a relatively inflexible standard the fisherman is not tempted to stretch his lobsters in order to put them into the "counter" class, and to sell animals which are likely to die from injuries thus received.

The weight is subject to considerable variation in consequence of molting, when a dense armor is exchanged for a much lighter though larger one. In the soft lobster the specific gravity of the solids and fluids of the body is considerably reduced, but on the whole the weight is chiefly affected by disparity in the size of the big claws.

The male is heavier than a female of the same length, at least after passing the 8-inch mark. The 10-inch males are about an ounce heavier than females of corresponding length. From this stage onward the balance in favor of the male becomes most pronounced. Thus the 11-inch male exceeds the female of this length by a full quarter of a pound. In a lobster 12.5 inches long there is a difference in favor of the male of 7.5 ounces.

It is evident from the data earlier presented (see *149*, table 31) that the greater size of the male, which is a sexual characteristic, does not appear until the animal has passed the 8-inch limit. At this period the sexes are of about equal weight, but from this point the male surpasses the female in weight, owing chiefly to the greater development of the large claws.

The average weight of females without and with eggs proves that females with spawn are in a poorer condition or weigh relatively less than females without eggs attached to the body. In one-third of the cases recorded the weight of females with eggs was actually less than that of females of the same length without eggs. In the 10-inch series 184 females were examined; 36 of them had eggs and weighed on the average but one-tenth ounce more than those without eggs. The average quantity of eggs borne by a 10-inch lobster is 1.73 fluid ounces, and since a fluid ounce of lobster eggs weighs very nearly an ounce avoirdupois, the average weight of the 10-inch female deprived of her eggs is 22.13 ounces, as compared with 23.76 ounces, the average weight of nonegg-bearing females of this size. There is thus a difference of 1.63 ounces in favor of the female without eggs. In the case of the 9.5-inch female lobsters, where 169 in all and 24 bearing eggs were examined, the average weight of the spawners was less by 0.09 ounce than that of the corresponding females without eggs.

The facts which have just been stated do not support the conclusion of Buckland and his associates on the fisheries work in Great Britain that "the lobster, when berried, is in the very best possible condition for food."

The average weight of the 10.5-inch male lobster (the present legalized length limit in Maine, New Hampshire, and certain districts of Canada) is about 1.75 pounds, a corresponding female without eggs weighing about an ounce less. At 9 inches (legalized in New York, Rhode Island, Connecticut, in Massachusetts since 1907, and in certain parts of Canada) the average for both sexes is nearly 1.25 pounds. The lobster 8 inches long (the present legal gauge for England, Norway, and parts of Canada) of either sex, has an average weight a little short of a pound, or 15.16 ounces. At the 12-inch length the male weighs approximately 2 pounds 12 ounces, the females being about 2 ounces lighter, while lobsters 15 inches long will weigh on the average 4.25 to possibly 4.5 pounds.

A lobster 17.75 inches long weighed nearly 10 pounds (though in this case the cutting claw was undersized), and the mammoth specimens recorded in table 1, weighing from 19 to 34 pounds, varied only from 19.5 to 23.75 inches in length. Indeed between the 18-inch and 20-inch length, as well as beyond this limit, great variation is seen in the weight of normal individuals of either sex of the same length as in the case of smaller lobsters, and due to the same causes, namely, variations in the size and the corresponding weight of the large claws or to the condition of the shell with respect to molting. Beyond the 20-inch size a slight increase in length may imply a great addition to the weight.

PROPORTION OF WASTE TO EDIBLE PARTS IN THE LOBSTER.

Atwater (*11*), in his chemical analysis of the flesh of the lobster, gives the proportion of the edible parts and shell as follows:

	Per cent.
Total edible portion	39.77
Shell	57.47
Loss in cleaning	2.76
	100.00

The proportions of water and dry substance in the edible portion are estimated as follows:

Water	82.73
Dry substance	17.27
	100.00

In this relation the analysis given in table 3 will be of interest. These data were obtained from a 13-inch (boiled) female lobster, with shell of medium hardness. Literally all of the soft and edible parts were carefully removed from the skeleton and weighed. This, without doubt, accounts for the higher percentage of "edible" parts obtained when compared with the result quoted above, it being assumed that all of the soft tissues of this animal are edible and wholesome excepting the stomach and intestine.

The flesh of the lobster is rich in nitrogenous or proteid substances and contains a considerable amount of phosphorus and sulphur. Its nutritive value as compared with beef taken as a standard is 61.97 per cent (*11*).

TABLE 3.—SHOWING RELATION OF EDIBLE TO WASTE PARTS IN THE LOBSTER.

Edible parts.	Pounds.	Ounces.	Waste.	Pounds.	Ounces.
(1) Tail muscles	0	8¾	(8) Shell and "lady" or stomach sac	1	3¼
(2) Meat of great claws, including joints of great chelipeds	0	7¼	(9) Liquids	0	6
(3) "Cream," or clotted blood from great chelipeds	0	1¼			
(4) Fine picked-out meat from linkwork of body and smaller appendages, including gastric, mandibular muscles, and green glands	0	7¾			
(5) "Cream," or clotted blood from body under shell	0	2⅞			
(6) "Coral," or ovaries	0	2			
(7) "Tomally," or liver	0	2⅛			
Total weight of claw and tail meat, 1 pound.					
Total weight of edible parts	2		Total waste	1	9¼

Total weight of items 1 to 9, 3 pounds 9¼ ounces.
Estimated living weight, 4 pounds 4½ ounces.
Dead weight, 3 pounds 9¼ ounces.
Percentage of clear meat in claws and tail, 27.
Percentage of all clear meat and edible parts, 55.
Total cost at current retail-market price, at 25 cents per pound, at Tilton, N. H., June 27, 1903, 90 cents.
Cost per pound of clear meat of big claws and tail (items 1 and 2), 90 cents.
Cost per pound of clear meat and other edible parts not usually saved (items 4–7), 45 cents.

Chapter V.—ENEMIES OF THE LOBSTER.

PREDACEOUS ENEMIES.

The adult lobster, whether with eggs attached to its body or not, is the prey of numerous fish which feed upon the sea bottom, like the sharks, skates, and rays. When of considerable size or in soft condition it is also devoured by the cod, pollock, striped bass, sea bass, tautog, and probably by many other species. In fact every predaceous fish which feeds upon the bottom may be looked upon in general as an enemy of the lobster.

Next to man with his traps, the codfish is probably the most destructive enemy of the lobster, for it not only takes in the soft and hard shell animals alike up to 8 inches or more in length, but is very partial to the young from 2 to 4 inches long.

If the lobster is thus attacked and destroyed in large numbers by fish after it has acquired the habits of the adult and has many devices to avoid its enemies, what shall we say of the destruction which is wrought on the young during the first three or four weeks of their life? From the time of hatching up to from the fourth or fifth stage the young lobster swims at the surface and becomes an easy prey to all surface-feeding fish, some of which, like the menhaden, roam about in vast schools, straining the water as effectively as the towing net. When lobsters settle in relatively shallow water the greedy cunners or even fish of smaller size would doubtless prove vastly more destructive. During this period the lobster measures from one-third to three-fifths of an inch in length, and is not only helpless in the hands of its animate enemies, but is subject to a vast amount of indiscriminate destruction from the forces of inanimate nature.

PARASITES AND MESSMATES.

But two parasites in the strict sense have thus far been known to infest the lobster, although it is probable that others will he discovered. One of these, a trematode worm (*Stichocotyle nephropis*) first noticed in the intestine of the Norwegian lobster, was later detected in the American form, about 2 per cent of these animals being infested by it. Its final host is probably some species of fish which preys upon the lobster, but the adult trematode is unknown.

The only other strict parasite which has been found to trouble the adult lobster is the large gregarine (*Gregarina gigantea*), discovered in the intestine of the European lobster by Van Beneden (*269*).

The European lobster is commonly infested with a small colorless worm, *Histriobdella homari*, of remarkable habits and doubtful relationship. Discovered in 1853 by Van Beneden on this lobster's eggs at Ostend, it was regarded as a larval serpulid, but later (*108*) shown by him to be an adult and placed among the leeches. An account of its anatomy was given by Foettinger (*108*) in 1884, but for the most exact anatomical analysis of this curious semiparasite or commensal we are indebted to Shearer (*324*), whose work has but recently appeared. He found that it not only lived among the eggs of the berried lobster, but took up its abode in the branchial chamber and on the gills of both sexes also, passing readily back and forth when its host was a female in berry. It crawls slowly, but is more active among the lobster's eggs, to which it attaches its own ova freely, as well as to the carapace side of the branchial chamber. It is very sensitive to changes in the sea water, and its selection of such lodgings seems to indicate clearly the need of an abundant supply of oxygen. Development is direct, there being no larval stage, and little is known of its distribution or the means by which this is effected. Though possessing toothed jaws, and though seen to bite one another, these parasites are not known to molest either the gills or eggs of their host, and since they often devour diatoms in quantity they may be the lobster's bosom friend rather than its enemy. So far as known at present, *Histriobdella* is not attached to the American lobster.

But although parasites are rare, the lobster is encumbered with a great variety of messmates, which attach themselves to the external shell. Whenever the lobster is confined in inclosures, or compelled for any reason to lead a sluggish life, the common barnacle fixes itself to the arched carapace and begins to secrete its tent-like covering as securely as it might upon a stone; mussels of various kinds insinuate themselves in convenient angles of the shell and joints, and small tunicates sometimes become attached firmly to the underside of the shell between the legs. Tube-forming annelids, lace-like bryozoa, form incrustations in various parts, and red, brown, and green algæ often decorate the antennæ and carapace with long streamers which are waved with every movement of the animal. At each molt the lobster of course frees itself completely from these troublesome companions. (For fuller account of parasites and messmates see *149*, p. 122–124.)

When young lobsters are hatched and reared in confinement they are apt to be troubled with a variety of parasitic fungi and algæ, including many species of diatoms, as well as stalked protozoans. Young lobsters captured at sea seem to be peculiarly free from foreign matters of every kind, but when the young of almost any crustacean are confined they are liable to become clogged with solid organic and inorganic particles of many kinds, including living bacteria, spores of fungi, and diatoms. The hairs which garnish the body and appendages of crustacean larvæ serve to gather up and hold particles from the water, so that one of the first considerations in the artificial rearing of these animals is to give them as clean a water supply as possible. Old lobsters, in which the molting periods have become very infrequent, are the worst sufferers from enemies of this kind, but the physiological condition of the animal is a most important consideration.

DISEASES AND FATALITIES OF THE LOBSTER.

There are few specific diseases to which adult lobsters are subject so far as known, yet they sometimes die off so rapidly as to lead one to suspect that they have fallen a prey to infectious disease.

Mr. N. F. Trefethen, of Portland, Me., relates the following experience: In May, 1893, he placed 100,000 lobsters in a pound at South Bristol, the area of which is about 3 acres. Very soon they began to die, and in a few days all of them were dead. There was from 12 to 13 feet of water in this pound at flood tide and not less than 9 feet at low tide. The pound was probably very much overstocked, but it is difficult to understand why these lobsters should have all died so suddenly, unless they were either poisoned or attacked by disease.

In the summer of 1889 a lobster with a large bunch on the side of the carapace was captured in Vineyard Sound. On the top of this tumoid growth was a crater-like depression covered with a membrane. This was probably a sore resulting from a wound which the animal had received in the back, and which failed to heal. A similar case is mentioned by Rathbun. Further, according to Prince (*218*), Professor M'Intosh has described a tumor-like growth in a large lobster which originated in the wall of the stomach sac, finally perforated the carapace and caused its death.

In another place I have alluded to the experience of the U. S. Bureau of Fisheries at Woods Hole in feeding the young lobsters with shredded menhaden. The larvæ became infected with a fungus, which spread to all parts of their tissues and was soon fatal.

To paraphrase the words of Hardy,[a] the lobster, like many other aquatic animals, is confronted by the same problem that has so long puzzled the shipbuilding world. Larvæ and spores are constantly settling upon the exposed surfaces of its body, where they tend to develop growths which would interfere with their movements unless some method of destroying or removing them were adopted. Hardy believes that "the presence of a film of soluble slime on the surface of an animal immersed in water would, like the copper sheathing of ships, mechanically prevent the occurrence of parasitic growths by continually forming a fresh surface," and further that this slime may in some cases have a specific poisonous power, directed chiefly against vegetable parasites.

The lobster apparently secretes no slime, but its shell is studded with the openings of the tegumental glands, the exact function and rôle of which is still in doubt. At all events it will do no harm to raise the question whether these bodies may not help to free the animal from such pests. That molting alone is not able to do this and that some additional aid is often needed is amply proved by the great variety of messmates or semiparasites which we have described.

Lobsters from a few inches in length up to the greatest size are sometimes driven ashore and stranded on the beach, where, stunned or crushed by the force of the waves,

[a] Hardy, W. B. The protective functions of the skin of certain animals. Journal of Physiology, vol. XIII, no. 3 and 4. London, 1894.

they are often left to perish. Well-nigh incredible accounts of the "windrows" of dead lobsters left by fierce storms on the shores of New Brunswick and of other maritime provinces were current in the earlier days of the fishery. Thus Prince (*218*) speaks of a memorable storm along the Shippegan shore, Gloucester County, New Brunswick, in 1873, and states that as many as 2,000 dead lobsters were counted in the distance of 2 rods.

The writer quoted above also speaks of the fish crow (*Corvus frugilevus*) as very destructive to lobsters on parts of the coast of Nova Scotia, where he says "when the tide goes down these birds destroy the lobsters left amongst the seaweed. They pierce the shield of the lobster where the heart and main blood vessels are situated, and the crustacean is at once rendered helpless and is devoured by its assailant." I have seldom known the lobster to be stranded in this way in calm weather. The adolescent lobsters, which alone remain in near the shores, ordinarily go deep down among the loose stones, where neither crow nor any other bird could possibly dislodge them.

Chapter VI.—ANATOMY OF THE LOBSTER, WITH EMBRYOLOGICAL AND PHYSIOLOGICAL NOTES.

Both the lobster and the crayfish have long been regarded as classical exponents of a zoological type and have figured so prominently in text-books that the elementary facts of the anatomy of few invertebrates are better known; yet there is still a wide field for more exact research in nearly every direction, as we have found whenever it was possible to dip below the surface. In the present chapter it will be necessary to restate certain elementary facts, but my embarrassment would be greater were this work intended solely for professional zoologists, who will probably find more that is new in the chapter which follows.

In attempting to give a fairly consistent account of the lobster's anatomy I shall not hesitate to enter into details, but shall endeavor to emphasize those parts of most zoological interest from the standpoint of morphology, physiology, and development. Numerous anatomical drawings are given, including the entire series of adult appendages, which may serve for more exact comparisons with the larval stages than have been possible hitherto.

THE BODY.

The lobster's body (pl. XXXIII and table 4), which the fisherman compares to a pistol in shape, but holds by the "barrel," is made up of a series of 21 somites or body segments (or of 18, omitting 3 of doubtful value), all but the last of which bear paired and jointed appendages. The first 14 are united into one piece called the cephalo-thorax or "barrel," while the last 7 form the flexible abdomen or tail. This primitive segmentation which is expressed chiefly in the exoskeleton or the hard and soft skin extends also to the nervous system, as well as to certain muscles and blood vessels, but does not involve the soft parts of the body as a whole. A cuticle, which is strengthened with lime and other minerals to form a hard crust wherever greater protection or rigidity is needed, follows every inward fold of the skin and covers every part of the body down to a microscopical hair.

The skeletal parts of head and thorax are fused on the upper and lateral surfaces to form a large cephalo-thoracic shield or carapace, often called simply the "shell," which is "buttoned" on to the tail by small overlapping pleura of the first small somite of this part. The carapace is marked and sculptured in a very definite manner by symmetrical folds or grooves, tendon marks, and absorption areas, not to speak of protective spines, and smaller tubercles, fringing sensory hairs, and the very minute depressions with which it is stippled all over, the hair pores to be later described. The light median stripe which runs, as if drawn with pen and rule, from the rostrum to the hinder border of the carapace represents an absorption area of the greatest importance to the molting lobster. A prominent fold known as the cervical groove crosses the carapace at a point about midway on the back to a triangular depression, representing a tendon mark, and is thence continued forward on either side as a groove, which ends between the antennæ and the mandibles. In a soft lobster a penknife can be readily inserted into this fold on the midline. Inwardly the pocket is continued into three divergent endotergites, which give attachment to parts of the posterior gastric muscles, but are absorbed previous to molting. Immediately below the forward end of the groove is seen the "grater," a peculiar roughened area of the shell at the outlet of the branchial cavity; just before reaching this place the groove rises slightly, as if to avoid a prominent swelling, which marks the position of the ball of the outer hinge of the mandible, to be seen upon opening the branchial cavity. A branchio-cardiac line passes backward from each tendon mark toward the hinder border of the carapace, and with the cervical groove divides it into cardiac, gastric, and branchial regions. These lines are obscure in young animals, but become prominent grooves later, and deep furrows in lobsters of mammoth size. The gastric mill underlies the shell immediately in front of the cervical groove; a puncture behind this fold draws blood from the pericardium or the heart, while one below the branchio-cardiac line pierces the gill cover to the branchial chamber. The meaning of other tendon marks and muscle impressions on the carapace is given in a later section. Of the last 10 thoracic legs in the decapod, the first pair bear the big claws in the lobster and are its largest and most characteristic appendages. Its smaller and slenderer legs are chiefly ambulatory and sensory. The tail carries at either side on its under surface a bank of elastic oar-like feet of simple type, the swimmerets or pleopods for forward swimming, while the greatly enlarged and displaced sixth pair, or uropods, make with the telson the tail-fan already referred to.

INTERNAL SKELETON AND HEAD.

If we examine a well-prepared skeleton of a lobster we see that besides the outer hard crust there is a delicate internal skeleton, consisting not only of hard strap-shaped tendons at the joints of the limbs, but of a complicated linkwork of very thin plates or apodemes (pl. XXXIII and XLI). These unite to form partitions between successive sterna and their appendages in the cephalo-thorax, and form an internal or endophragmal skeleton. This intricate structure is produced by infoldings of the epidermal layer of the skin in the sternal and epimeral parts of the cephalo-thorax. The apodemes of which it is composed, are formed like the rest of the exoskeleton from matter secreted by the epidermis. Each plate or rod is thus double in origin, being formed in a flattened pocket like the tendons of the legs (*tp*, fig. 1, pl. XLIII).

According to Huxley [a] four apodemes are originally developed as ventral folds of the skin between any two successive somites of the body, the anterior wall of each pertaining to the somite in front, and the posterior wall to the somite behind. These four apodemes thus form a single transverse series, the two nearer the middle line being called the endosternites, and the two farther removed the endopleurites. The linkwork

[a] Huxley, T. H. The Crayfish, p. 158. New York, 1880.

which thus arises by the repetition of simple units on the ventral side of the thorax becomes more complex through the divergence and coalescence of both endosternites and endopleurites at a higher level to form an archway for the sternal sinus. The roof of this passage is discontinuous, being formed by the fusion on the midline of the inner processes or mesophragms of the endosternites of each side, while their outer processes or paraphragms unite with corresponding horizontal plates of the endopleurites.

The endophragmal skeleton greatly increases the area for the attachment of muscles, and serves to bind the somites of the cephalo-thorax together with greater rigidity, as well as to protect important organs, for not only does the archway securely lodge the large blood sinus, but it also gives passage to the nerve-cord, access to which from above can not be had without cutting through its roof (pl. XXXIII and XXXIV). Since, as is well known, this linkwork is shed in one piece, how do the central nervous system and the parts adjacent to it escape unharmed? I have never heard this simple question raised, but the answer is given by the molted shell, in which it will be seen that the roof of the archway is completely absorbed as well as a large part of the intersegmental and dividing partitions of the bulkheads referred to above, so that the whole under surface of this part of the body with the delicate gills can be withdrawn with impunity.

The endophragmal skeleton bears the hinges for the articulation of the limbs, the arrangement of which is peculiar (pl. XXXVII and XXXVIII). The central hinges which lie close to the mid-line are all cups and are borne on the sterna and close to the endosternites, while the outer or peripheral hinges are all balls and are borne on the epimeral surface of the branchial cavity, close to the endopleurites. The transverse partitions are parallel with the axes of articulation of the appendages in successive somites.

The hard skeleton of the lobster's head immediately in front of the mouth, representing apparently the sterna of somites ii to iv, consists of a conspicuous plate shaped like an Indian arrowhead or spear, with the point drawn out into a sharp spine lying between the first segments of the lesser antennæ, while its broad base, raised into a ridge, bears the soft upper lip or labrum; immediately in front of the ridge this triangular plate is traversed by a deep furrow, in the midst of which lies a small closed pit, most obvious in a soft-shelled animal. This marks the position of a median endosternite to which are attached certain small muscles leading ventrally to the esophagus and dorsally to the membranous covering of the brain.

Upon examining the skeleton of the head from the inside, it is seen that the epimeral and tergal parts are fused to form a ring into which the eye stalks open, close to the brain. On the upper side at the base of the rostrum the ring forms a solid bar, which Professor Huxley thought might represent the tergum of the antennulary somite in the crayfish, and from either side of this bar spring two large leaf-like divergent plates, the procephalic processes, to which the anterior gastric muscles are attached. Below the ring the calcified epimeral surface surrounds the large paired openings for the antennules and antennæ, and is continued to form the wall of the branchial chamber on either hand.

APPENDAGES.

The 20 pairs of appendages of the lobster are developed as tubular folds or outgrowths of the body wall, and consist of ectoderm with mesodermic cores, a rule which seems to be broken only in the case of regenerating limbs, where ectoderm appears to contribute to the renewal of both muscles and nerves. The order of embryonic development is: (1) Antennules, (2) mandibles, (3) antennæ, (4) maxillæ and the thoracic limbs in regular succession. Four pairs of swimmerets (somites XVI–XIX) are released together in the second larval stage (fig. 41); the uropods in the third stage (fig. 42) and the first pair of pleopods, which are the last to appear, are not usually recognizable until the sixth molt or later.

The eyestalks, which are omitted from the enumeration given above, and the antennules are prostomial in origin, while the originally postoral antennæ reach a position in front of the mouth by the twentieth day, when the compound eyes are distinctly lobate. Segmentation in the limbs is a gradual process, constrictions early marking future joints, while the division into outer and inner branches begins at the apex of the appendage except in the antennules, as noted below. Most parts of the adult appendages are recognizable in the first larva, and all, excepting those of the XV somite, in the lobsterling. From the fourth stage on through the adolescent period the changes are gradual and relatively slight, excepting only those which involve the

TABLE 4.—THE BODY SEGMENTS AND APPENDAGES OF THE LOBSTER

Divisions of body.	No. of somite.	Name of somite.	Name of appendage.	Functions of appendage.
Head (6)	I	Opthalmic	Eyestalk	Visual
	II	First antennal	Antennule	Olfactory or chemical, chiefly through outer branch, and static.
	III	Second antennal	Antenna	Tactile chiefly, and probably chemical
	IV	Mandibular	Mandible	Crushing and triturating small, hard parts of food.
	V	First maxillary	First maxilla	Masticatory and chemical, but chiefly for passing the food.
	VI	Second maxillary	Second maxilla	Respiratory chiefly; also chemical, masticatory, and for passing on the food.
Thorax (8)	VII	First thoracic	First maxilliped	For passing, and like the maxillæ possibly subserving the chemical sense.
	VIII	Second thoracic	Second maxilliped	For transference of food, the chemical sense, and respiration.
	IX	Third thoracic	Third maxilliped	Chiefly masticatory, with brushes for cleaning.
	X	Fourth thoracic	Great cheliped, or first pereiopod.	Chelate; big claws adapted on one side for crushing and on other for seizing and rending prey; respiratory, tactile, and possibly olfactory.
	XI	Fifth thoracic	Second pereiopod	Chelate; ambulatory, tactile, and possibly with chemical sense, for seizing, testing, and transference of food; respiratory.
	XII	Sixth thoracic	Third pereiopod	The same
	XIII	Seventh thoracic	Fourth pereiopod	Nonchelate; the same
	XIV	Eighth thoracic	Fifth pereiopod	Nonchelate; the same, and for cleaning swimmerets.

great chelipeds and the first pair of swimmerets. The complex and varied relations of the successive somites and appendages of the lobster in the larval and adult state are outlined in table 4.

In their type form (fig. 2 and pl. XXXVI, fig. 5) the appendages consist of an inner and outer branch borne on a basal stem, known respectively as endopodite, exopodite, and protopodite. The protopodite is composed of two segments, a proximal coxa, or coxopodite, and distal basis or basipodite. The coxa of each limb from the maxillæ to the fourth pair of pereiopods (somites V–XIII) bears a hairy respiratory plate or epipodite, from which rises a gill or podobranchia on all but the first two of these somites. The primitive type of crustacean limb was probably biramous, since in the course of development we frequently find the uniramous condition produced by loss of the more transitory exopodite, and further, since the foliaceous form of appendage of the lower branchiopod crustacea is secondarily assumed by certain of the mouth parts of the lobster and other decapods. The undivided form of limb is permanently preserved in metameres I and X–XV, in the last of which the appendage is modified in the two sexes to perform distinct functions. The origin of the two-branched antennules will be considered presently. The exopodite is frequently abortive, or multiarticulate and elastic, as in the swimmeret, a condition which the endopodite has also preeminently assumed in the long whips of the antennæ.

WITH THEIR CHIEF FUNCTIONS AND MODIFICATIONS IN LARVA AND ADULT.

Relation of appendage to type form.	Relation of adult to embryonic and larval appendage.	Apertures of body.
Doubtful; stalk in two segments	Transitory ocellus in first larva; compound eye relatively large, and stalks short.	
Doubtful. Basal segment lodges statocyst sac	Bifid, and later uniramous in embryo; finally biramous in first larva; inner flagellum a secondary outgrowth from primary stalk. Prostomial.	Pore of statocyst on upper surface of basal segment.
Exopodite wanting; exopodite reduced to scale, and endopodite irregularly segmented.	Bifid, and later completely biramous in embryo; poststomial in origin, but later advance in front of mouth.	Papilla for opening of renal organ on coxa.
Biramous; two distal segments of palp supposed to represent the endopodite.	Body and palp at comparatively late embryonic stage.	Mouth, screened by labrum, between mandibles.
Foliaceous; exopodite wanting; endopodite of two modified segments.	Early larval condition similar to adult, but endopodite unsegmented.	
Biramous and foliaceous; respiratory fan formed by fusion of exopodite and epipodite.	First larval condition similar to adult	
Biramous and foliaceous, and like maxillæ, with protopodite modified for testing and passing the food. Endopodite 2-jointed.	The same, but epipodite without fold for "bailer".	
In type form; endopodite 5-jointed, and epipodite with rudimentary gill.	First larval state similar to adult	
In type form, modified for mastication, and cleaning; second and third podomeres fused, and exopodite reduced. Epipodite with functional podobranchia in IX–XIII.	In first larva with long swimming exopodite, lost at fourth stage, and third joint free; no cleaning brushes, and no teeth on ischium.	
Uniramous through loss of exopodite in fourth stage. Second and third podomeres modified for autotomy, and fused "breaking joint" between them.	Biramous to fourth stage. Big claws nonprehensile in first larva; of toothed type in fourth, and symmetrical up to sixth or seventh stage. Torsion of limb completed at fourth stage, after which big claws are horizontal, and dactyls face, opening toward mid-line of body.	
Uniramous through loss of exopodite in fourth stage.	Swimming exopodite shed at fourth stage	
The same	The same	Oviduct opens on coxa.
The same	The same	Seminal receptacle.
The same, without epipodite and podobranchia	The same; torsion of terminal segments away from mid-line of body completed at fourth stage, when limb is directed backward.	Vas deferens opens on coxa.

TABLE 4.—THE BODY SEGMENTS AND APPENDAGES OF THE LOBSTER WITH

Divisions of body.	No. of somite.	Name of somite.	Name of appendage.	Functions of appendage.
Abdomen (7)	XV	First abdominal	First pleopod	Modified in male for copulation, and reduced in female to prevent attachment of eggs. Uniramous.
	XVI	Second abdominal	Second pleopod	Biramous; for forward swimming; in female for holding and aerating the eggs, and possibly for secreting the glue by which they are fastened to certain of the setæ; tactile, with chemical sense in doubt.
	XVII	Third abdominal	Third pleopod	The same
	XVIII	Fourth abdominal	Fourth pleopod	The same
	XIX	Fifth abdominal	Fifth pleopod	The same
	XX	Sixth abdominal	Sixth pleopod, or uropod.	Enlarged and modified for forming with telson, the tail-fan, for backward swimming; tactile.
	XXI	Telson	Wanting	

In the typical thoracic leg (pl. XXXVIII) the endopodite is divided into 5 segments, which, with the two divisions of the protopodite, give the limb 7 podomeres, numbered[a] and named from base to apex as follows: (1) Coxa or coxopodite, (2) basis or basipodite, (3) ischium or ischiopodite, (4) meros or meropodite, (5) carpus or carpodite, (6) propodus or propodite, and (7) dactyl or dactylopodite. These successive segments are articulated to the body and to one another by soft membrane and usually by hinge joints which limit the movements of each to a single plane at right angles to the articular axis, or to the line joining the two hinges; each segment, with the exceptions to be noted later, is actuated by opposing muscles, a larger flexor and a smaller extensor, the fibers of which are implanted over the hard shell of their respective segments and are inserted on strap-shaped tendons which react on the distal podomere (fig. 1, pl. XLI). The tendon is derived from an ingrowth or flattened pocket of interarticular membrane (fig. 2, *mb.*, pl. XLI, and fig. 1, *tp.*, pl. XLIII), and is sometimes closely united to the shell of the distal segment. Each joint or articulation is therefore crossed by tendons which belong to the proximal podomere and pull on the distal one.

In the successive somites of the tail the axes of articulation are all parallel, and at right angles to the longitudinal axis of the body so that movement is limited to the vertical plane. In the appendage, on the other hand, the direction of the axis of articulation varies in successive podomeres (see figs. 6 and 7); moreover the initial direction of movement of the base of each limb, which depends upon the angle which its articular axis makes with the long axis of the body, varies greatly from head to tail (135° in the mandibles, about 55° in the great chelipeds, and 90° in the swimmerets). Accordingly each segment acts as a lever of the third order, and the successive thoracic limbs are capable of universal movement, and in a variable field. By reference to figures

[a] This order seems preferable to the reverse, which is sometimes adopted, since the protopodite has been less modified than either of its branches, and we thus avoid the ambiguity of speaking of the seventh segment of a pleopod or of an antenna

THEIR CHIEF FUNCTIONS AND MODIFICATIONS IN LARVA AND ADULT—Continued.

Relation of appendage to type form.	Relation of adult to embryonic and larval appendage.	Apertures of body.
Uniramous, presumably through loss of exopodite.	Appear as buds in fifth to eighth stage, and sexually differentiated in eighth to tenth stage.	
In type form, with endopodital spur in male.....	Appear as bifid buds beneath cuticle of first larva; released as rudimentary limbs in second larva; fully functional at fourth stage.	
In type form..................................	The same..	
The same......................................	The same..	
The same......................................	The same..	
The same, with protopodite undivided, and 2-jointed endopodite underlying exopodite.	Appear as buds at base of telson in second stage; released in third and completely functional in fourth.	
....	Bifid in an embryo of two weeks; later elongated and forked; released in larva as a triangular swimming plate, with terminal fringe of large spines and small setæ, which are more distinctly plumose and greatly elongated at the fourth and later stages.	Anus on lower side at base.

1 and 2, plate XLI, the working of this effective mechanism is readily understood. In the sectional view of the big claw and walking leg the tendons of the terminal joint lie in the plane of the paper, and the axis of articulation is at right angles to it; a contraction of the large flexor muscle (*fl.* 6) pulls on the large inner tendon and thus closes the claw, while an impulse sent into the extensor (*ex.* 6) draws on the opposite tendon (*t.* 6), which springs from the opposite side of the dactyl, and thus opens the claw. Contraction of the flexor of the next segment (*fl.* 5) would raise the whole claw toward the eye, and so on. In this case, where considerable power is required, there is a double or divided tendon for this muscle. Owing largely to the variation in the field of movement of the successive pereiopods, referred to above, the lobster is able to cover a wide front in defense, move forward, sideways, or backward, reach every part about the mouth, and scratch the underside of its tail.

Whether the stalked eyes of decapods are metameric appendages or not is a question upon which zoologists are not agreed. In the lobster the eye-stalk (fig. 1, pl. XXXV) is composed of two segments, the basal of which is minute, and imperfectly calcified, as in the protopodite of the swimmeret, and that flagella-like outgrowths occasionally follow partial excision or injury of the eye is well known. "I think," says Professor Brooks in his monograph on Lucifer, "that the presence of a distinct ocular segment in *Squilla* compels us to recognize an homology between the stalked eyes and an ordinary appendage, although it is no doubt true that all the groups in which stalked eyes occur can not be traced back to a common ancestor, and also true that the stalked eyes themselves can not be traced back to ordinary appendages."

The first antenna (fig. 4, pl. XXXV), as we have seen, is first in the order of embryonic development, arising on about the ninth day, just behind the thickenings which form the optic diks, and before the mouth invagination is formed. The latter appears a few hours later than the antennules, and on a line drawn through their posterior margins, so that these appendages are essentially prostomial. The mandibles come next in order, followed in a few hours by the second antennæ, both arising as simple buds, and all three pairs become concentrated about the mouth in the early egg nauplius stage, which is thus reached at the tenth or eleventh day. Both pairs of antennæ are then distinctly divided at the tips, as if about to branch, but the second pair only becomes biramous, the first remaining as single constricted stalks up to near the end of embryonic life. When the larva emerges, what is to be the inner and slenderer branch of this appendage is seen arising as a small bud from the base of what becomes the outer and thicker flagellum (fig. 34). The inner branch of the antennule is therefore probably not homologous with an endopodite. The outer branch develops its club-shaped "olfactory" setæ in the second larval stage, and remains very short and stout up to the fourth or fifth stages, when it rapidly lengthens.

It should be noticed that the lower or sternal part of the head faces forward instead of downward, as a result of cephalic flexure which arises in the course of embryonic development; in consequence of this the anterior sterna are bent upward through nearly a right angle, so that the eyestalks and both pairs of antennæ are directed forward, and their originally anterior faces have become their upper sides. (Pl. XXXIII.)

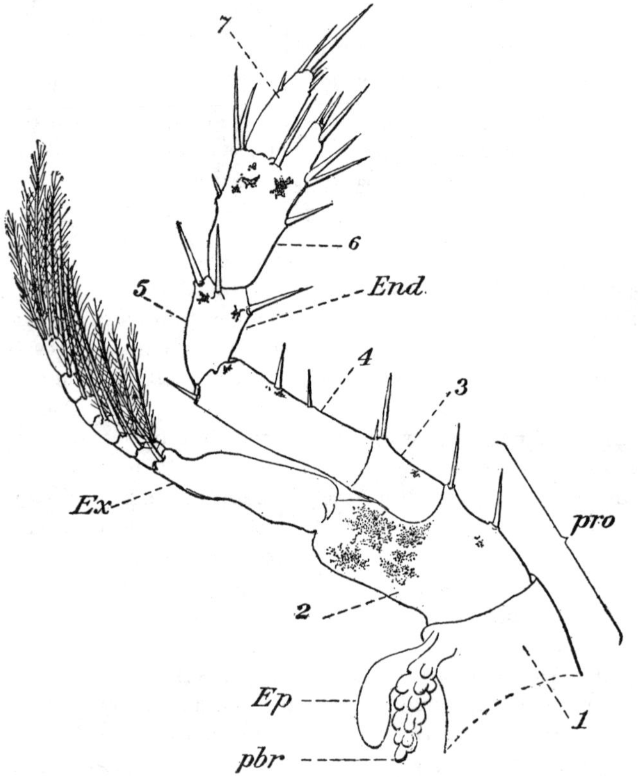

FIG. 2.—Left second pereiopod of first larva of lobster, showing the primitive divided form of the limb, with successive segments or podomeres of protopodite (*pro*, segments 1–2), and permanent inner branch or endopodite (*End*, 3–7). *Ex*, deciduous swimming branch or exopodite; *Ep*, epipodite or gill separator, with its gill or podobranch (*pbr*).

Assuming that neither the eyes nor antennules are metameric appendages, and that the telson is not a true somite, the body would consist of a prostomium bearing the two pairs of articulated processes named, eighteen metameres, and a terminal telson, the first four somites being fused with the prostomium to form the head, with appendicular antennæ, mandibles, and maxillæ.

Since it will be necessary to examine the swimmerets, the compound eyes, and statocysts in relation to other organs, the account which immediately follows will be limited to the mouth parts and certain adaptations found in the walking legs, further details being given in table 4. The history of the big claws is reserved for the following chapter.

MOUTH PARTS.

In addition to labrum and metastoma, we designate as mouth parts the six pairs of limbs which are concentrated about the mouth opening, and which are modified in some degree for dealing with the food.

When the mandibles open, a conspicuous pink fold of fleshy tissue is revealed overhanging the median V-shaped fissure which is the lobster's mouth (pl. XXXIII). The labrum is shield-shaped and compressed in a peculiar manner, being keeled above and below on the middle line, with a thin free edge or border, so that it presents two upper and two lower concave surfaces. The lower keel by fitting into the slit of the mouth forms the upper bound of this opening as it passes into the dorsal wall of the esophagus; the fissure is limited below by a soft, round papilla, from the sides of which spring a bifurcated "lower lip," or metastoma. The metastoma on either side consists of a short strap-shaped blade fitting closely over the convex body of the mandible; it is slightly ridged on the outer side and sparingly sprinkled with setæ. Both labrum and metastoma are richly supplied with organs which there is reason for regarding as sensory buds. The sides of the mouth are formed by rounded swellings of the esophageal wall, and are directly continuous with the metastoma below. When the jaws are closed and their outer masticatory ridges meet on midline just over the mouth fissure, the concave sides of the labrum fit into deep grooves which traverse the opposing mandibular surfaces, and since the groove of each mandible lies below the level of its cutting ridge, it is impossible for the lobster to "bite its lip." The V-shaped mouth described leads through a very short esophagus directly to the large stomach sac. All of the mouth parts which succeed the mandibles are thin and leaf-like up to the somite VII; and all conform to their outer convex surfaces.

The six pairs of appendages which are concentrated about the mouth are abundantly supplied with sense organs, and are charged with a variety of functions, the most obvious of which are handing the food along to the mouth and mincing it in the course of passage; that they further serve as organs of the chemical sense and of touch more or less completely is not to be doubted.

The mandibles of the adult lobster (fig. 7, pl. XXXV) are in form like hinged double doors set in front of the mouth, and so hung to the cephalothorax of the animal that they are capable of swinging only a little way in or out, or toward and away from the middle line. The body of the mandible, which probably represents the coxa of a typical limb, is a triangular convex bar, with a very oblique axis of articulation corresponding to its long anterior side; the opening tendon of the abductor mandibuli muscle is inserted on the anterior border, near the outer socket and exerts a pull sufficient to open the "door." The posterior border bears at a more favorable point near its middle, a long tendon, from which fan out the fibers of the powerful adductor mandibuli muscle (see p. 242). These muscles arise from the inner surface of the carapace on either side in front of the cervical groove, and between two white tendon marks; when they work the "doors" are swung to with force.

The masticatory surface of each jaw is represented by the short side of the triangle which meets its fellow on the midline in front of the labrum. It is divided by a deep groove into an outer cutting ridge, capped with a dense mass of yellow chitin, and a lower and flatter surface, which appears to be available for mastication in but a slight degree, if at all. The groove (*g*, fig. 7, pl. XXXV) not only protects the fleshy upper lip, but gives play to a 3-jointed hairy palp, the two distal segments of which are supposed to represent the endopodite. The palp is actuated by muscles lodged in the body of the mandible itself, and possibly serves to direct food particles to the mouth, below the level of the groove, and just beneath the tip of the labrum.

The lobster's mandibles work essentially on the principle of the modern stone-crushing machine; little or no lateral motion being possible in an animal with a hard shell, they can serve only by repeated closing movements to divide and triturate the larger particles of food, which, having resisted the preceding mouth parts, get pinched between the meeting edges of the swinging "doors."

The leaf-like first pair of maxillæ, the smallest of the mouth parts (fig. 1, pl. XXXVI), bear on their first segment a fringe of stiff hairs and on their second a comb of bristles, which help to pass up the food or mince it when soft. The second maxilla serves chiefly as a "bailer," or rather as a fan for driving water out of the respiratory cavity in front. (Fig. 2, pl. XXXVI.) This thin elastic plate lies nearly horizontal, the divided protopodite and rudimentary endopodite closely fitting over the mandible and the conforming first maxilla, and is formed by the fusion of an anterior exopodite and posterior epipodite, the upper side of the former and lower side of the latter, when not in rythmic movement, resting against the sides of the respiratory cavity. (For action of fan see p. 247.) The "masticatory ridges," or setigerous coxa, and basis of the second maxilla are partially cleft and distinctly separated by a superficial fold.

The first pair of maxillipeds (fig. 3, pl. XXXVI), except for one or two particulars, are modified only in minor details from the condition seen in the first larva. The parts are all rather soft, flattened, and curved to fit over the swelling mandibles and one another; the setæ of the meeting borders of the bases and coxæ are soft and useless for mastication; the exopodite lies against a shallow groove on the outer side of the two-jointed endopodite, the groove being marked by independent rows of setæ and the branch presenting a modified four-sided appearance. There is a long respiratory epipodite which carries no gill, but a part of its outer border is folded or turned under so as to form a trough, *fd* in which plays the posterior blade of the "bailer," or scaphognathite.

In the slender, outwardly swelling second maxilliped (fig. 4, pl. XXXVI) there is a fused joint (x) between the ischium and reduced basis. The brushes of setæ which fringe the inner border of this compound segment and the long curved meros are all soft, and on the small knob of the dactyl only do we find short stiff spines which can in any way effectively react on the food in mastication. Both epipodite and podobranchia are rudimentary.

The third and last pair of maxillipeds are similarly curved and conform perfectly to the typical limb, with the exception of a fused third joint between ischium and basis. (Fig. 5, pl. XXXVI, *x*.) The three terminal segments of this appendage are flattened and, as commonly carried, crooked downward upon the longer and more modified meros and ischium. The latter podomeres are curved upward and outward, are three-sided, and, like the former, bear double fringes of dense setæ which are used, among other purposes, as cleaning brushes (see p. 179). In place of the upper or inner fringe, however, the trihedral ischium is provided with a serrate crest, or row of about twenty closely set "incisor" teeth. These tooth-like spines increase in size distally and end over the joint in a strong curved fang. They work on the principle of an old-fashioned nutcracker, but in this case with toothed jaws which are very effective in cutting the coarser pieces of food delivered by the slender claw feet before they are passed on to the smaller mouth parts. The first three segments of this limb are closely appressed and quite flat where they meet on the midline, the coxa bearing two flat and hairy spurs.

The third pair of maxillipeds are the only really effective "jaw feet," and with the mandibles the only appendages which play an important part in reducing the food. Of the other mouth parts, the maxillæ, especially the smaller first pair, and the second maxillipeds without doubt help in the mincing process to which the food is subjected, but their chief function, as in the first maxillipeds, is without doubt sensory and for passing the food up to the mandibles. When the latter have finished their work the "grist" is ready for the gastric mill.

THE SLENDER LEGS.

The ten thoracic legs, which are designated as the pereiopods in the higher Crustacea, consist of the great chelipeds and four pairs of slender walking legs (pl. XXXVIII), the first two of which bear weak compound or double claws and the last two end in simple dactyls.

The successive segments of these limbs move on hinges, a description of which is given in chapter VII, and are actuated by opposing muscles in the typical way with the exception of basis and ischium, in each of which a flexor is absent. (Fig. 1, pl. XLI.) The basis has but one ventral or posterior extensor, with movement limited to a few degrees of arc, and the ischium two posterior extensors inserted upon two tendons, which are set close together on the margin of the shell at the opening of the meros. Accordingly these limbs can not be flexed at the fourth joint. There are no fused joints in the slender legs, which commonly break between basis and ischium, and are regenerated from this plane.

Aside from their direct use in locomotion, the smaller pereiopods present a variety of functions, the last pair possessing brushes for cleaning the abdomen (see p. 303), and incidentally serve as picks to steady the animal as it crawls over the bottom. Far more significant, however, are the clusters of sensory setæ (*s. s.*, pl. XXXVIII) arranged in symmetrical rows on the last two segments of the slender legs. One can count a hundred brushes upon a single leg, and each brush contains from 50 to 100 setæ, the bundles themselves being gradually concentrated toward the tip. In other words, each limb is furnished about its apex with from 5,000 to 10,000 sensory hairs, each of which is supplied with at least one nerve element. With such sensitive feet the lobster can feel its way securely at every step, whether by night or by day, as well as test every object before handing it up to the mouth.

THE CENTRAL NERVOUS SYSTEM.

The nervous system, the coordinating and regulating mechanism of the body, is composed of a complex series of distinct but closely related nerve elements, and each element consists of a ganglion cell and one or more outgrowing processes, the principal of which in certain cases is termed the nerve fiber. Three kinds of nerve elements or neurons have been described, as follows: (1) Coordinating elements, which lie wholly within the central system, the probable function of which is to coordinate the action of its parts; (2) motor nerve elements, which consist of a ganglion cell in the central mass and of a fiber process which passes out to a muscle or gland; and (3) sensory elements, composed of specially modified cells of the outer layer of the skin and of sensory fibers which enter the ganglia of the nervous system proper. Certain nerve fibers which pass out to the skin or its immediate neighborhood end in close relation with sensory cells and serve to convey impulses from them to the centers, while others conduct motor impulses from the centers to the muscles or glands. The epidermic cells of the skin may be regarded as the simplest sensory cells, or as the direct ancestors of such, and all the specialized sense organs, such as the eye or statocyst, are essentially modified patches or pockets of the outer skin layer.

The most primitive sense being that of touch, it is not surprising to find in an animal like the lobster that virtually every part of the skin is capable of receiving and distributing either tactile or chemical sense impressions. The proper sense organs, however deep their final position in the skin or tissues, come into close relation with the nerve fibers with which each is abundantly supplied. The sense organs are thus a primary means by which any form of energy to which they are able to respond starts a series of changes which are finally translated into what are known to us as sensations, feelings, and other mental states.

The lobster has a nervous system of the relatively simple "ladder" or "chain" type characteristic of the higher invertebrates (pl. XXXIII), in which segmentation, begun at a lower level in the animal scale, is the dominant character of its structure and instinct the ruling method of its response. Its reflexes and instincts are very precise and very stable, but not necessarily invariable, and, as we shall see at a later page, the lobster even at the fourth stage is able to modify its actions in relation to experience and to form habits, and thus is gifted with a certain degree of what is usually defined as intelligence in vertebrates. The uprights of the ladder are the long commissures of the chain, the rungs the transverse commissures, while the paired ganglia for each somite lie at the junctions of these parts. In addition to this cord with the appendicular and other nerves which spring from it, the lobster has certain stomatogastric nerves and ganglia which have been described as a rudimentary sympathetic nervous system.

The brain or compound supra-esophageal ganglion (pl. XXXIII) is united, by means of a ring-commissure which embraces the esophagus, to the chain of paired ganglia; this traverses the mid-ventral portion of the body and is protected by an archway of the internal skeleton in the thorax. The brain, which is thus the only ganglionic part of the central nervous system dorsal to the alimentary tract, appears as a small whitish mass at the base of the rostrum and between the stalks of the compound eyes. It gives origin to the following paired nerves: (*a*) The large optic nerves, which terminate in the optic ganglia and the compound eyes of the eyestalks; (*b*) the antennular nerves supplying the first pair of antennæ, and (*c*) antennal nerves which innervate chiefly the second pair of antennæ. The brain thus represents the fused ganglia of the first three somites and is connected by esophageal commissures with the central cord.

The subesophageal ganglion, or first ventral link of the chain, lies below the mouth and is composed of the ganglia of the mandibles, the maxillæ, and the maxillipeds (segments IV–IX), more or less intimately fused together, the ganglia of the large maxillipeds being nearly or quite independent.

Then follow five pairs of thoracic ganglia, which supply the legs and body wall, and six abdominal ganglia, the last of which sends nerves into the terminal telson. The longitudinal commissures between the twelfth and thirteenth somites diverge to admit the sternal artery, which thereupon divides, one of its branches passing forward and the other backward immediately under the nerve cord. (For nerves of cheliped, see ch. VII, p. 265).

In the embryo and larva the nervous system is much more concentrated than in the adult, and according to Allen (2) the thoracic ganglia are fused into one mass, which is united by short commissures to the brain. The hinder part of the embryonic brain is connected by a bridge commissure, which in the adult lies immediately behind the esophagus.

The nervous system is composed of a central "Punkt-Substanz" or neuropile, which, though granular in appearance, is in reality a felt work of fibers running in all directions, and an outer covering of ganglion cells. According to Allen the posterior ganglia of the chain give off two pairs of nerves, an anterior and posterior division; the anterior nerve becomes a double branch in the adult lobster and supplies the limbs, while the posterior division innervates the body wall.

THE PERIPHERAL STOMATO-GASTRIC SYSTEM.

In passing down the esophageal commissures, at a distance of about two-thirds of their course from the brain, a small commissural ganglion is seen upon either side lying against the wall of the esophagus. The delicate bridge commissure, which indirectly unites both sides of the brain, lies immediately behind these small ganglia and toward the lower side of the gullet, as already seen. Each commissural ganglion gives off two nerves, a dorsal medio-lateral and a ventral or antero-lateral nerve of Huxley, which send branches to a diffuse esophageal ganglion to be seen resting against the upper anterior wall of the esophagus (pl. XXXIII); from this ganglion, moreover, a median bundle, the anterior visceral or azygos nerve, runs up the wall of the stomach sac, to end in a minute gastric ganglion lying between the origins of the anterior gastric muscles. A smaller anterior median nerve also joins the esophageal ganglion to the brain.

The stomato-gastric system thus consists of four peripheral ganglia, two of which form a pair, and of peripheral nerves, which spring from them, in addition to a smaller ganglion belonging to the labrum, to be mentioned presently. The dorsal or medio-lateral nerve gives off two branches to the wall of the esophagus and bifurcates, a dorsal division going to the esophageal ganglion and a ventral forming the labral nerve, which has hitherto escaped notice. I have found that the two labral nerves end in a small labral ganglion embedded in the fleshy mass of this organ; from it issue fibers which presumably supply the sense organs of this part (see p. 237). The ventral nerve gives off a small branch to the esophagus and divides, one section going to the esophageal ganglion and the other passing to a plexus of fibers on the lower border of the mouth; from this plexus a very diminutive median nerve is sent to the esophageal ganglion.

Allen has traced with great skill the origin and course of the fibers in various nerves. Many of these fibers, which have bipolar cells in their course and which terminate on the walls of the esophagus, are possibly concerned with sensory cells.

SENSE ORGANS.

Special-sense organs, in so far as they are definitely known to exist in the lobster, are (1) the eyes, and (2) the sensory hairs or setæ, distributed over the body and appendages, if we omit from this category those organs of equally wide distribution which have the appearance of sensory buds and have received the general designation of tegumental glands. The hairs embrace (a) tactile setæ, which, though apparently aimlessly scattered over the appendages, are really distributed in a definite manner, including the setæ of the statocysts, and (b) chemical setæ, which abound on the antennules and where for a long time they have been supposed to possess an olfactory function, as well as on the mouth parts, to which a gustatory sense has been ascribed, and indeed upon the surface of virtually the whole body, where experiment seems to prove that chemical sense organs of some sort exist.

EYES.

At the time of hatching, the lobster possesses three visual organs, a median cyclopean ocellus, a mere rudiment of the simple type of eye which proved useful to its ancestors and is still retained in the lower orders of Crustacea, and the paired lateral or compound eyes. The latter, so conspicuous at all later stages of life, appear very early, and at the close of the fourth week their black pigment can be detected as a dark crescent-shaped line on either side of the head of the embryo. The eye is first disk shaped, then lobate, and finally stalked. In the first larva the stalks are immobile but very large, being relatively four times longer than in the adult. From the fourth stage the faceted eye is typically borne at the apex of a cylindrical movable stalk, which projects from either side of the base of the rostrum. Each stalk (fig. 1, pl. XXXV) is capped with a hemispherical surface, over which the cuticle has become modified into a thin flexible membrane as transparent as glass. Through it is seen the black pigment which defines the retinal area. This window-like cornea is interrupted by a process which juts in like a peninsula from the opaque shell at a point where the field of vision seems to be interrupted by the rostrum.

After the first larval stage the eyestalks recede somewhat until the lobster attains a length of from 1½ to 3½ inches, when their prominence is again very marked. In short, they now assume the form and relative size of certain fossil Crustacea from which the modern lobsters have probably descended.

The structure of the compound eye of the crustacean appears to be extremely complicated, because it is composed of units repeated many thousands of times. As was shown in 1889,[a] it is wholly derived by differential growth from a single plate of columnar ectodermic cells, the optic disk, which arises very early in development on either side in front of the future mouth and before the buds of the antennules are formed.

When the lobster's eye is examined with a hand lens, its clear corneal membrane has the appearance of a glass mosaic, composed of minute square disks of great uniformity both in size and arrangement, especially in its central parts (fig. 2 and 3, pl. XXXV). Each disk is the facet of an eyelet or ommatidium of the compound eye, and each supplies a part of the mosaic image produced in vision when the light is sufficiently strong. Each eyelet is developed from a cell cluster of the optic disk and this in turn from a single columnar cell of the primary optic plate.

The axial part of the ommatidium consists of (1) the corneal lens secreted by 2 underlying cells, (2) the refractive cone derived from 4 cone cells, and (3) a long striated and sensitive rod, the rhabdom, secreted and sheathed by 7 retinular cells, in addition to 2 peripheral pigment cells which surround the crystalline cone; in this rod also a nerve fiber terminates at the level of a basement membrane which divides the proper eye from the complex optic ganglia, muscles, and other tissues contained in the rest of the stalk. In ordinary daylight each eyelet is completely isolated by its sheath of black pigment cells, all of which display ameboid movement, but which respond differently to the intensity of the light stimulus.

In 1890, while working at the laboratory of the U. S. Fish Commission at Woods Hole, Mass., I showed by experiments upon the prawn *Palæmonetes vulgaris* that when this animal was placed in total darkness there was an immediate adjustment of the pigment cells of the ommatidium, in consequence of which the whole eye became intensely black and prominent, and that when returned to the light the eye began to lighten in a few minutes and in a relatively short time assumed its normal daylight appearance. It was shown that the blackening was due to a forward movement of processes of the

[a] The development of the compound eye of Alpheus. Zoologischer Anzeiger, bd. XII, p. 164–169, fig. 1–5. Leipzig, 1889.

distal pigment cells. One shrimp was kept in darkness 38 days, but the change was the same whether the interval was one of a few hours or weeks.[a] The true significance of this response was clearly established by Exner in his remarkable work on the physiology of faceted eyes in insects and crabs, published in 1891.[b] It was shown that the distal and proximal pigment cells or the "iris" and "retina" pigment moved in opposite directions in response to waning light, the former in its "night position" moving up to the cornea and leaving the refractive cone exposed and the latter crowding down upon the basement membrane, thus exposing the sensitive tip of the rhabdom. In the "day position" the converse movement takes place when the eyelet is completely isolated, and only those rays which are parallel to its long axis can enter and reach the rhabdom.[c] When the pigment screens are separated and drawn wide apart at night, on the other hand, light rays of any angle can pass freely from one ommatidium to another to be refracted by the exposed cones upon the upper ends of the exposed sensitive rods. The response is thus an adjustment to economize light, though at the expense of clearness of image. At dusk the lobster can presumably distinguish moving objects, but only dimly, since the eye at this time can produce no clear mosaic images.

The compound eye of the house fly is said to have about 4,000 facets, that of a dragon fly 20,000, while in a 12-inch lobster I estimated the number to be 14,000. Assuming that the ommatidia are equally well isolated and equally sensitive in each case, the relative efficiency of mosaic vision in insect and crustacean would be proportional to the number of facets. Upon this showing the lobster has a rather poor eye when we consider the unfavorable medium in which its visual powers must be exercised. The image produced by this organ, as Exner showed by a photograph made through the medium of the faceted insect eye itself, is single and upright; sight is attended by great loss of light, and must be very imperfect except for short distances and when the animal is moving in shallow water strongly lighted. The fact that the lobster is most active at night, that it is abundantly supplied with tactile organs for feeling its way about, and that the greater part of its life is spent at depths where clear vision is impossible for lack of light, show us further that its visual organs can play but a subordinate part in the activities of its daily life.

SENSORY HAIRS.

Certainly the most numerous and probably the most important sense organs of crustaceans generally are the sensory hairs or setæ, which are all of epidermic origin. Each hair consists of a hollow, conical, or nearly cylindrical shaft of chitin, continuous with the general cuticular basis of the shell, and is associated with one or more sensory nerve elements connected with the central nervous system.

---

[a] Memoirs of the National Academy of Sciences, vol. v, 4th mem., p. 454. Washington, 1893.

[b] Exner, Sigm. Die Physiologie der facettierten Augen von Krebsen und Insecten. Leipzig, Wien, 1891.

[c] It has been found by Congdon that increased temperatures cause movements in the pigment cells, which are probably of a non-adaptive character and are reverse in direction to those caused by light. See Congdon, E. D: The effect of temperature on the migration of the retinal pigment in decapod crustaceans. Journal of Experimental Zoology, vol. IV, p. 539–548. 1897.

The exact analysis of the sense organs of the higher Crustacea is still a vexed problem, and the literature of the subject far from satisfactory.[a] In the description to be given I shall follow in the main the account of Prentiss (*217*), who worked upon the common prawns, *Palæmonetes* and *Crangon vulgaris*, with which the lobster undoubtedly agrees in these particulars. The sensory bristles of decapods have been found to conform to two types: (1) The tactile, and (2) the olfactory, or better, the chemical setæ which are sensitive to chemical stimuli. The former have straight, long, and often plume-like shafts, and at the base of each a spherical enlargement is formed, which, owing to its thin wall, permits the hair to swing freely as upon a joint. Bristles of this type occur all over the body and appendages, and the "auditory hairs" of what has been called the "ear-sac" or otocyst (fig. 2 and 4, pl. XXXV) are of this form. According to Prentiss, each is supplied with a single nerve element. The "olfactory" or "chemical" bristles are shorter, more cylindrical, or less tapering chitinous tubes, with no marked basal swelling. Their tips are either perforated or possess so thin a wall as to permit the ready diffusion of chemical substances from the water to the inside of the shaft. Each bristle is supplied with a cluster of nerve elements, which may be very numerous, their fibers ending free in the shaft, but not penetrating to its apex. Such setæ are apparently more highly specialized and are restricted to the small antennæ, where they are called olfactory hairs, or to the mouth parts, where they are often spoken of as gustatory bristles, though it is probable that their functions are the same wherever found.

RELATION OF THE SETÆ TO HATCHING AND TO MOLTING.

The way in which these sensory hairs are formed and renewed at each molt is very interesting. The subject has been investigated by a number of naturalists, but in the brief account which follows we shall depend mainly upon the observations of Prentiss. Each hair is secreted by a number of matrix cells which send their processes up into its shaft. In preparation for the molt the protoplasm recedes from the shaft of the hair and its matrix cells sink into the tissues and with other cells form a "papilla" around the nerve fiber and begin to secrete a new hair. This condition lasts for a long time in an adult animal, but for a few days only in the larva, which often passes several molts in the course of a week. The cuticle which is to form the new shell and hair is secreted under the old which is soon to be cast off, but the new hair is invaginated, so that below the level of the skin its wall is double, while its tip only projects into the hollow shaft of the old hair above it. The walls of the double hair tube are thus continuous with each other and with the general cuticle which is to form the new shell.

In this condition the hairs may be compared to the fingers of a glove which have been pushed in or telescoped, so that their tips only project from the surface. When the lobster is ready to molt every new hair on its body is in this condition. Now at each molt we always find between the old and new cuticle a sticky, homogeneous substance which adheres both to the old shell and to the tips of the new hairs. Molting

[a] For a review of this subject, see Bell: The reactions of crayfish to chemical stimuli. Journal Comparative Neurology and Psychology, vol. XVI, p. 299–326. 1906.

thus becomes a means of drawing out or evaginating every microscopical haii of the newly-formed armor.

This adjustment is even more complicated in the young lobster about to hatch. Its "swaddling clothes" are so pinned together that all come off as one piece; the animal hatches and molts at the same time. The outer egg membrane splits lengthwise like the skin of a pea; it is glued in certain places to the inner membrane or true egg shell; this adheres to the outer deciduous cuticle, which in turn sticks at innumerable points to the hairs; by the time the animal has kicked off its covers it is thus ready to swim, for every hair is drawn out to its full length.

In hatching the eggs of lobsters by artificial means in jars or boxes, this delicate adjustment often fails at one point, and the little animal is doomed. The egg membranes fail to stick, and thus to pull out the swimming hairs, so that the young lobster is hatched in a helpless condition. It struggles in vain, a prisoner inside of its own skin, which it is unable to shed.

Blood pressure is another factor which enters into this important process of evaginating the setæ, and in all adult lobsters withdrawal of the blood from the great claws is an essential condition of the molt. As a consequence, when the animal escapes from the old shell, the hair clusters on the deformed plastic flesh of the great claws are scarcely visible, while they are prominent in other parts. With returning blood pressure the hairs of the toothed claw are fully evaginated. It seems evident that when once the shell has become hard no further evagination of the hairs is possible.

From the method of formation of new hairs it follows that at each molt, as Prentiss has shown, the nerve fibers lose their connection with the old hairs and enter into relations with the new ones.

TOUCH, TASTE, AND SMELL.

As long ago as 1868 Lemoine (*179*) suggested that the senses of taste and smell in higher Crustacea might be blended with that of touch, and while many able workers have since attacked this problem and produced far better results, we are still unable to speak with much exactness upon the subject. As I have shown by earlier experiments, nearly every part of the lobster's body is subject to tactile or chemical stimulation, and must therefore be supplied with sense organs of some sort. (See *149*, p. 129.) We found that the parts most richly supplied with setæ, with the exception to be noted below, were most sensitive, and it seemed evident that all the soft setæ, whether fringing and protective or not, were sensory. It was further observed that the greater sensitiveness was lodged in the antennules, and especially in their outer whips, which bear the peculiar club-shaped setæ, the antennæ, the tips of the slender legs, and in younger animals, at least, in the fingers of the big claws. Stimulation with various gases and liquids, injected with a pipette upon a given part, gave more or less prompt reflexes either in the limb itself or in the appendages nearest the part affected. If any stimulus, whether electrical, tactile, or chemical, be applied to the right second maxilla or right first maxilliped, vigorous chewing movements are immediately started in the affected appendage of that side, and may spread to the side opposite.

The swimmerets of the lobster were also proved to be quite sensitive under most conditions, as well as the thoracic sterna, the wings of the seminal receptacle of the female, and even the hard carapace, which was nearly as responsive to weak acids as is the soft skin of the frog, and the scratching movements made by the legs in the direction of the stimulated part are essentially the same in each case. We concluded that the sense organs were the setæ, reenforced by sensory buds, which lie in the tissues beneath the hard shell, but open upon it by capillary ducts. For other reasons these perplexing structures were given the name of tegumental glands. We have found no reason to alter this conclusion, and can still point to the upper lip as a supporting case. The labrum while possessing no true setæ is highly responsive to chemical stimuli, and is full of the organs in question, which open by ducts all over it in the lobster, but are most abundant on the under concave surfaces, to which a greater sensitiveness was attributed in the crayfish by Lemoine; here the ducts are clustered in large sieve-like plates bearing 60 to 70 holes each. We have further shown (see p. 232) that the labrum is not only well supplied with nerves, but possesses an independent ganglion of its own. That these labral organs are not glandular in function might be also indicated by the fact that the upper lip is always clean in the lobster, and free from anything suggesting a glandular secretion.

Experiments on the crayfish by Bell and others have shown conditions essentially similar in most respects. In getting food, sight plays little part, the blinded crab or crayfish going unerringly to the bait. This is certainly true of the lobster, as the experience of fishermen amply proves. Apparently through their chemical sense organs, for we do not seem warranted in using either the word "smell" or "taste," they become aware of the presence of food, and are attracted to it, while in the crayfish accuracy in the localization and in the seizure of the food seems to be secured through the medium of touch.

Bethe, who performed some strking experiments with the common green crab, *Carcinus mœnas*, found that the chemical reaction was the most important in its search for food.

> The mouth parts, says Bell, in summarizing Bethe's results, seem to be more sensitive to chemical stimulation than the antennæ or the antennules, since the animals react when the latter are removed. The threshold of chemical stimulation is extremely low, for the animals react most vigorously to the trail left in the water by a finger that has been in contact with meat, and greedily devour filter paper which has barely touched meat, but to really clean filter paper they pay no attention.

Holmes and Homuth[a] have repeated Bell's experiments on the crayfish and tested its reactions to chemical stimuli after removal of the antennules and antennæ, and after destruction of the brain and a section of the ventral nerve-chain. They confirmed the old opinion that the olfactory sense was lodged chiefly in the outer branches of the antennules, but found it exercised in a lesser degree by the antennæ, the mouth parts, great chelipeds, and the slender legs. Destruction of the brain or nerve cord tended if anything to slow down the reactions, but did not put an end to response.

---

[a] Holmes, S. J., and Homuth, E. S.: The seat of smell in the crayfish. Biological Bulletin, vol. XVIII, p. 155–160. Boston, 1910.

The lobster feels its way in the dark or gropes about in twilight by the aid of the sensory hairs with which it is abundantly supplied. From 50,000 to 100,000 of these organs are present on the big claws and slender legs alone. In most cases we do not find it possible to discriminate between hairs which are solely tactile or for the chemical sense alone. The lobster finds its way, however, to the fisherman's baited trap after dark or in dim light by the aid of all those setæ which respond to the chemical stimulus, and chiefly no doubt by those on the anterior appendages, the hairs which project from the lower sides of the outer whips of the antennules being probably the most sensitive. Fine particles of the bait which diffuse through the water from all sides of the trap, or are carried by currents, furnish the stimulus which draws this animal to their source.

BALANCING ORGANS OR STATOCYSTS.

It is commonly observed that while a living fish swims with its body erect and poised, a dead one floats on its side, and that the former position is one of unstable, and the latter one of relatively stable equilibrium. The upright unstable position is maintained in life by compensating movements which are automatically called into play by aid of special sensory bodies called static organs. This is true of the lobster, and of all animals which carry themselves upright, in opposition to the force of gravity.

There is now considerable evidence to show that what were formerly regarded as true "otocysts," or ear sacs, in the basal segments of the first pair of antennæ, are static rather than auditory in function, and accordingly they have been more appropriately called statocysts or organs of equilibration. The sac of either side (fig. 2) fills nearly the entire segment, and is open to the outside by a fine pore barely large enough to allow a minute grain of sand to pass, or to admit the point of a pin. The membrane overlying this sac is thin and taut (fig. 4, pl. XXXV, *mm.*); long setæ encircle it, and also surround the mouth of the sac.

The sac originates as a shallow pit of the skin, sinks into the tissues, becomes horizontally flattened, and remains attached to the cuticle along its transverse front, the opening being gradually constricted to a minute pore on the inner side of the thin membrane. Upon dissection and examination of the sac from within, we see on its floor a semicircular or horseshoe-shaped sensory ridge (*s. r.*, fig. 3), studded with a median row of about 75 plume-like hairs and four times as many shorter setæ arranged on either side or crowded about its mouth. Three hundred and seventy-five hairs were present in a single case examined, but the number may be considerably greater. Some of the hairs have bent shafts; some are thread-like, and scattered among them and often glued to their tips are numerous fine sand grains, the "ear-stones" or otoliths, as they have been called. In one of the sacs examined there were several hundred grains, ranging from one-fortieth to one-six-hundredth inch in diameter, the smaller being far too minute to be picked up with the points of the finest forceps. Each hair of the sac is supplied with a nerve-element, and as Prentiss has shown, with but a single one, as is the case with all tactile setæ.

From the foregoing account it will be seen that in the water-filled sacs just described, with their rich supply of sensory hairs, many of which, having little weights in the form of sand grains glued to their tips, and all being subject to the impact of free particles with the least displacement of the body, we have what would seem to be an admirable apparatus for enabling the animal to carry itself erect in walking or swimming. Any swaying of the whole body would sway the little hairs, or rattle the sand over them, and the stimulus thus given, would act as a sign to which the nervous system of the animal could respond in an adaptive and useful manner.

The study of development throws some light on the probable use of these peculiar sense organs. As shown by my earlier studies but first carefully worked out with histological definiteness by Prentiss, the sacs are developed in the free-swimming stages. They are barely visible as shallow depressions in the second and third larvæ, but in the fourth stage sensory hairs and sand grains are present, and closure of the sacs, which has now begun, is gradually effected with each successive molt. As Prentiss has shown, this "sudden leap" in the appearance of the sacs at the fourth stage is probably related to the abrupt change in form and method of swimming exhibited at the fourth molt.

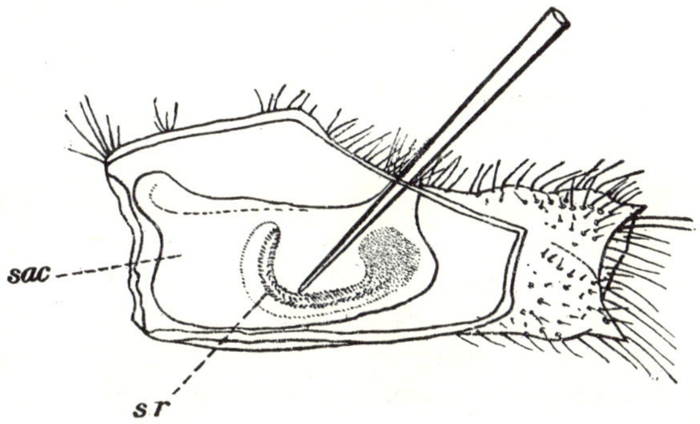

FIG. 3.—Sectional view of antennal segment to show statocyst, with needle inserted in pore at surface and pointed to sensory ridge, *s r*.

Every one who has watched the swimming movements of the young lobsters up to the fourth stage (fig. 34 and 42) has noticed how unsteady they become whenever the water is in the least degree disturbed. In ordinary swimming, when their equilibrium is not upset, the thorax is horizontal and the abdomen bent; in rising the head is inclined downward, but at best they are very unstable, and frequently pitch and reel to and fro, swimming now on their backs, now with their heads directed up or down. (See fig. 40.) It should be added, however, that under certain conditions, as in dull light, the young larva, as Hadley observes (*131*), swims with grace and precision, and there is no doubt that the eyes act before the statocysts as organs of orientation.

At the fourth stage (pl. XXXI) the little animals uniformly bear themselves erect like an adult and move about with great speed and definiteness. Prentiss has pointed out that when the young at this stage are unable to get sand for the statocysts, their movements again become uncertain, like those of an adult animal from which the sac has been removed. It is thus evident that while other organs, such as the eyes and antennæ, may help a crustacean to maintain its erect attitude, the sacs are indispensable for this purpose, at least after the larval stages.

It seems to be established that the supposed response of aquatic animals to atmospheric sounds of ordinary intensity is a myth, for sound waves propagated in air are almost totally reflected from the surface of water, but since sound vibrations are transmitted by water it does not follow that aquatic animals are necessarily deaf. An animal so abundantly supplied with tactile organs as a lobster has little need of ears, since sounds transmitted through the water would be perceived or felt by means of the sensory hairs. "The range of the average auditory organ in mammals," to quote from the work of Prentiss, referred to above, "is from 30 to 16,000 vibrations per second; waves of less than thirty vibrations per second do not usually produce auditory sensations, but are appreciable to the tactile sense. It is important to note that decapods respond most vigorously to low notes, and not at all to high notes or sounds produced by very rapid vibrations. This fact would seem to be good evidence that the vibrations imparted to the water and perceived by decapods correspond to those which produce tactile rather than auditory sensations in vertebrates." It has been noticed that the so-called "auditory" hairs of certain crustaceans will vibrate to different musical notes, as will the hairs on the back of one's hand or the strings of a violin, but they are not auditory, as Prentiss remarks.

It is only natural to find that the senses of touch and hearing grade into each other, and in either case it is the effect of a vibration which is perceived. While it is a matter of convention how these sensations are described, it is evident that an aquatic animal like the lobster has no organ strictly comparable to a vertebrate ear or even to the auditory or chordotonal organ of insects, and that if possessed of such an instrument it would have little occasion to use it. The basal segments of the large antennæ of *Palinurus* possess a peculiar structure often called a "stridulating organ," but nothing seems to be known of the real uses which it serves. (See p. 160.)

To return for a moment to the sacs, which have the form of a narrow-necked bottle, and are carried in the antennulæ, how do the sand grains find their way through their minute openings, guarded with hairs? Professor Brooks has seen the megalops larva of the crab, *Callinectes*, pick up the grains and place them in the sac with its claws. As an illustration of animal instinct, this is truly remarkable, for it is peculiar to the larvæ alone, the adult crab having no sand grains or otoliths of any kind in its sacs. The lobster at the fourth stage nearly corresponds to the crab megalops, but it has never been seen to behave in this manner. Whatever method the young may adopt to replenish their stock of sand after each molt, it is evident from the microscopical proportions of the grains that adults behave in a different manner. The animal in all probability thrusts its head in the sand, while the smaller grains, selected by the one opening of the "strainer," gradually sift into the sac by the force of gravity. The spiny lobster (*Palinurus*), which also keeps its antennal sacs well supplied with sand, has no claws with which to pick up anything, and must have recourse to a similar method. In reference to this peculiar need of the animal, it is interesting to notice that molting lobsters often burrow in the sand, where they remain for some time after casting the shell.

THE MUSCLES.

The muscles of the lobster's body are of two kinds, the striped or striated and the non-striated, distinguished in higher animals as the voluntary and involuntary muscles. The involuntary muscular tissue is inconsiderable in quantity, excepting the "fine meat" at the tips of the claws, being mainly confined to the walls of the alimentary canal, the blood vessels, and sexual organs. The heart and powerful skeletal muscles are composed of distinctly striated fibers.

The skeletal muscles, of which the large adductor of the mandibles is a good example, are attached to the hard shell on the one hand, and to tendinous ingrowths of the softer cuticle on the other. Just how the union with the shell is effected is a somewhat vexed question. In the first larval stage of the lobster the prominent muscle just referred to is distinctly striated up to the basement membrane. (Fig. 2, pl. XLVI, *bm.*) At this level its fibrillæ are directly continuous with attaching fibers within the cells of the epidermis; the basement membrane is accordingly penetrated at this point. Examination of earlier embryonic stages shows essentially the same conditions. The epidermis of the shell in the area of attachment (*fb. ep.*) is modified in a characteristic manner; its cells are columnar and elongated, and their cytoplasm develops fibers which appear to fuse with those of the muscle-fibrillæ; moreover, their nuclei are eventually reduced and spindle-shaped, though this was not the case in the specimen figured. The basement membrane in this region is a distinct cuticular sheet, to which blood cells and other elements (*ms.*) presumably of mesoblastic origin also attach themselves, with long axes parallel with the surface, thus making a distinct lamella. The horizontally placed lamellar cells can be detected beneath the modified epiblast, where the cuticular portion of the membrane appears to be reduced or absent. In some cases the epiblastic fibrils brush out perceptibly at their periphery against a concavo-convex layer of chitin, upon which the outermost stratum of the shell is molded. Since the clearer inner chitinous layer frequently peels off in preparations, it may represent a renewal of the shell at this point previous to molting.

In his study of regenerating limbs in the lobster, Emmel (97) has found that the striated muscles are regenerated from ectoderm, and that the outer ends of the myofibrillæ are differentiated as tensile elements, which pass between the proper epidermic cells, are frequently spread out in branches, and are fused directly to the chitin of the shell.

The muscles of the tail, which form a great part of the edible flesh of the lobster (pl. XXXIII) consist of two paired masses, the dorsal extensors, by the contraction of which the abdomen is straightened, and a much larger pair of ventral muscles, mainly flexor in function, which form the principal source of power for locomotion. As we have seen, the segments of the shell in this region are united by flexible membrane, and move over articular surfaces as well as upon double hinges of the typical ball-and-socket form, and that the parallel and horizontal arrangement of their articular axes limits the flexion of the tail to the vertical plane. The ventral muscles are very complex, being composed of external bundles attached to the side walls of successive segments, and of interlooping or enveloping strands, which are fixed to the lower or sternal parts of the skeleton. A twisted rope-like mass is thus formed, the forward strands of which are attached to the linkwork of hard tendons in the thorax. There are also in the thorax, rotator abdominis, ventral thoracico-abdominis and tergo-epimeral muscles, as well as flexors of the telson and tail fan in the abdomen.

The weaker dorsal muscles (pl. XXXIII) form a pair of segmented strands overlying the alimentary canal and dorsal blood vessel. They are inserted into the anterior border of each abdominal somite and diverge as extensor abdominis muscles in front, where they are attached to the walls of the thorax below the cervical groove. When the ventral muscles suddenly contract at the command of the nervous system, the combined pulls on successive joints bring the tail with expanded tail fan quickly and violently down upon the thorax, and the animal shoots backward through the water. By the contraction of the weaker extensor muscles the body is again brought into a horizontal position, and ready for another downward stroke. Raising the abdomen tends to send the animal forward, but owing to the obliquity and slowness of the stroke after closure of the tail fan the speed is but little checked. The muscular equipment of the great claws and legs are described in chapter VII.

Two prominent light spots are conspicuous on either side of the carapace of an adult lobster, one at a point about an inch behind the base of the large "feelers," and the other about as far behind the first, close to the irregular depression known as the cervical groove. (See p. 220.) The first, which is large and very conspicuous at the sixth stage, when the animal is barely five-eighths inch long, is the mark of a straight rod-like tendon which binds the carapace firmly to the internal skeleton below. The latter was without doubt originally a tendon-mark also, but in place of a distinct tendon, short muscle fibers issue from its margin, and from the groove in front, to be attached to the wall of the gill chamber. The scar-like impression conforming to the groove and immediately in front of it marks the attachment to the shell of the posterior suspensory muscles of the stomach sac. The powerful adductor of the jaws, by the contraction of which their cutting surfaces are brought to bear on the food, divides to give passage to this gastric muscle, one section of which is attached to the carapace in front of the groove, and the other just behind it on the endotergites, which as stated above are tendinous ingrowths from the fold itself. The anterior gastric muscles are inserted on the procephalic plates.

Some fourteen pairs of extrinsic and intrinsic gastric muscles have been described by Williams (279). These serve either to suspend the stomach sac to the inner wall of the carapace (anterior gastric, anterior dilators, and posterior and lateral gastrics) or to move its nicely articulated framework, bring the food to mill, work the grinding teeth, and to effect in some measure the sorting and straining of the comminuted food particles.

THE BLOOD AND ORGANS OF CIRCULATION.

The blood of the lobster when freshly drawn is quite colorless, leucocytes or white blood cells being the only corpuscles present, but after exposure to the air for a few minutes it becomes tinged with blue, and thickens or coagulates. The bluish color is imparted by a respiratory pigment called hæmocyanin, which like the haemoglobin of red blood becomes deeper in color as it takes up oxygen. The bluish tint of the larval lobster is probably due in part to the hæmocyanin of its blood. The blood is also regarded as the bearer of other pigments, the lipochromogens, which are probably elaborated in the digestive gland, transmitted by the blood, and laid down in the pigment cells and the shell.

The heart begins to pulsate rythmically when the lobster is an embryo, between 4 and 5 weeks old,[a] at a time when the black pigment spots of the compound eyes have begun to show, but when the nervous system has been only roughly blocked out and long before any nerves are developed. The heart, although later brought under nervous subjection and control, is at first quite automatic and independent in its movements.

The circulatory system of the lobster (see pl. XXXIII) consists (1) of a muscular heart for driving the blood, (2) of arteries or definite channels for conveying it to the tissues, and (3) a system of irregular channels called sinuses or lacunæ, besides certain well defined vessels, the veins for leading it back to the pericardial chamber and heart. The arteries end in microscopic capillaries which open directly into the lacunar system.

The freshly aerated blood of the lobster is driven from the gills to the pericardial sinus, enters the heart through the ostia, is pumped thence by the rhythmical contractions of its walls into the arteries, and by their subdivisions is distributed over the entire body. Having performed its physiological work of giving up to the tissue cells dissolved oxygen and food materials, and having received from them carbon dioxide and other waste products, it returns by the lacunar system to the large ventral sinus, which surrounds the ventral nerve-chain; thence the venous blood is driven to the gills, where aeration is effected by the absorption of oxygen from the fresh streams of sea water in which they are constantly bathed. More simply expressed, the path traversed is heart, body, gills, heart. The gills are placed in the returning blood stream, so that the vessels which both supply the gills with venous blood (afferent branchial vessels) and which conduct arterial blood from the gills to the heart (efferent branchial and branchio-cardiac vessels) may be described as veins.

THE HEART.

Examining the heart more closely, it appears as a boat-shaped or somewhat hexagonal body, rounded below, flattened above, and broader in front. It is pierced by three pairs of openings, the dorsal, ventral, and lateral ostia, which admit blood from the pericardial sinus. Each ostium is provided with valves which open inward, so that the blood once admitted to the heart can not be regurgitated to the sinus.

The heart gives off a series of arteries, five in front and two behind; these are also supplied with valves (or at least in the largest of them, the sternal), so that the heart can empty only into the arteries, while it can fill only from the sinus.

THE PERICARDIAL SINUS.

The chamber in which the heart is suspended, called the pericardial sinus, lies at the extreme upper and hinder part of the carapace; it is lined with connective tissue

---

[a] The beating of the embryo lobster's heart has been noted in winter (December 14) at 100 times per minute.

and muscle fibers,[a] and has an arched roof and floor, with sloping sides. This chamber lies close to the back, so that if the shell is perforated anywhere in the cardiac area the animal will quickly bleed to death. The convex floor of the sinus covers the sexual organs and the digestive gland, while at the sides only the thin shell of the body wall (inner epimeral surface) separates the sinus from the upper part of the branchial cavity. Moreover, the extensor muscles of the tail virtually pass through the sinus and are inclosed between its sides and floor.

The heart beats rythmically and heat accelerates its action. Plateau (*214*) found that the isolated lobster's heart, when placed in a moist chamber, would beat for nearly an hour; according to this investigator the movements of the decapod heart are governed as follows: (1) By a cardiac nerve which arises in the stomato-gastric ganglion and ends in the heart muscle; (2) by ganglion cells within the tissue of the heart itself, by means of which its automatic movements are maintained, and (3) by depressor nerve fibers which moderate the heart's action, but the real courses of which are not known. The brain is found to have no direct influence upon the action of the heart.

THE ARTERIES.

Of the five anterior arteries the ophthalmic or cephalic runs along the middle line just beneath the shell, and makes straight for the brain, which it supplies, together with the eye stalks (upper side), giving off a few twigs to the stomach sac in its course. The paired antennal arteries issue from the side of the ophthalmic, and in passing forward along the surface of the gastric glands they give off numerous small branches to the following organs: The glands themselves, the gastric muscles and walls of the stomach, the sexual organs, the thoracic muscles, and the body wall, or the integument of the carapace and the inner epimeral wall of the branchial cavity; finally the same vessel sends twigs into the eyestalk, the antennule, the adductor mandibuli muscles, the antenna, and the green gland which lies at its base. The paired hepatic arteries supply the gastric glands. Both ophthalmic and antennary arteries are subject to considerable variation in both the lobster and crayfish. (See fig. 1, pl. XLIV.)

Two arteries issue from the hinder end of the heart, where it swells into a bulb, namely the sternal artery, which passes straight down and penetrates the nerve cord, and the superior abdominal artery, which supplies the greater part of the tail. The sternal gives off twigs to the sexual ducts before it swerves to pass the intestine, and entering the ring formed by the long commissures between the fourth and fifth ganglia of the ventral chain (somites XII and XIII), gains the ventral side, where it divides or gives off a posterior branch, the inferior abdominal artery, which supplies a small part of the ventral surface of the abdomen, but none of the appendages. The main branch of the sternal, the inferior thoracic artery, runs forward under the nervous system, and supplies the slender legs, the great forceps, and the mouth parts.

---

[a] According to Dogiel (*72*), the pericardium also contains blood vessels, which can be injected from the superior abdominal artery, as well as nerves supplied by a trunk (nerve of Dogiel) which is given off from the ganglion of somite XII. The valves of the heart are further regarded as properly sphincters, rather than of the bilabial or semilunar form. On the other hand the sternal artery, of which the superior abdominal may be considered a branch, is provided with true valves of the bilabial type.

ARTERIAL SUPPLY OF THE SWIMMERETS.

The dorsal or superior abdominal artery passes backward just above the intestine and gives off six pairs of segmental lateral vessels, which, besides supplying the intestine itself, send arterial blood into the great muscles of the tail, the posterior lobes of the gastric glands, and the sexual organs. To complete the statement, however, it must be added that the main branches of the lateral segmental vessels are curiously continued around the sides of the body to the swimmerets or pleopods, which they feed with arterial blood.[a]

The swimmerets have been invariably described as receiving their blood from the inferior abdominal artery, both in the lobster and crayfish, an error which may have arisen in the first instance from failure to inject the vessels or from inference, probability favoring the inferior vessel, on the principle that organs as a rule draw their blood supply from the nearest source. The error, started in some such way, has escaped the scrutiny of such keen observers as Professors Huxley, T. J. Parker, and Howes, and is to be found in all the text-books and literature dealing with these forms. It can be seen, however, without recourse to much dissection, that the inferior abdominal artery is too diminutive and passes altogether too small a quantity of blood to supply the swimmerets, which are the most active of all the appendages, excepting only the respiratory plate or "bailer" of the second maxilla.

The superior abdominal artery divides at the hinder border of the fifth somite into two branches, which embrace the intestine where it gives off a short cæcum on its upper side, and which run backward and diverge to supply the sixth somite and tail fan.

The principal artery of the big claw (pl. XL) traverses the lower side of the limb and gives off numerous branches to the muscles of the segments. In the fifth podomere it sends off a shoot which enters the big claw, passes to the abductor muscle along the inner border of the big tendon, and ends in the fine meat of the dactyl. The main artery, upon entering the claw, again divides, giving rise to four branches, three of which supply the big adductor muscle and the fine meat of the propodus, while the other passes to the adductor muscle and divides, sending a branch to both dactyl and propodus. The division to the dactyl is united by a cross branch to the vessel which supplies the abductor and enters the propodus from the fifth joint. In the index and dactyl the arteries ramify in tree fashion, and apparently break up into a lacunar system of irregular spaces in the fine meat. From this situation the blood returns by a large irregular channel and enters the sternal sinus, whence it reaches the gills.

It has been shown by Emmel (97) that as the returning sinus of the great cheliped passes the ischium or third podomere it is divided into two channels by a septum of connective tissue. These dorsal and ventral sinuses, moreover, possess valves which originate as folds from the septum and become operative to staunch the flow of blood from the breaking joint the moment a claw is shot off (see p. 282).

---

[a] I am indebted to Prof. Carl B. James for first directing my attention to this fact, which must have been noticed by other teachers in the laboratory.

THE GILLS.

The adult lobster is provided with 20 pairs of gills, 1 of which, belonging to the second pair of maxillipeds, is rudimentary. Of these, 6 are podobranchiæ, 10 arthrobranchiæ, and 4 pleurobranchiæ, distributed according to the following table:

TABLE 5.—BRANCHIAL FORMULA OF THE LOBSTER.

Thoracic segments and appendages.	Podo-branchiæ.	Arthrobranchiæ.		Pleuro-branchiæ.	Totals.
		Anterior.	Posterior.		
VII, first maxilliped	0 (ep.).	0	0	0	0 (ep.).
VIII, second maxilliped	1 rud. (ep.).	0	0	0	1 rud. (ep.).
IX, third maxilliped	1 (ep.).	1	1	0	3 (ep.).
X, first pereiopod	1 (ep.).	1	1	0	3 (ep.).
XI, second pereiopod	1 (ep.).	1	1	1	4 (ep.).
XII, third pereiopod	1 (ep.).	1	1	1	4 (ep.).
XIII, fourth pereiopod	1 (ep.).	1	1	1	4 (ep.).
XIV, fifth pereiopod	0	0	0	1	1
Total	6 (ep.).	5	5	4	20 (1 rud.).

ep.= epipodite. rud=rudimentary.

The first larva has no rudiment of a podobranchia in the eighth somite, but all the other branchiæ are represented. The podobranchiæ of the following segments are very small and are partially exposed, together with their reniform epipodites (fig. 34). In the second larva the podobranchiæ are covered by the carapace (fig. 41) and the branchial formula is complete.

The gills are developed in the embryo as simple folds or pouches in the body wall. (fig. 8, *g. fil.*) They belong to the trichobranchiate type, the respiratory surface being gradually increased by growth of multiserial branchial filaments.

In the fourth larva the podobranchia carries four rows of filaments, and the mastigobranchia, or epipodite proper, is a long, tapering, hairy plate.

The adult gill (pl. XXXVIII), suggesting by its form a bottle brush, is a pyramidal tuft, consisting of a central stem and numerous longitudinal rows of branchial filaments, which enormously increase the area of the surface exposed to the water. The number of rows of gill filaments gradually increases with the size of the animal and with its need of a greater respiratory surface, until it reaches between 30 and 40 in an adult 10½ inches long, while the total number of filaments in such a gill is between 3,000 and 4,000. The filaments are "parted" into two groups by a median longitudinal furrow and in the larger posterior section tend by transverse partings to separate into quadrangular masses. The filaments gradually lengthen in passing forward or backward on either side of the "part" and terminate in several rows of short filaments next the efferent division of the stem, opposite the body wall. Further, the filaments are so regularly spaced that they come to assume an arrangement in circular rows from base to apex of the branchia, corresponding to the circular efferent vessels (fig. 2, pl. XLVII *c v*) with which they communicate.

THE BRANCHIAL CAVITY AND RESPIRATION.

The branchiæ are lodged in a cavity of peculiar form upon either side of the body, where they are securely protected by the broad sides of the curving carapace. The gills (pl. XXXIV) arch upward in pyramidal form from the bases of the limbs and the sides of the body to which they conform, those of successive somites being divided by the gill separators or epipodites, which are hairy respiratory plates, springing from the basal segments of the limbs. Currents of water set upward and forward from under the free edges of the carapace, pass over the myriads of fine filamentous processes of branchiæ, and are led into a trough or groove at the forward end of this curved narrow passageway on either side of the body. From this trough the water is fanned out by the rythmic beating movements of the "bailer" or respiratory plate of the modified second maxilla (see p. 228). The fan or respiratory paddle thus works with up-and-down strokes in a narrow passageway,[a] which is horizontal in front, and behind curves upward abruptly to the pyramidal apices of the gills. The lower bound of this passage is formed mainly by the epipodite of the first pair of maxillipeds, which is folded over so as to form a sort of trough in the part where the free inner division or epipodite of the bailer plays (pl. XXXVI, fig. 3 *fd.*). This fold presses against the side of the carapace and keeps water from entering the trough until it has passed over the lower half of the gills. The outgoing stream is thus essentially limited to the forward upper part of the gill cavity.

By the alternate beating of the hinder (epipodite) and anterior (exopodite) divisions of the bailer the water is driven forward and out of the cavity.

At the extreme hinder end of this chamber the carapace overlaps a small hairy leaf-like plate belonging to the fourteenth somite and bearing a small oval lacuna in its chitinous cuticle, just behind the pleurobranchia of this segment and above the hinge joint of the limb. This corresponds to similar lacunæ for the four pleurobranchiæ in front and without doubt represents the position of a former gill, every other vestige of which has now disappeared.

As blood slowly passes through the 20 pairs of gills and their protective plates the act of respiration is accomplished. Carbon dioxide diffuses from the blood through the thin walls of the filament, and from the air dissolved in the sea water the oxygen supply of the blood is renewed. The water in the respiratory chamber is kept stirred up by the legs, to the bases of which 10 of the gills are attached, while the incessant beating of the fan at the front end of the cavity (marked by the frothing which commonly occurs when the animals are taken from the water) causes an active forward flow through the chamber and over the gills as described above. If the motion of the fan is stopped the animal soon becomes asphyxiated. The lobster will live for a long time out of water, in some cases for upward of two weeks, provided the branchiæ are kept moist, and even in hot weather when the air is cooled by ice.

From the filaments the aerated blood is conducted down one of the efferent branchial veins on the inner side of the stem in each gill, and thence through a distinct channel, one of the branchio-cardiac veins, to the heart.

---

[a] The "fan" has been noticed to beat at the rate of 95 to 178 strokes per minute in summer, in lobsters which had been out of the water long enough to become quiet.

COURSE OF THE BLOOD IN THE GILL.

The description of the course of blood through the gill given above usually suffices for the text-books of zoology, but the physiologist wishes to know how the blood circulates in the gill filaments, for if these were simple capillary tubes it would tend to flow past rather than through them. The gill in reality is a complicated structure, and the actual course of the blood is not easy to follow.[a]

Each filament, like the stem of the branchia, is a double tube or vascular loop, consisting of outer afferent and inner efferent divisions (fig. 2, pl. XLVII.) All the blood must pass from the afferent branchial vein (*af. v.*) to the afferent divisions of the loops, thence to the efferent divisions, and then to the main efferent of the stem (*ef. v.*). The wall of the branchial afferent vein which carries unaerated blood to the filament suggests a cylindrical sieve or grater, with fine holes arranged in regular transverse rows. As the blood enters one of these holes it is conducted by a short passage to the afferent division of the loop or filament, but, as Dahlgren and Kepner have shown, the course by which the efferent half of the filament is reached is indirect. The venous blood in the afferent section enters a plexus of fine channels or capillaries, by which it is conducted around the filament and into the efferent loop. In the course of this passage the venous blood is brought close to the cuticular surface, but never quite touches it, there being always a cytoplasmic layer of the true epidermis of the filament, from which the cuticular covering is supplied at each successive molt. Thus, in passing through the filament the blood is kept in close relation to its surface, a condition which tends to promote the most active exchange of gases essential to respiration. These capillaries do not, apparently, have definite walls, but worm their way between or through the cells. The connective-tissue cells of the central core of the filament are described by Dahlgren and Kepner as being essentially peculiar and characteristic in possessing loosely branched protoplasmic processes. The efferent channel of each filament empties into a circular vessel (fig. 1, pl. XLVII, *c. v.*) which runs around the main afferent of the stem, and thus conveys the arterialized blood to the efferent vein (*ef. v.*).

The course of the blood through the gill is thus, in brief, as follows: Stem afferent to filament afferent, through filament capillaries to filament efferent, to circular vessel in wall of stem afferent, to stem efferent, to branchio-cardiac vein, to pericardium and heart.

This system of vessels is filled with blood, which, owing to the rhythmic contractions of the heart and the dispositions of its valves, is kept moving in the same direction, from heart to tissues, from tissues to gills, and from gills to heart again. The heart is "arterial," and the breathing organs of the crustacean are thus introduced into the returning stream of venous blood, the converse of the conditions found in fishes, where the heart is "venous" and the gills participate in the arterial system which leaves it.

---

[a] The account of the circulation of blood in the gill given in this section was written six years ago, when the drawings illustrating it were made. Certain details concerning the capillary plexus have been added since reading the work of Dahlgren and Kepner, who, so far as we are aware, were the first to describe the histology of the filament and the course of the blood through it.

THE ALIMENTARY TRACT.

The alimentary tract (pl. XXXIII), extending from mouth to anus, consists of three parts, which are quite distinct in origin, namely: (1) The foregut (stomodæum of the embryo), formed by a tubular invagination of ectodermic epithelium; this remains distinct until late in embryonic life, and gives rise to the epithelial lining of the esophagus and grinding stomach; (2) the midgut (mesenteron of the embryo), lined with endodermic epithelium, and formed by the walling in of the great mass of the yolk by endodermic cells; paired outgrowths or folds of the endodermic sac arise early in embryonic life and eventually form the liver of the adult; aside from the liver or gastric glands, the mesenteron appears to take no part in the formation of the alimentary tract; (3) the hindgut (proctodæum of the embryo), formed by a solid ingrowth of ectodermic epithelium which subsequently becomes hollowed out, its walls merging with those of the mesenteron; it gives rise to the lining of the intestine and cæcum.

The foregut and hindgut, being infolded parts of the outer surface of the body, are covered with a cuticle which is continuous with the chitinous exoskeleton, and is cast off in the molt.

THE GRINDING STOMACH.

The higher Crustacea are the only animals which grind the food after it reaches the stomach as well as before it enters the mouth. Granivorous birds swallow their food whole, and with the aid of gravel stones or other hard bodies pulverize it in a muscular gizzard; in a number of gasteropod mollusks analogous organs occur, but the stomach mill of a decapod crustacean is a much more complicated machine.

When a bit of fish or clam is offered to a hungry lobster, it seizes the food with the claws of the slender forward legs and passes it up to the mouth, where it is held by the large maxillipeds. The cutting teeth and spines of the mouth parts, especially the maxillæ and mandibles, are successively brought to bear upon it, and chop it into mince-meat, while it slowly enters the mouth in a stream of fine particles.

The stomach of the lobster is truly a complicated mechanism, and could not be fully described without entering into great detail. In the brief account which follows I shall rely mainly upon a study of this subject by Williams (*279*), which is by far the best that has appeared.

The stomach sac (pl. XXXIII and XXXIV) serves for storing, grinding, sorting, and straining the food, as well as for delivering the finest particles in liquid streams along definite channels to the intestine and to ducts of the liver; for, as Jordan has shown, the huge gastric glands serve also for the direct digestion and absorption of food. Further, the coarser particles of the food may be sent to mill time and again to be reground, while the indigestible parts are regurgitated. Again, it should be added that newly molted lobsters instinctively devour their own cast, and I have found soft lobsters with their stomachs stuffed full of the shells of mollusks and other calcareous fragments (see *149*, p. 89), actions which point clearly to the need of the animals at such times to obtain a supply of lime as quickly as possible.

The stomach is divided into a larger forward, or cardiac division, for storage chiefly, and a smaller hinder, or pyloric section (pl. XXXIII and XXXIV), mainly for sorting and straining the food. Between the two lies the gastric mill, the grinding "stones" of which consist of a single dorsal median tooth and of two large lateral grinders. The wall of the stomach is composed of two layers of connective tissue, in the inner and looser of which are lodged the blood vessels and muscles, a gastric epithelium, and a chitinous lining. The lining of the stomach is thickened in certain areas and hardened by deposits of lime, to form the calcareous plates or ossicles which make up the framework of the gastric mill; the largest and strongest ossicles culminate in the "millstones," or teeth, just mentioned. The lining of this organ is further thrown into various permanent folds, pads, ridges, or bands, between which lie definite canals for the circulation of liquids containing the comminuted food. Most of these parts are thickly studded with short setæ, which in general point toward the gastric mill, and serve to direct the food mechanically into its proper channels, whether to or from the mill, whether into the pyloric strainer or from this to the intestine and liver.

Aside from the grinding mechanism, the most essential parts of the stomach, according to Williams, are the distributing and circulating canals (the upper and lower cardiac and the lower pyloric canals) and the five food gates or valves, namely, the cardio-pyloric valve between the two main divisions of the stomach and the four pyloric valves which guard the passage of food to the intestine and the liver. There is a small intestinal cæcum, which extends forward over the dorsal wall of the stomach, and the short duct of the liver or gastric gland opens into the intestine between the ventral and lateral pyloric valves on either side. The conspicuous horn-shaped processes at the base of the pyloric sac and in front of the intestinal cæcum are the lateral pyloric pouches, where the finer particles of food are sifted out for delivery to the liver. In addition to the canals mentioned there are also a pair which traverse the median section of the pyloric sac. A small rudimentary tooth (infero-lateral tooth) is seen projecting from between folds of the stomach wall immediately below the anterior end of the lateral tooth, on either side (pl. XXXIII).

Upon each side of the stomach sac, at its forward end, a large ovate plate (pl. XXXIII) is to be seen, called the gastrolithic plate (lying immediately above a small gastrolithic bar). This plate is composed of a modified epithelium, which between the molts secretes the rounded mass of snow-white prisms known as the stomach stones or gastroliths. Williams has found that the gastroliths make their first appearance in the fourth stage, when for the first time the skeleton abounds in lime.

Over thirty distinct plates, ossicles, and bars enter into the complex framework of this organ, governed by some fourteen pairs of intrinsic and extrinsic muscles, some of these serving to suspend the sac to the dorsal wall of the carapace (such as the anterior, posterior, and lateral gastric muscles), for "turning the wheels" of the gastric mill and feeding the "hopper," as well as for dilating or constricting the cardiac and pyloric chambers.

From the mouth the food passes into the short esophagus, through an esophageal valve, and into the cardiac chamber of the stomach sac. Thence it is delivered through the cardio-pyloric valve to the mill to be ground. The contraction of the anterior and posterior gastric muscles reacts upon the articulated plates of the elastic frame in such a way as to bring the lateral grinders together and to draw the median tooth forward with great force. This upper middle tooth, or prepyloric ossicle, is shaped like a bird's beak and has brown indurated surfaces, while the lateral teeth, or surfaces of the zygocardiac ossicles, the principal grinders, are divided by parallel transverse furrows into a series of yellowish-brown hardened tubercles. According to Williams the forward and downward movements of the median tooth tend to drive much of the food back into the cardiac sac, so that it is reground again and again. Some of it, however, enters the pyloric division of the stomach, and filters back and forth in its chambers and canals. Here it is sorted and strained; the finer parts, suspended in fluids, are delivered by the canals to the intestine in four streams, while the coarser elements are swept up by bristles of the cardio-pyloric valve and sent to mill again. Two streams from the dorsal pyloric canal pass into the intestinal cæcum; a stream from the middle pyloric canal also delivers food to the intestine, while finally a current from the lower pyloric canal conducts food particles to the lateral pouch, where a final sifting occurs, the finest parts, suspended in fluids, entering the liver by the "bile ducts," and the coarser by way of the middle pyloric canal reaching the intestine.

When the muscles of the gastric mill relax, the elasticity of the framework is sufficient to separate the parts. While it is not possible to see these movements in the living animal, they can be roughly imitated by concerted pulls upon the anterior and posterior gastric muscles. Undoubtedly the clashing movements of the teeth go on for hours after a full meal until all of the food has been thoroughly stirred up, brought to mill, ground, and reground. After the soft and semiliquid parts have been filtered and delivered to the intestine and gastric glands, the indigestible residue is regurgitated through the mouth, as is the habit with many birds.

The intestine is a delicate tube of small caliber, and since there are no coils it is quite short. This suggests the need of a gastric mill, and the absorptive function of the glands, for the area of the intestinal surface being limited, the digestive process must be conducted as rapidly and efficiently as possible. As already seen, there is a cæcal enlargement on the dorsal side of the pyloric sac of the stomach. The intestine suddenly enlarges at the beginning of the sixth segment of the tail, where it gives off from its dorsal side another slender blind pouch or cæcum, which is apparently a rudimentary structure. (Pl. XXXIII.) From this point to the vent, which is closed by a sphincter muscle, and from the mouth to the beginning of the intestine, the canal is lined with cuticle which is continuous with that over the body and is accordingly renewed at each molt. The embryology of the animal shows that the inner wall of the intestine is primarily due to an ingrowth from the outside skin and in the early larvæ an intestinal cuticle can be detected, but if the latter is present in the adult it is reduced to a layer of extreme thinness.

THE LIVER.

The "liver" (pl. XXXIII and XXXIV), called also the gastric glands, hepatopancreas, and by the chefs "tomally," is the largest single organ in the body. It is paired of a green, bright yellow or yellowish green or yellowish brown color, and lies along the sides and partly below the alimentary tract of which it is a part.

The liver is a soft, lobulated mass, divisible on either side into three parts—a thick anterior lobe, a long posterior lobe, and a less clearly marked dorsal or lateral lobe. Each lobe is composed of many lobules, and each lobule of a multitude of short aggregated tubes called the cæca. The lobules are covered by a delicate transparent membrane, and when this is broken can be shaken out in water like tassels.

A part of the secretions of the cæca is gathered by a system of converging tubes and is finally admitted to the pyloric division of the grinding stomach, near the junction of the latter with the intestine. These ducts also serve to admit streams of food particles (see p. 249) to the glands themselves, where they are acted on by ferments and are directly absorbed.

THE KIDNEYS OR GREEN GLANDS.

The direct excretion of nitrogenous waste products is effected by a pair of glands which open at either side by a prominent papilla on the lower side of the basal segment of the first pair of antennæ. (Pl. XXXIII and fig. 6, pl. XXXV, *g. gl.*) In their fundamental relations these organs agree with the segmental nephridia of worms and vertebrates.

When unraveled, the entire organ has been found to consist of the following parts: A large, thin-walled peripheral vesicle or bladder, and closely applied to this, in front or below, the proper excretory organ or gland. Together these parts form a rounded or flattened body of a light green color, closely fitting in the convex depression over the articulation of the antenna on either hand and just in front of the stomach sac.

The bladder empties to the outside by a short duct, the opening of which on the papilla is guarded by a valve. The kidney proper is composed of a central saccule or end sac, and of a convoluted tubule, both of which are glandular. According to Dahlgren and Kepner (67) the tubule is lined throughout with nonciliated epithelial cells, and is covered by a tunic of connective tissue, it being in this section only that a cuticle is secreted. Upon taking a lobster in hand a fine jet of liquid is sometimes thrown from the papilla to a height of an inch or more. Inasmuch as water does not apparently have access to the bladder, the walls of which are contractile, the liquid is probably a true secretion. This fountain display of the green glands has been noticed but two or three times.

Chapter VII.—THE GREAT FORCEPS OR BIG CLAWS.

THE CRUSTACEAN CLAW.

The last ten thoracic legs of higher Crustacea all end in hard-pointed segments technically known as dactyls. In the account which follows, when not thus designated, they will be called "single claws," "nails," or "digits," the original meaning of the word. In *Palinurus*, the spiny lobster, all of the thoracic legs end in talon-like claws of this simple type; but in the true lobsters, crayfishes, crabs, and many other decapods a unique organ is developed in certain of the forward legs by the extension of an opposable finger-like process of the subterminal segment, the propodus, which is often large and powerful. In the great cheliped of the lobster (pl. XXXIII and XXXVII) this division is also called "the hand" and the terminal part of it the "index," as distinguished from the opposed "thumb" or dactyl. Thus is formed the admirable forceps, commonly known as the "claw" or chela.[a]

Those legs ending in forceps are described as chelate and the others as nonchelate, and the technical use of these terms is unobjectional. This, however, need not lead to the ambiguity of saying that the last two pairs of legs in a lobster or crayfish have no "claws." To avoid this absurdity, we may adopt Huxley's terms, "double claws" and "single claws" for the forceps of the first three and the nails of the last two pairs of legs, respectively, since they describe the conditions met with in both lobsters and crayfish exactly. The chelate legs all pass through the simple claw stage in either the egg or early larval state.

The big claws of the lobster are remarkable organs whether considered in the light of their structure, their development, or the process of their renewal, and the more we study them the more remarkable they appear.

In most of the higher Crustacea the great claws are the chief weapons for both attack and defense and very efficient means for seizing and rending the prey, as well as for grasping and holding the female in the act of pairing, when the spermatophores are transferred to her seminal receptacle or to some other part of her body.

While three pairs of pereiopods in this animal bear double claws or forceps, in the first pair alone are they entitled to be called "great." In many crabs, as well as in the lobsters and crayfish, the great claws are weapons whose grip is not to be despised.

In some of the crayfishes the great chelipeds are equal to about one-quarter of the weight of the entire animal, while in lobsters above medium size their proportionate weight sometimes reaches one-half, and tends to increase with age. Moreover, the disproportion between the big claws of either side, which are normally asymmetrical,

---

[a] Latinized from the Greek word for any armed appendage; in plural form chelæ, corrupted from chele.

tends also to increase with age and in favor of the "crusher," which in old males reaches an extraordinary size (fig. 1). Many crayfish when incautiously handled readily draw blood, and there can be little doubt that a lobster weighing upward of 30 pounds could easily crush a man's arm at the wrist.

The differentiation of the large claws is often very marked in crabs, and all degrees are represented. The character of the adaptation is equally varied, as may be seen in the common green crab (*Carcinus mænas*), the fiddler (*Gelasimus pugilator*), and in the "king crab" of the West Indies (*Caleppa marmorata*). In *Carcinus* the slightly larger claw is of the "knobbed" or crushing type. A singular differentiation has apparently been started in the same direction in the more remarkable *Caleppa*, where the great chelipeds have been modified in a different manner for the protection of the animal. The great trihedral claws of this singular species swing in and out in front of the head like double doors, and when these are closed or folded in, the crab is as secure as the tortoise in its shell.

In many of the small shrimps belonging to the *Alpheus* family, the huge "hammer" claw, which is usually largest in the males, is most interesting, whether considered as a "snapper" or popgun, as a saber for delivering a slashing blow, or as a means of controlling the development of its fellow in regeneration (see p. 277).

But of all the crustaceans known to me the shrimp-like *Jousseaumea* which I found at Nassau, Bahama Islands, in 1887, but did not describe, presents the most singular differentiation of the claws. When viewed from above this animal presents a very deceitful appearance, no formidable weapons of any kind being visible. In reality, it possesses a huge and ugly looking claw, which in rest is completely concealed, being nicely folded like a pocket rule and tucked under the grooved cephalothorax, ready at any moment to be shot out and to strike an unsuspecting victim. The fellow to this "pocket" weapon is very diminutive. Were this little shrimp as large as the common lobster it would be justly regarded as one of the most remarkable animals in the sea.

While in *Cambarus* and in crayfishes generally right and left claws may be more or less unequal in size, they are often very similar in structure and function, suggesting the primitive toothed type seen in the lobster, but not approaching it with any degree of detail. There is no lock spine in *Cambarus*, but the hooked tips cross, the dactyl underlapping the propodus. The armature consists of small rounded tubercules, set like a row of corn on a cob. When this claw is closed a large gap is left at the proximal end where the teeth are most numerous, and the fingers touch only at their tips.

THE GREAT CHELIPEDS.

The legs which carry the big claws consist of the 7 typical segments already enumerated (pl. XXXVII), united to the body and to each other by articular membranes, and moving in the way described on double hinges of variable form, excepting only the basis and ischium, or second and third segments, which after the fourth or fifth stage fuse into a single piece. This limb in the adult state therefore possesses 6 free podomeres and 6 free joints. The suture of the stiff joint (x in all figures) marks the "breaking plane," since whenever the lobster "shoots a claw," the limb always breaks at the suture of this joint.

The musculature of the great chelipeds is essentially normal and like that of the slender legs, with the exception of the basis or second segment, which has no muscles in the adult state, a condition to be considered in relation to autotomy and the breaking joint; as in the smaller pereiopods the ischium carries two posterior extensors only.

The hinges of this limb are quite peculiar, and suggest possible adaptations to the "breaking joint," and "interlock," considered in a later section. In place of anterior balls working in posterior sockets, as in the tail, we have proximal balls moving in distal cups,[a] with the exception of the first, fifth, and sixth podomeres, for the hinges between the carpus and big claw are so peculiar that they merit special attention. As we have seen, the order in the hinges of the basal joints of all the thoracic appendages is socket and ball of limb, united to ball and socket of the body.

LOCK HINGES OF BIG CLAWS.

By far the most peculiar joint and one of the most unique mechanical devices in the lobster's skeleton are the concealed, sliding hinges, by means of which the great forceps are securely locked and articulated to the rest of the limb. By referring to plate (XXXVII and text fig. 4) it will be seen that the great claw swings between flattened processes of the carpus, which embrace the upper and lower sides of its proximal end near the joint. These two processes (*u* and *l h p*) conceal the joint in question, and lock the claw firmly to the carpus, upon which it is free to move in the horizontal plane through an arc of about 135°, but from which it can not be removed without breaking either segment.

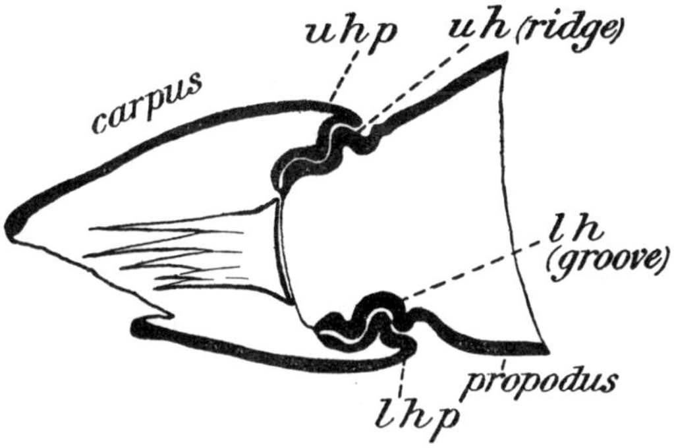

FIG. 4.—Locked sliding joint of big claw of lobster. Sectional view of left chela seen from side towards median plane of body, showing reversed grooves and ridges of upper and lower hinges. This locked joint is strengthened by the overgrowth of upper and lower hinge processes (*u h p* and *l h p*), which arise from the carpus.

When the hard shell is broken at this joint the upper hinge on the claw side is seen to consist of a prominent semicircular ridge, which fits into a corresponding carpal groove, but of greater length. Further, on the inner or proximal side of this groove rises a ridge of lesser arc, which runs in a corresponding groove under the curved ridge of the claw; in brief, circular ridge and groove of claw work on corresponding groove and ridge of fifth segment. To complete this adjustment there is an outgrowth from the hinge process of the carpus, which is outwardly curved, and runs in a corresponding groove distal to the articular ridge on the claw; this serves as an additional lock to the joint, but the proper articular surfaces are those described above. Turning now to the lower or originally anterior side of the claw, we find the conditions completely reversed, and instead of ridge groove we have groove ridge, with corresponding ridge groove on

[a] These terms are used for the successive segments of the limbs in reference to the median plane of the body. The dacty possesses proximal balls only.

the lower hinge process of the carpus. It follows from these relations that the articular surfaces of the carpus face, while those of the claw look in opposite directions.

This remarkable joint suggests the hinge of an ordinary folding pocket rule, but with a different locking device. It is neither a true pivot, tenon-and-groove, or ball-and-socket joint, and so far as I am aware its principle is not found embodied in any of the common mechanical devices. We find it well developed at the fourth stage, with little later change except in the further overgrowth of the hinge processes. (Fig. 9.) Such a joint works with great precision in its prescribed plane, with little or no appreciable lost motion, and would seem to be an adjustment by means of which the big claw is firmly secured to the supporting carpus, and the voluminous flexors of this segment can react upon the great weight of the claw to the best advantage.

In the crayfish (*Cambarus*) the big claw is not locked to the carpus, but moves loosely on double hinges of the typical ball-and-socket order, each hinge consisting of carpal ball, and propodal socket mounted on a round tubercle. In *Callinectes* and certain other Brachyura examined (text fig. 5) the great cheliped has suffered little or no torsion, and the dactyls open upward as in the larval lobster. The claws move on modified ball-and-socket hinges, which are firmly locked to the claw but in quite a different manner from that of the lobster. The propodus in this case bears cups (*l h* (socket) fig. 5) on both upper and lower sides, which are locked over the balls by processes (*u* and *l h p*) growing out from this segment and not from the carpus. The crab's claw thus swings vertically in and out through an angle of upwards of 90°.

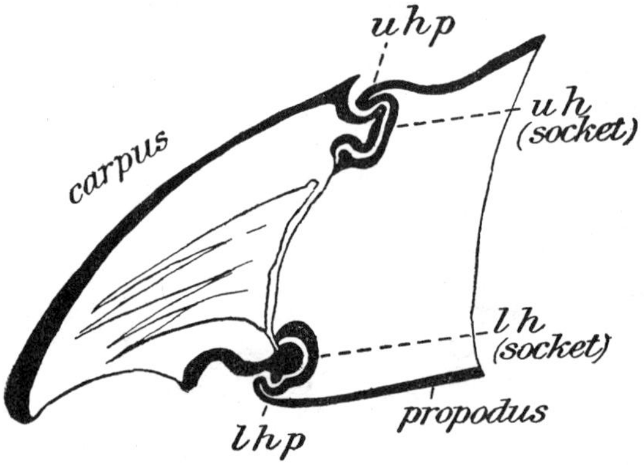

FIG. 5.—Locked sliding joint of big claw of crab (*Callinectes hastatus*); in same plane as represented in figure 3, showing modified balls and sockets, but with no reversal on upper and lower sides; hinge processes (*u* and *l h p*) here arise from the propodus of claw.

While the locked, sliding joint of the lobster, particularly in the reversal of its hinges, suggests the ordinary ball-and-socket device of the other limb segments, and even more that of the crab's chela, it would be difficult to decide whether one was better from a mechanical standpoint than the other, or to imagine how either could have arisen from the simpler type upon any principle of selection.

ASYMMETRY IN THE BIG CLAWS OF THE LOBSTER.

The marked dissimilarity of the big claws (pl. XXXVII) in regard to both their structure and chief functions in all lobsters above an inch or an inch and one-half long, has led to various distinctive names on both sides of the Atlantic. Fishermen often speak of the "knobbed" and "quick" claws. The larger is adapted for crushing the food, and to emphasize the function, we shall call it the cracker, crusher, or crushing claw; the smaller and slenderer, which suggests a patent lock forceps with serrated jaws, is used for seizing, holding, piercing, tearing, and slashing the prey. We shall call it the lock forceps or toothed claw, in preference to the phrase "cutting claw" formerly used. In young animals from 2 to 5 inches long the teeth of this weapon are completely concealed by dense clusters of sensory hairs, which though seldom absent become less conspicuous with advancing age. It is therefore evident that the toothed claw is highly sensitive and "feels" the blows it gives as well as those it takes.

Przibram (*223*), who classifies the higher Crustacea according to the similarity or differentiation of the big claws into the "Homoiochelie," and the "Heterochelie," calls the larger claw the "Knoten" or "Knackschere," and the smaller the "Zähnchen" or "Zwickschere," in view of their form and function respectively. Stahr (*257*), who uses the terms "Zahnchenschere" (toothed claw), and "Knotenschere" (knobbed claw), as descriptive of their structure, after a discussion of their probable functions, says that he is justified in designating the claws of *Homarus gammarus* as follows, "the beautiful, regular, elegantly formed, thin-walled forceps, provided with periodic teeth and sensory hairs as the ornamental ("Schmuck-") and sensory claw ("Spürschere"), and the other, plump, oval, thick-walled form, provided with tubercles, as the crushing ("Knack-") and grasping claw ("Greifschere"). As will later appear, the development of these organs affords no warrant for regarding the toothed claw as an ornament, not to speak of the psychological difficulties involved.

TORSION OF THE LIMB.

Of greater interest than the difference in size and structure of the big claws is the complete change in their position on either side which takes place after birth, due to a twisting of the limb and mainly of the fifth joint or carpus or the third podomere reckoned from the distal end.

This curious torsion of the crustacean leg is of very ancient origin, dating from as early as the Cretaceous period, and is shared by many of the higher crustacea decapods (for first account of torsion and fuller discussion see *153*). It further affords a good illustration of how a very obvious fact may long escape the notice of naturalists, my own attention not having called to it until 1905, although drawings of the larval and adult stages had been repeatedly made.

In the adult lobster or crayfish the free dactyls of the smaller chelate legs all open upward and outward in a plane which is nearly vertical, while in the big claws the dactyls of opposite sides face and open inward or in a nearly horizontal plane. In the lobster at birth, on the other hand, and up to the fourth stage, all the chelæ have the same relative positions; all open vertically upward with an outward inclination. (Compare fig. 1, 6, and 7 with pl. XXVIII.)

It is thus evident that the position of the great forceps in an adult animal has been reversed through a rotation of either claw through an angle of 90°, toward the median plane of the body, in consequence of which their inner or anterior faces have become their under sides. This rotation is completely effected at the fourth stage (pl. XXXI) and with the molt which registers so many other marked changes in the structure and habits of this animal. It is responsible for the torsion or twist to be clearly seen in the carpus of the limb. In conformity with this change in position, the claw has undergone a change in coloring, for the deep green chromogen pigments which cover the present upper surfaces are completely lacking from their pale red under sides.

It would appear in the highest degree improbable that this condition in the big claws could have been produced through the inheritance of slight variations leading to a greater and greater degree of torsion, and finally extending through so great an arc, although it is conceivable that such a variation may have been correlated with others which were of so favorable a character as to be of selective value and to have been "dragged" along with them.

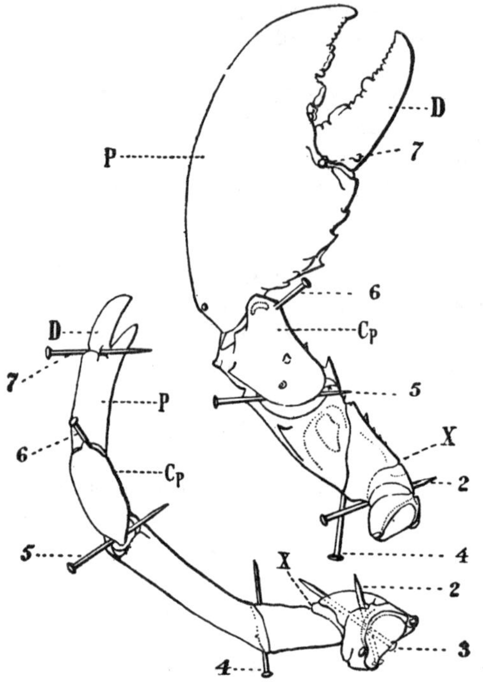

FIG. 6 and 7.—Great first and small left third claw feet of adult lobster with pins (no. 2–7) inserted in the axes of articulation of successive podomeres, to indicate normal torsion in the great cheliped. Position of the big claw up to the fourth stage is identical with that of the little claw of the slender leg. Compare plates XXVIII and XXXI, with figure 14 of text. Cp carpus; D, dactyl, and X, breaking joint. Podomeres or segments of permanent limb numbered, as in all succeeding figures, in Arabic numerals, from base to apex.

Again, it is even more difficult to regard this torsion of the crustacean limb as the resultant effect of use through inheritance. The carpal podomere has but one flexor and one extensor muscle, both of which react on the claw at points outside of the joint itself; at the same time the muscles, of course, pull on the shell of this part at their points of origin, but no conceivable position or strain of these fibers can convert the pull into a twist. If the increasing weight of the claws in the growing animal had any effect upon their ultimate position it should tend to turn them outward. In other words, their modification is just the reverse of what we should expect were the effects of strain or use inherited.

If we examine other crustaceans we find that the big claws open inward, upward, or outward, irrespective of their relative size or weight. In the Alphei, which usually have one claw of enormous size and of peculiar structure, the dactyls open outward, while in the fiddler crabs (*Gelasimus pugnax*) they incline inward, as in the lobster. This is true not only of the single huge claw of the male fiddler but of its diminutive fellow and of the small, almost rudimentary chela of the female. In the common crabs (*Carcinus*, *Callinectes*) the claws open obliquely outward. It therefore appears that in the rotation of the crustacean limb we have an illustration of an adaptive variation, which in origin and the extent to which the process may be carried is independent of use and the mechanical strains to which the organ may be subjected.

Apart from their crushing or piercing teeth and sharp indurated tips, the large claws are armed along their facing edges by stout tooth-like spines, while the exposed surfaces and angles of the lower segments of the limb are similarly protected. These spines are generally directed forward and mostly upward and tend to guard the space about the head which the outstretched claws inclose (see, p. 273).

The terminal segments of the last pair of slender legs have undergone torsion but of a different character, as described in chapter IX, page 304.

BREAKING PLANE AND INTERLOCK.

We have seen that both of the large chelipeds have a stiff or breaking joint in the compound segment at their base, as well as peculiar hinges, which are not only adapted to the ordinary uses of such limbs, but possibly to the resources of the animal in sacrificing them for its own preservation. There has also been developed in relation to the breaking plane an interesting interlocking mechanism, which seems to have escaped notice up to the present, although its importance in the life of this animal would appear to be great.

This interlock (fig. 1, 3, and 4, pl. XXXVII) is a simple but effective adjustment by means of which it is impossible for an enemy to pull out or twist off one of the chelipeds, as may be done in a cooked lobster, without bringing autotomy into play, to which process it seems to form a sort of emergency "brake."

Turning the body of the lobster over and working the chelipeds by hand, we perceive that they move freely forward and backward, the striking or thrust movement, at the junction of coxa with basis. In such movements the lobster's most powerful blows are dealt, whether in attack or defense. We observe further that any lateral movement of this joint would be serious, and that is guarded against by huge interlocking spurs (s^1, s^3) on the first and third podomeres respectively. This condition seems to be related to the fact that the breaking joint (x) lies between these points, or peripheral to a free joint, so that when the strain upon this articulation and the interlocking spurs is too great or, in other words, sufficient, the limb is reflexly cast off in the breaking plane.

This mechanism, moreover, together with the complete fusion of the joint, is not developed until after the fourth stage, when there is probably less need of strengthening the hinges between these particular segments. Yet autotomy occurs at this stage, and we find the hinges strengthened in a degree by the interlock of distinct but different spines (fig. 8–10, s^4, and s^2), although this early adjustment is not quite so marked as in the adult animal. At all events in the lobsterling there is an interlock between the second and third podomeres, which evidently increases the resistance of the limb at its base during this period. These spurs of the fourth stage lobster become later reduced to rudiments, and new interlocking processes are developed in the adult animal between the first and third segments. The principal spur at the fourth stage (fig. 8, s^2) is still to be seen in its rudimentary state in the adult lobster immediately in front of the large functional spurs already described. (Pl. XXXVII, fig. 1, s^2 *rud.*)

THE TOOTHED CLAW OR LOCK FORCEPS AND ITS PERIODIC TEETH.

If the armature of the smaller claw is closely examined, the teeth or spines are seen to be arranged in periodic sequence, a fact first noticed by the German naturalist, Stahr (257). Stahr's description is correct, so far as it goes, but we can not adopt his remarkable conclusions that this should be called the "ornamental" or "beauty claw," and that the æsthetic sense of this self-admiring crustacean is aroused as its eye wanders over the dentate margin of its "hand." We should fail, however, to do justice to the imagination of this writer without quoting directly from his work, in which he concludes "That it is not a far-fetched idea to recognize in the periodic teeth or rows of points of the ornamental and sensory forceps an embellishment—an architectural and artistic ornament. We may mention their close relation to music, poetry, and dancing, where we have to do with rhythm, time, measure, composition, everywhere with periodic sequences. * * * Thus it is only natural to suppose that the beauty sense of a crustacean would receive an agreeable impression as its eye wanders over the periodic points of its claw." [a]

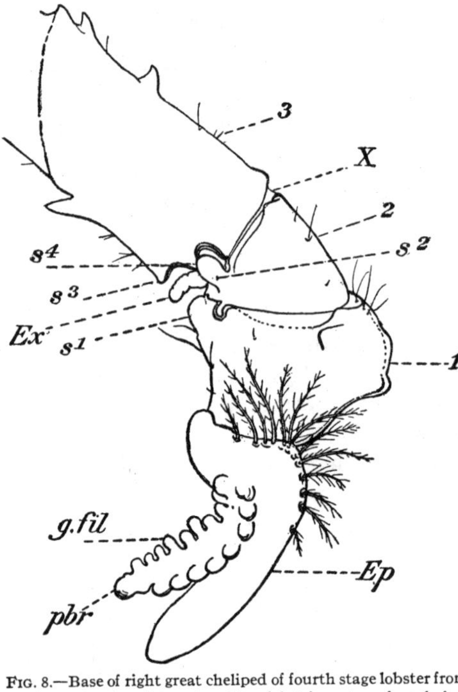

FIG. 8.—Base of right great cheliped of fourth stage lobster from below, showing future breaking joint free at surface, before complete fusion has occurred, a temporary interlock at this stage by spurs (s^2 and s^4) of the second and third podomeres, as well as rudiments of the spurs (s^3 and s^1) of the first and third segments, which form the permanent interlock of the adult limb. Compare with figure 13. Swimming branch or exopodite (*Ex*) functional up to this stage is reduced to a rudiment. Gill filaments (*g. fil*) are developed as secondary outgrowths of the primary filament, which is a fold of the body wall.

We have worked out the history of development of both types of claw, in the light of which their peculiar structure becomes more intelligible. The arrangement of the teeth or spines on the smaller claw may be expressed by a diagram (fig. 11), in which they appear as a linear series, made up typically of periods of eight. In respect to size and age, or order of development, the eight teeth of each period are symmetrically distributed and fall into four orders or series, of which the first and second contain one each, the third two, and the fourth four. On this basis the formula for each perfect period or sequence would

[a] All quotations from foreign languages in this work are freely rendered into English.

be: $1+1+2+4=8$, or, designating each spine by its serial number as in table 6, $1:4:3:4:2:4:3:4=8$.

About midway on the dentate margin of the "hand" (fig. 12 and 13) or propodus one finds a stout spur which I shall call the "lock spine" (*L* in all the figures). As we shall see, it is really a displaced spine of the first order. It fits into a shallow groove of the dactyl, which is often slight or wanting, and forms the lock of the claw. Upon closing, the dactyl falls on this spur, and, its teeth sliding under those of the opposed jaw, it is firmly locked in this position, so that no lateral motion is possible. (Fig. 1, pl. XXXVI.) To complete this adjustment, the tips of the forceps are bent like the mandibles of a crossbill, the dactyl underlapping. The spines of the propodus are bent upward, those of the dactyl downward so that in the claws of some individuals they make an angle of 45° with the lock spine, which is nearly vertical. Moreover, the spines are aligned very accurately, and in a peculiar manner. The spines of the "upper jaw" or propodus are all tangent to a line traversing its lower border, while those of the dactyl or underlapping jaw meet a line drawn along its upper margin. This reversal of the alignment it will be observed makes it possible completely to close and at the same time to lock fast the jaws of an instrument having this structure. It follows that the teeth do not interlock but overlap (fig. 12 and 29).

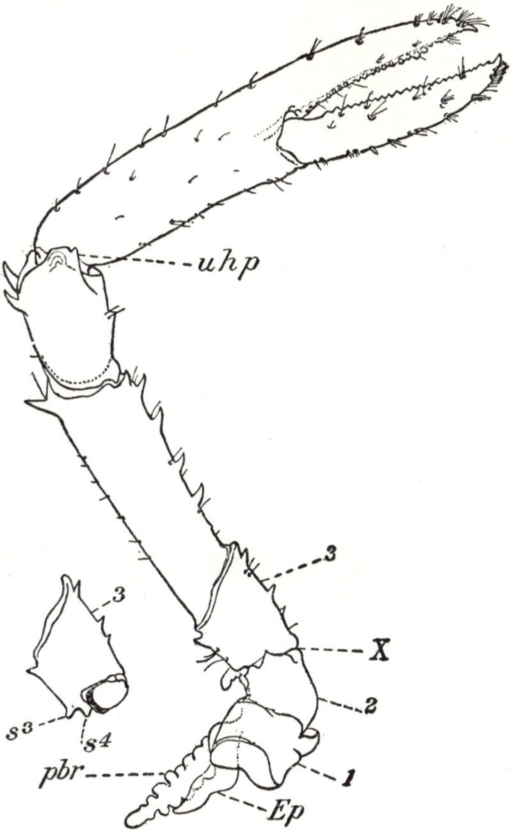

FIGS. 9 and 10.—Right great cheliped of fourth stage lobster, from above, showing upper hinge process (*u h p*) of carpus, and disarticulated ischium with interlocking process (s^4), and future interlocking spur (s^3), which is rudimentary. Compare text figure 8, and plate XXXIV, figures 1, 3, and 4.

The tendency of the spines to increase in geometrical ratio is often present and if effective would in the next progression give a period of 16 spines. Under these conditions the periods are generally incomplete, seldom yielding over 13 spines.

The formula given above seldom holds good for more than two or three periods, and in many claws no period is quite perfect. At both proximal and distal ends of the series the periods become irregular and the identity of the spines is lost. Some means of identifying the principal periods, however, is necessary, if we are to follow the course of development and the changes which attend the molt. Fortunately two guideposts are always present at either end of the series, the lock spine (fig. 12 *L*) and a distal spur or tubercle on the lower side of the propodus near its tip (*Sp.*) For convenience of description we assume, then, that the first period lies proximal to the spur, and that the "lock" spine is the primary member of a hypothetical fifth period. Between these boundaries lie three, four, or exceptionally five, periods, of which the fourth is rarely perfect. This leaves three or at most four periods (numbered in all the figures I–IV) for special consideration.

Counting the tip of the claw as a primary spine (though it really is not, since it develops as a seta), we should have from five to seven periods between it and the lock

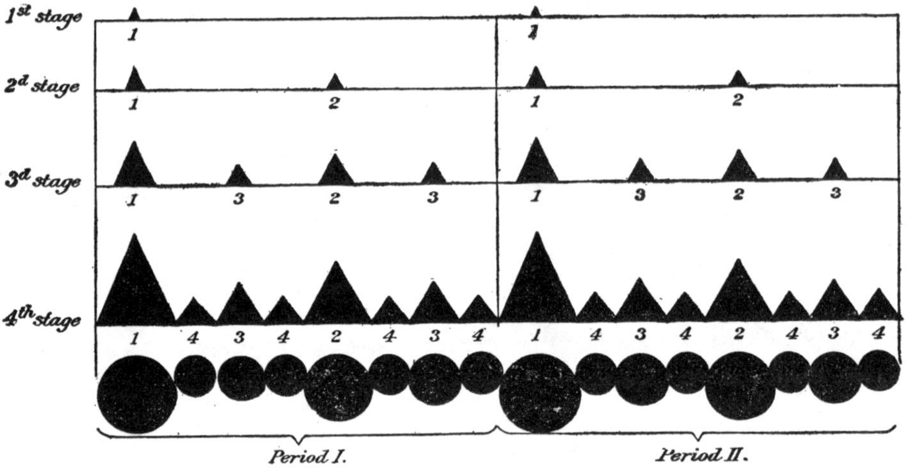

FIG. 11.—Diagram to show the serial arrangement of the spines in the toothed forceps of the lobster in periods of 8, and the development of these spines by interpolation from the first to the fourth stages. Arabic numerals indicate orders of teeth (here reading from left to right).

spine. Proximal to the lock spine, the linear series is completed by from three to five primary teeth, with small secondary spines among them, which like similar spines elsewhere are a fluctuating quantity. Consequently in the propodus there are from 8 to 12 primary spines which represent periods, of which never more than 3 or 4 are complete, or in eights. (Compare fig. 29.)

In order to set these relations in clearer light as well as to illustrate individual variation I append a table of formulæ for the teeth in the large segment of the toothed claw of 10 lobsters taken at random (table 6), and of the teeth before and after the molt in the claw of an adolescent (no. 11*a*, 11*b*, stages VII and VIII) and an adult animal (no. 12*a* and 12*b*).

TABLE 6.—SEQUENCE OF SPINES IN PERIODS I–IV OF TOOTHED CLAWS OF ADULT, AND IN PERIODS I–III OF MOLTING ADULT AND ADOLESCENT LOBSTERS.

No.	Period I.																Period II.															
1a	1	5	4	5	3	5	4	5	2	5	4	5	3	5	4	5	1	5	4	5	3	5	4	5	2	5	4	5	3	5	4	5
2a	1		4		3		4		2		4		3		4		1		4		3		4		2		4		3		4	
1	1		4		3		4	5	2		4		3		4	5	1	5	4		3		4		2		4		3		4	5
2	1		4		3		4		2		4		3		4	5	1		4		3		4	5	2	5	4		3		4	5
3	1		4		3		4	5	2		4		3		4	5	1	5	4	5	3		4	5	2	5	4		3		4	
4	1		4		3		4		2	5	4		3		4		1		4		3	5	4		2		4		3		4	
5	1		4		3		4		2	5	4		3		4		1	5	4		3		4		2		4		3		4	
6	1		4		3		4		2		4		3		4		1	5	4		3		4		2	5	4		3		4	5
7	1		4		3				2				3		4		1		4		3				2		4		3		4	
8	1				3		4		2		4		3		4		1		4		3		4		2		4		3		4	
9	1				3				2				3		4		1		4		3		4	5	2		4		3		4	
10	1		4		3		4		2	5	4		3		4		1		4		3		2						3		4	
11a (VII)	1				3		4		2		4		3		4	5	1		4	5	3		4	5	2		4	5	3		4	
11b (VIII)	1		4		3		4		2		4		3		4		1		4		3		4	5	2		4		3		4	5
12a	1		4		3		4		2		4		3		4		1	5	4		3		4		2		4		3		4	5
12b	1		4		3		4		2		4		3		4	5	1	5	4		3		4		2	5	4	5	3		4	5

No.	Period III.																	Period IV.																Summation of periods.
1a	1		5	4	5	3	5	4	5	2	5	4	5	3	5	4	5	1	5	4	5	3	5	4	5	2	5	4	5	3	5	4	5	16+16+16+16=64
2a	1			4		3		4		2		4		3		4		1		4		3		4		2		4		3		4		8+8+8+8=32
1	1			4		3		4		3		4		3		4		1		4		3												10+10+8+3=31
2	1		5	4		3		4		2		4		3		4		1		4		3		4		2		4		3		4		9+11+9+8=37
3	1		5	4		3		4	5	2		4		3		4		1		4														10+12+10+2=34
4	1		5	4		3		4		2		4		3		4		1		4		3		4		2		4		3		4		9+9+9+8=35
5	1		5	4	5	3		4		2		4		3		4		1	5	4		3		4		2		4		3	5	4		9+9+10+10=38
6	1			4		3		4		2		4		3		4		1		4		3		4		2				3		4		8+11+8+7=34
7	1			4		3		4		2		4		3				1	5	4		3		4		2		4		3				6+7+7+8=28
8	1	6	5	4		3		4		2		4		3		4		1	5	4		3		4	5	2		4		3		4		7+8+10+10=35
9	1			4		3		4		2		4		3				1			5													5+9+7+2=23
10	1			4		3		4		2				3				1		3				4										9+6+6+3=24
11a (VII)	1		5	4		3		4		2		4		3																				8+11+8
11b (VIII)	1	6	5	4	5	3		4		2		4		3		4																		8+10+11
12a	1			4		3				2		4		3																				8+10+6
12b	1			4		3		4	5	2		4		3																				9+12+8

It will be observed that four periods usually occur between the spur and lock spines; that in ten individuals only seven regular 8-tooth sequences occur; in one there are two, and in four cases none. The disturbances arise from the interpolation of exceedingly small spines, or the tendency to advance to the next progression, which if complete would give 16 spines to the period. The largest number of spines to the single period given in the table is 12, but I have seen a case in which the third period contained 15 spines.

A fairly regular claw of large size is represented in profile and horizontal projection in figures 12 and 13, the formula of which for the four principal periods is 31 (table 6, no. 1), only one of the sequences being in eights, and the spines of the entire armature totaling 48.

The serration of the dactyl of the toothed claw is more regular than that of the propodus and similar except for the disturbance introduced by the "lock spine" of the latter. Three or four 8-tooth periods usually occur and the sequences are often perfect.

The toothed claw, as already remarked, is richly supplied with tufts of sensory hairs above and below the line of teeth and also along the margin of the claw near its tip. These are specially abundant on the underside, and with them the animal is constantly feeling the bottom when it assumes the common alert attitude with the tips of the claws bent down. These tactile setæ are arranged in bundles of 200 to 300 or more short, stiff bristles which, like little scrubbing brushes, project from depressions in the shell. The floor of each depression is a sieve plate, the perforations of which correspond to the number of setæ as well as to the number of nerve fibers supplying the bundle. In the adolescent stage, when the lobster has attained a length of 3 or 4 inches, the setæ of the lock forceps become large matted tufts which sometimes completely conceal the teeth. (Compare fig. 15 and 16).

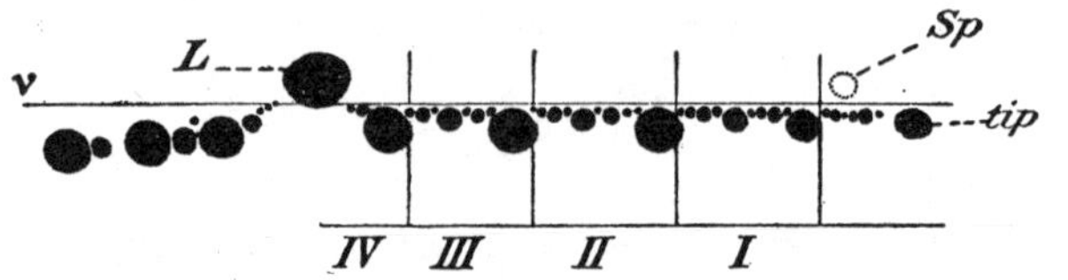

FIG. 12.—Projection of serial teeth in segment of big claw of large adult lobster represented in figure 13, showing alignment to lower or ventral (*v*), originally the anterior, side, the position of the *tip*, spur (*Sp*), and the large displaced lock spine (*L*), the two last serving as guide posts for identification of the periods i-iv. In this and following figures the periods are enumerated from the distal to the proximal end of the claw.

THE CRACKER OR CRUSHING CLAW.

In place of tooth-like spines the great crushing claw presents a number of rounded tubercles, both large and small, single or double, and arranged in a characteristic manner (fig. 2 and 3, pl. XLIII). These crushing tubercles are very dense, and in old hard-shell

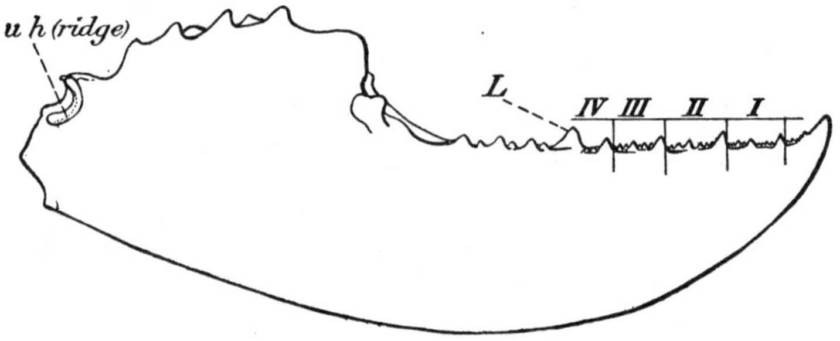

FIG. 13.—Large segment of right toothed claw from above, to show the periodic teeth; compare projection in figure 12; *u h* (*ridge*), upper ridge of sliding lock joint.

lobsters the pigment and enamel is completely worn away from long and rough usage. The tips overlap slightly, but the dactyl is curved, and not straight as in the toothed claw, consequently when closed there is often a wide gap between the jaws, the tubercles touching at but one or two points only. (Fig. 2, pl. XXXVII.)

The crushing claw, as shown in the drawing (pl. XL), has a far more powerful musculature than its fellow, and is accordingly richer in its supply of blood vessels and nerves. Two tendons (fig. 2, pl. XLI) spring from opposite sides of the proximal end of the free dactyl and afford a surface for the attachment of the huge flexor and smaller extensor muscles. Each tendon is a keeled plate which is developed in a flattened pocket of the skin, but the closing muscle of the great claw being the largest and the strongest in the body requires the largest tendons. The tendon of the flexor (*t. fl.*[6]) is a broad leaf-shaped plate, keeled above and below, while that of the weaker opening muscle is narrow and strap-shaped.

At the time of molt these huge tendons, like all others in the body, are drawn out, attached to the cast-off shell, and leave deep open pockets into which in a large animal the little finger can be easily inserted. As soon, however, as the soft claw becomes tense with blood, the water is driven out and, the opposed surfaces of the pocket uniting, a new tendon is gradually formed. (Compare fig. 1, *t p*, pl. XLIII.)

The coarser flesh of the claws represents, as we have indicated, the characteristic flexor and extensor muscles, while the "fine meat" of the dactyl (fig. 3, pl. XLVI) and distal half of the propodus is composed of a sponge work of involuntary muscle fibers in addition to fine-blood vessels of the arterial system, nerves, glands, and connective tissue, the whole being enveloped by the soft pigmented skin (pl. XL). No special sense organs, aside from the setæ, have been detected in it. The meshes of the sponge work form a system of communicating sinuses into which the arteries appear to open through very small branches or capillaries.

During the molting process, when the fleshy mass of the claw is drawn through a series of narrow rings as if it were a piece of candy, the blood is of necessity withdrawn from these parts. The sponge work is an adjustment which meets this prime need of the molting period. At the time of molt the muscles are extremely tense and the flesh hard, and the contraction of the fibrous sponge work apparently keeps back the flow of blood until the animal escapes from its old shell, when it again becomes completely relaxed (see p. 206).

The abundant blood always found in the large claws, except when molting, is supplied by a large artery, which at the point of entry from the fifth segment divides into an inner and a smaller outer branch. The inner division passes between the two muscles, and gives off small twigs in its course; then as it curves outward over the distal end of the flexor muscle, it sends off somewhat irregularly a branch to the upper and lower division of each muscle, and to upper and lower parts of dactyl and propodus.

The nerves of the great cheliped (pl. XL) consist of two main bundles (n^1 and n^2), made up of a number of closely related strands. In the basal segments of the limb the larger and more complex bundle (n^2) is anterior while the smaller bundle (n^1), which is double, follows it closely on its posterior or outer side.

The nerves usually enter the claw in three closely related strands, one of which supplies chiefly the extensor, one the dactyl and flexor, while the outermost branch is distributed to the flexor and large "finger" of the claw. Both arteries and nerves regularly divide and subdivide in the terminal parts of the claw to form a very complicated system.

DEVELOPMENT OF THE GREAT FORCEPS.

How has the differentiation of the great claws been brought about? It is easy to follow the history of their development molt by molt from the first larval stage onward. This history clearly shows that the toothed claw represents an original or an older type, and that the crusher claw was later developed by a modification of this primitive pattern.

In the first larval stage of the lobster the future big claw (fig. 14) is distinctly of the embryonic type, relatively short and thick, and armed with few tactile bristles, its tips being drawn out, as it were, into long sharp-pointed spines. The dactyl, which bears the longer and straighter spine, is larger than the undeveloped index. This inequality is much more marked in the smaller chelipeds, where the index appears as a bud-like outgrowth, setate and bearing one or more stiff, barbed, or serrated bristles (fig. 2).

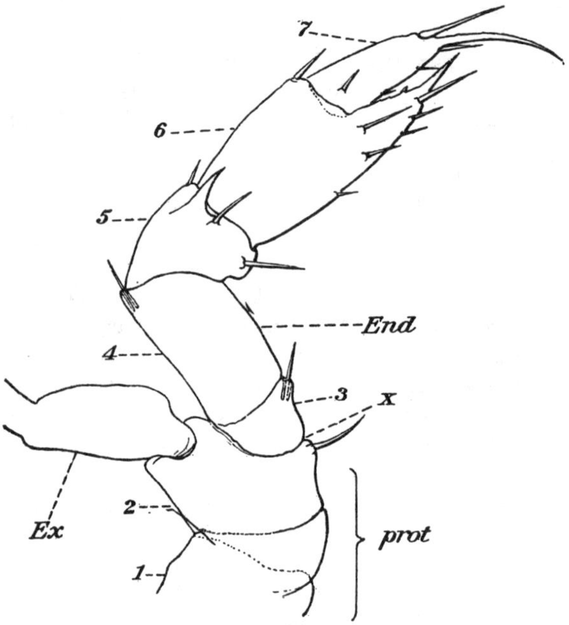

FIG. 14.—Left great claw-foot of first larva of the lobster, from above, showing outwardly opening dactyl, before any torsion of the limb has occurred, the short ischium (3), with free joint at future breaking plane (*x*), and base of swimming branch (*Ex*). Compare with text figures 6 and 9.

In the second and third larvæ (fig. 41 and 42) the claws become broader and more voluminous, while their spinous tips are reduced and both index and dactyl are curved.

In the fourth stage (fig. 9 and pl. XXXI) the great chelipeds suddenly become very conspicuous, bearing long slender forceps which now for the first time serve as organs of prehension with marked success. The jaws of the forceps are slender, dentate, and tufted with tactile hairs. The condition of symmetry, with this general structure, on right and left sides, continues through the fifth and in some cases up to the seventh or eighth stage, when the first traces of asymmetry begin to appear, though not necessarily apparent to the naked eye. (Fig. 15 and 16.) By the ninth stage, when a total length of about one and one-quarter inches has been reached, the differentiation of the crusher claw is easily recognizable, but the changes registered at each molt are slight. In the account which follows we shall consider in more detail the beginnings of asymmetry and the development of the teeth and tubercles which characterize the two types of big claw in the adult animal.

In the fourth stage the great claws are not only symmetrical, but of the toothed type. According to Emmel (*96*) the transition to the asymmetrical condition begins in the sixth stage, but in the material studied as a basis for this account it was impossible to detect any morphological differences until the seventh or succeeding stage. There is doubtless some variation in this respect. It is true that at preceding periods the big claws may differ in size or slightly in form as a consequence of molting or regeneration, but without implying the differentiation in question. Again at the seventh stage these

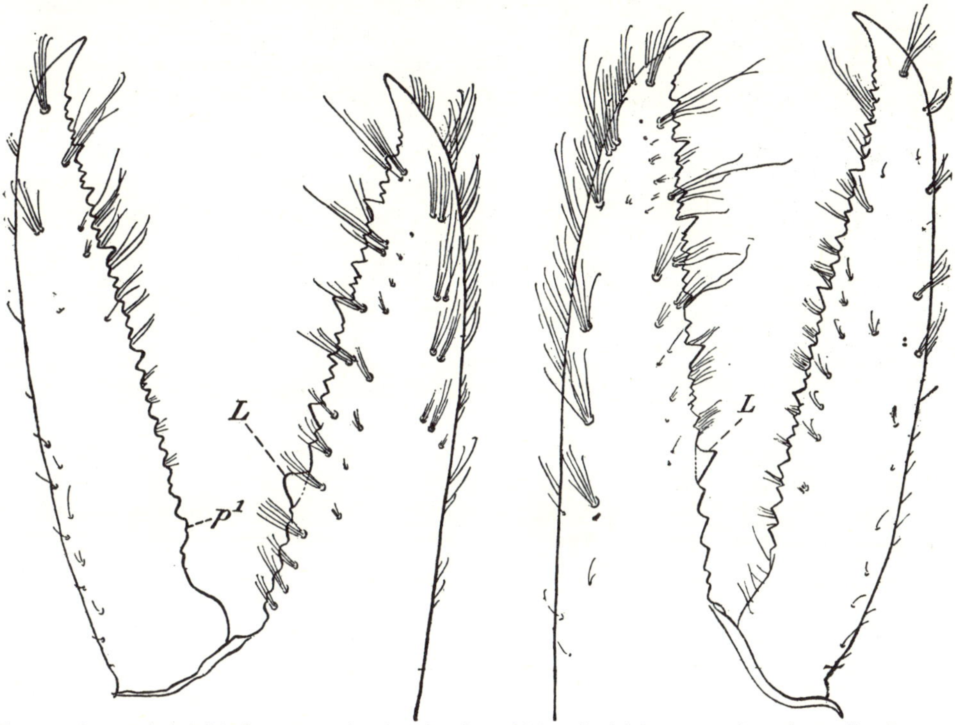

FIG. 15 and 16.—Left and right future toothed and crusher claws of lobster in eighth stage, seen from above. The claws are of equal length, and the differentiation of the crusher is not apparent to the naked eye. Compare with figures 21 and 22. *L*, lock spine; *p*, compound proximal tubercle of crusher claw. Enlarged about 34 times.

claws may appear to the naked eye essentially alike in form and size. Thus, to give a concrete example, a lobster in the eighth stage, measuring 19.75 milimeters, September 22, showed a rather striking similarity in the forceps, the dimensions of which were as follows:

	Millimeters.
Right claw (future crusher):	
Length	7
Breadth	1.7
Left claw (future toothed forceps):	
Length	7
Breadth	1.4

When these claws are magnified thirty or forty times (fig. 15 and 16) the first steps in the differentiation of the crushing from the primitive toothed type of claw become evident. They are expressed by a blunting or rounding off of the sharp points of the teeth, and a tendency to fusion among those situated at the proximal extremity of both divisions of the claw. (Compare fig. 21–24.)

We therefore conclude that during the fourth, fifth, and in some cases at least in the sixth or even seventh stages of the lobster, both of the big claws represent the older or phylogenetic type which is retained as the toothed or lock forceps of the adult. The chela destined to become the crusher is a little broader though not necessarily longer than its fellow, and its teeth which still show the periodic sequence are more rounded, as we have just seen, at the proximal end of the series. The tufts of sensory hairs are, moreover, less prominent on the future crushing claw, as apparent in all the later stages.

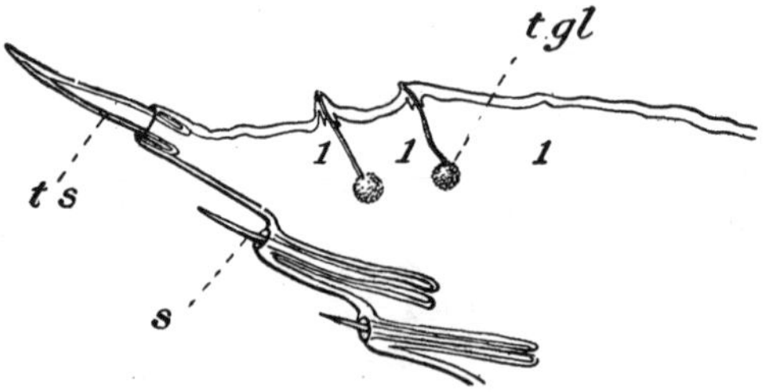

FIG. 17.—Outline of great claw tip, showing serrate margin of propodus, in first larval stage of the lobster, represented as a transparent object, from glycerine preparation. Note the invaginated sensory hairs or setæ (*s*), and claw-tip (*t s*), and three teeth of the first order (I, I, I) developing from apex to base, and representing the three primary periods (I, II, III), indicated in figures 11 and 18; also a tegumental gland opening at the tip of each of the two oldest spines.

The development of the toothed type of claw is represented by a series of drawings (fig. 17–25, and pl. XLII) from the first to the ninth or tenth stages, in which the orderly appearance of the spines can be followed with approximate accuracy up to stage 3, and with certainty beyond it. The large propodus only is represented in most of the figures.

The spines of the toothed claws are developed in a linear series, and the order in respect to size corresponds to that of age, or time of appearance. The larger teeth of the first order are the first to emerge. They are set at wide intervals and evenly spaced. From 2 to 3 are recognized in the chelæ of the first larva (fig. 17) and from 3 to 5 in the claw of the second stage (fig. 18). In the third stage the normal number of primary teeth are present (fig. 19), although some of them are very small, and in the intervals between them are interpolated rudiments of the teeth of the second order. In a single series the first trace of the third series of teeth may be detected also. At the fourth molt (fig. 20) a single period of eight may be completed by the intercalation of the four small teeth of the fourth order; but the process does not always stop here, and an attempt, so to speak, is often made at the seventh, eighth, or at some subsequent molt to introduce a fifth series of 8 teeth, which if completely

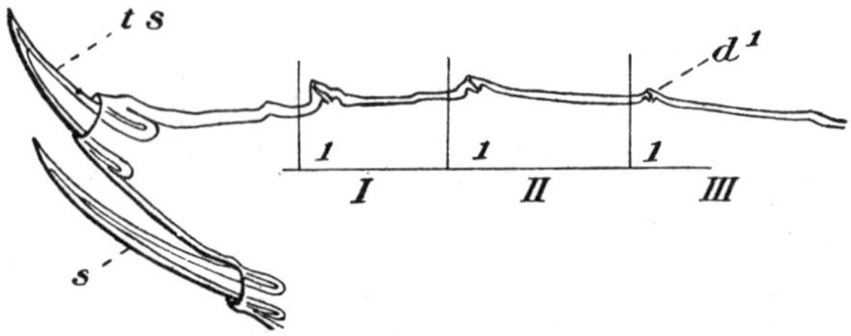

FIG. 18.—Outline of corresponding part of big claw shown in figure 17, but in second larval stage, showing the separated primary teeth, invaginated claw-tip, and setæ, as well as a new spine arising at either end of the series. Spine I of period III now bears the duct of a gland. See figure 11.

successful would increase the serial number to 16. A few cases are noted of the introduction of a tooth of the sixth series (table 6, no. 8, 11*b*). The process of interpolation is illustrated in the diagram (fig. 11) up to the usual 8-period stage, which is commonly attained at the fourth or fifth molt.

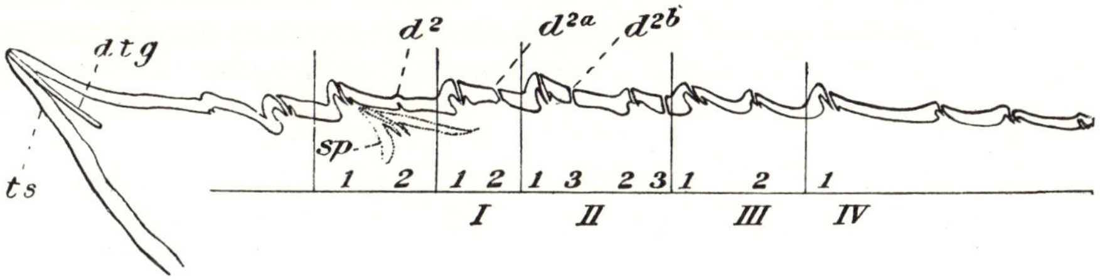

FIG. 19.—Outline of corresponding part of great claw shown in figures 17 and 18, but at third larval stage, showing spines of the second order, sometimes preceded by ducts of glands (*d 2a*, and *d 2b*), interpolated between those of the first, also spur (*sp*) and tip of claw (*t s*), both of which arise like the setæ, and like the teeth are provided with glands, the ducts (*d t g*) of which open at their summits. Compare figure 11.

The first teeth to appear apparently occupy the same plane, but at the seventh stage, or even before this, the alignment is similar to that of the adult claw, and the future "lock spine" or tooth (*L* in all the figures) is readily distinguished by its form and position.

It is interesting to notice that in all the early larval stages and up to at least the fifth or sixth molt, each serial tooth is regularly pierced by the canal of a single tegu-

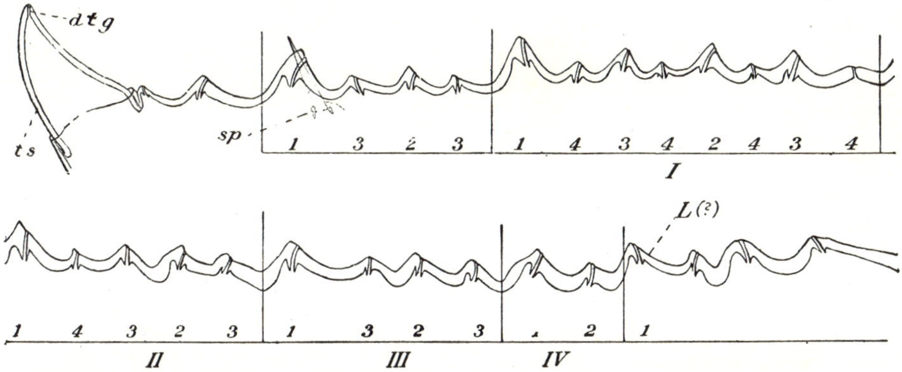

FIG. 20.—Outline of corresponding part of big claw represented in figures 17 to 19, but at fourth stage, showing spines of the third and fourth orders, and the establishment of a single period of 8, though the identity of the periods in this case can not be exactly defined. Invaginated claw tip still bears duct of gland (*d t g*), and the spur (*sp*) is still invaginated like a hair.

mental gland (fig. 17–20), which opens on its proximal side and just below the summit. In some cases the opening of the duct precedes the spine and marks its future position exactly (fig. 19 d^2 *b*). While the serial spines are always developed as outgrowths of the skin, the tips of the claw (fig. 17–20, *t. s.*) and peculiar tubercle or spur (*sp.* in all figures) originate like ordinary hairs, and like them are always invaginated previous to molting.[a] It is to be further noted that as early as the third larval stage and for some time thereafter the claw-tip, like the tooth, gives passage to the duct of a gland (*d. t. g.*, fig. 19–20). I have not found glands of this type in the spines of the adult claw, and if present in older adolescent lobsters they are successfully concealed by the opacity of the shell. The adult spines were sectioned, but in all the young stages glycerine preparations were relied upon. A single tooth sometimes bears the ducts of three independent glands, in which case it is probably compound, resulting from the fusion of a corresponding number of teeth. Rarely a bifurcated duct is seen (fig. 2 pl. XLII), each

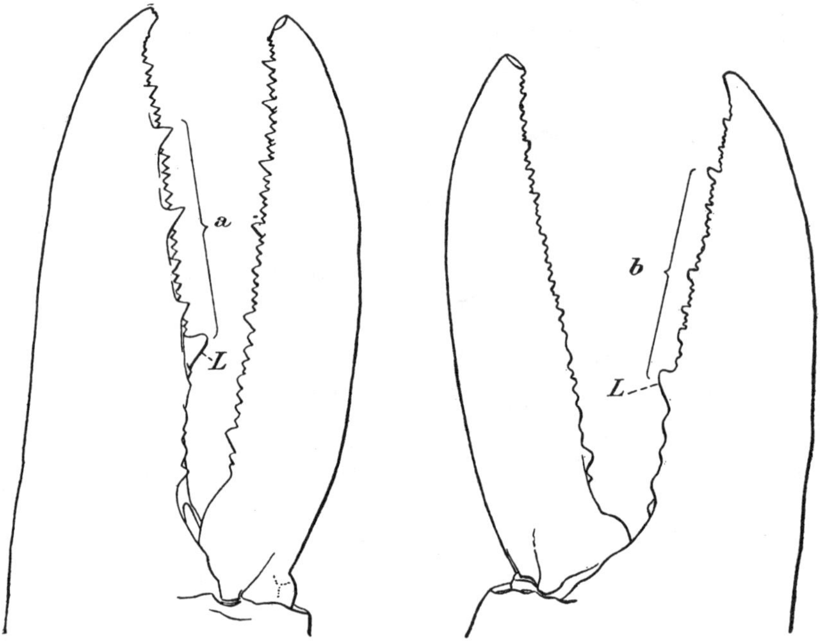

FIG. 21 and 22.—Right and left forceps of lobster 24 mm. long, reared in captivity, and 11 months old, in eighth or ninth stage; seen from above, showing early state in the differentiation of cracker (right) and toothed claws. Enlarged about 40 times. *L*, lock spine, as in all figures.

tube issuing from a separate gland, but with common opening at the summit of tooth. Whether these organs possess any special significance in these parts or not I am unable to say.

The first step in the differentiation of the cracker claw, as already remarked, is seen in the rounding or blunting of the teeth, particularly at the proximal end of the series (see fig. 22 and 24, and especially fig. 25). The teeth appear to be retarded in growth, and while these remain blunt and irregular, those of the toothed claw become

[a] The sensory hairs, as already stated, are derived solely from the epidermis, no mesoblast ever entering them, and they are invaginated with every molt. The claw teeth are tubular outgrowths of the wall of the appendage, and are never invaginated. The rostrum, as well as at least the tips and terminal spur or tubercle of the propodus, are seen to arise like the setæ, and like them are invaginated during the early molting periods, but they are eventually entered by mesoblast.

even sharper than before and retain their periodic character. The spines of the lock forceps are also noticeably larger for a time at least. Then follows a characteristic process of concentration and fusion in the spines of the future crusher claw (fig. 24, *c. s.*, and fig. 25, *d.*), which eventually leads to the reduction of their number. The crushing tubercle is thus formed by the fusion of a greater or lesser number of spines, like those of the toothed claw in the fourth to sixth stages.

In the light of this process are to be explained the "transition forms" which Przibram found to arise in the course of regeneration of the crusher claw, showing the knobs as fusing masses of teeth. The occurrence of such transitional stages has also been mentioned by Stahr and Emmel.

In the adult cracker claw (pl. XLIII, fig. 2 and 3) the propodus bears two large and six or more smaller tubercles. The big proximal tubercle (*p* (*L*), fig. 25) repre-

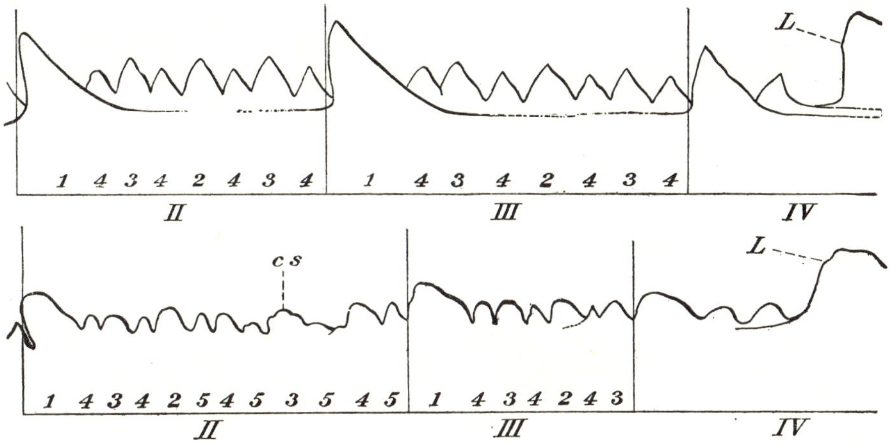

FIGS. 23 and 24.—Serrate margins of claws shown in figures 21 and 22, in regions marked a and b, and corresponding to periods ii-iv. Two perfect periods of eight sharp spines appear in the future lock forceps, and interpolations with fusions of teeth (*c s*) in the future crusher.

sents mainly the lock spine of the toothed claw, with the addition of lesser elements, while the great distal tubercle (*d.*) is composed of a fused mass of upward of thirteen spines, embracing the whole of the third and a part of the second periods. The dactyl of the crusher also possesses two tubercles of greater size, which close over the intervals between the "molars" of the propodus, besides a dozen or more small ones, resulting in each case from the fusion of several spines. There is also a small rounded tubercle on this segment at its proximal end and below the serial line.

The final differentiations established between the great crusher and lock forceps are illustrated by a perfect set of typical claws from a hard-shelled lobster which must have weighed approximately 12 pounds. In all measurements excepting length this crusher greatly exceeds its fellow, being one-third broader, weighing twice as much (in the dry shell), and having more than double the cubic capacity. In animals of adult size the slenderer claw has often a slight advantage in length over the more powerful cracker, as in this case, and in giants the difference is sometimes striking. The dry shell of this crusher is so dense and strong that it will bear the weight of a man of average size without giving way. The measurements of these claws are as follows:

Crushing claw:		
Length propodus	inches..	8½
Greatest breadth	do....	4$\frac{9}{16}$
Greatest girth	do....	11⅛
Contents	cubic centimeters..	680
Weight of shell (8¼ oz.)	grams..	235
Toothed forceps:		
Length propodus	inches..	8⅝
Greatest breadth	do....	3½
Greatest girth	do....	8$\frac{1}{16}$
Contents	cubic centimeters..	320
Weight of shell (4⅛ oz.)	grams..	116

The armature of this cracker claw (fig. 2 and 3, pl. XLIII) is typical and does not essentially differ from that found in giant lobsters weighing upward of 25 pounds.

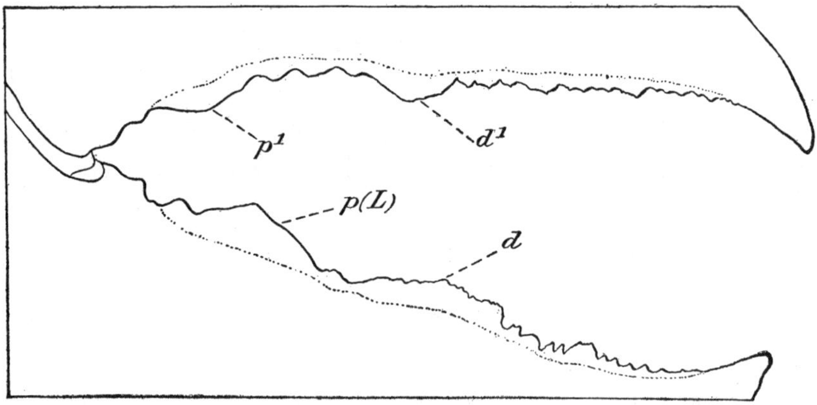

FIG. 25.—Armature of right crusher of female lobster 35 mm. long, and at approximately the tenth stage, showing origin of "molars" by fusion of spines. The proximal tubercle of the propodus (*p* (*L*)) is derived from the lock spine, while the distal (*d*) is composed of a fused mass of over a dozen teeth, embracing the whole of period II, and part of III. Length of claw 14.5 mm.

As in their case also the blunted end of the dactyl meets the big distal "molar" of the propodus, which, in the Belfast lobster, is worn flat and is 1¾ inches long by 1¼ inches broad. The dactyl in the slenderer claw is considerably longer, and as noticed above in mammoth lobsters the toothed forceps tends to surpass the crusher in length.

Since writing the preceding paragraph I have had the opportunity of reexamining the New Jersey lobster, which holds the record for size and weight (see no. 9, table 1), and find that the great claws which here reach the extreme known development of such organs, conform to the types already described and to conditions met with in mammoth lobsters generally. The cracker claw of this giant is remarkable for its swollen ovoidal form, its girth being 20½ inches, and for its worn and blunted tips; the blunt end of the "hand" is even recessive, the tubercular margin being convex as is frequently noticed in very large animals, and this in spite of the fact that the big molars are worn nearly flat. The worn-off end of the dactyl strikes about midway on the big distal tubercle, while the arrangement of the tubercles themselves is typical and essentially that given above; the propodus showing only two big "crushers," with one small intermediate and two paired or double proximal tubercles.

In the lock forceps of this specimen the hooked points are broken, rasped, and worn down, while its serrated margins are slightly convex, as is often the case in the fourth or fifth stage. The dactyl of this claw presents 7 to 8 primary spines. The huge, pyramidal lock spine of the propodus is much worn, and the first period distal to this bears 10 spines, having the formula: $1+1+2+4+2=10$. Then follows a long and probably compound period of 17 spines; then a primary spine and several smaller ones opposite the "spur." Thus, in this huge claw from lock to spur there are only three or at most four periods represented, as in all the younger stages hitherto discussed. This again illustrates the fact that while the procession of spines is constantly "on the move," the "dental formulæ" for the toothed claw never being identical for any two successive molts, the losses are so well balanced by the gains that the toothed claw, which attains its characteristic form from the fourth to the seventh molt, remains essentially unchanged throughout life.

We have seen how the toothed type of claw, which Stahr considers an ornament fitted to please the "æsthetic sense" of these animals, has arisen, but the wonder is not that the teeth are arranged in periods of eight, but that they are developed in order at all. The problem is similar to that of the orderly arrangement and appearance of the paired mesentaries of certain coral polyps, and fundamentally the same as that of the orderly development of the parts of all organic bodies, concerning the mechanics or the regulative control of which nothing is definitely known.

When we consider the known structure and development of the great claws in relation of the known habits of their possessor, we find no warrant in considering them as an "ornament" or in any other light than that of most efficient tools and weapons, chiefly for defense, for the capture of prey, for rending it in pieces, and afterwards for handing over the edible parts to the grinding mechanism which begins with the mouth parts and ends in the stomach. The developmental history of the lock forceps and its periodic teeth, as narrated above, renders any criticism of Stahr's fantastic theory, on the ground of comparative psychology, superfluous.

On the inner margins of the great claws appear certain prominent spines (fig. 2, pl. XXXVII *up. ser.*, and *l. ser.*), which are very regular in form and position, but vary somewhat in number. They consist of an upper series of 4 to 6 stout spurs curved upward and forward, and a lower of 1 to 3 teeth of lesser size, alternating with the first, and bent downward and forward. They probably originate from a single series, by displacement. They are eminently protective, while the proximal and often double spur on the upper side may act as a buffer when the claw is folded inward. Greater attention, however, is called to the serrated jaws of the forceps themselves, owing to the origin of their teeth by interpolation in the way described, and to the periodicity thus established, but the biological significance of one set of spines may be as great as that of the other.

VARIATION IN THE POSITION OF THE GREATER FORCEPS.

As was long ago remarked by Aristotle,[a] it seemed a matter of chance whether the crushing claw were on the right or left side of the body, but this is not altogether the case. The large claw occurs about as frequently upon the right side as upon the left, without distinction of sex, as shown by the following table, in which 2,433 individuals are recorded:

TABLE 7.—SHOWING VARIATION IN POSITION OF BIG CLAWS.

Sex.	Crushing claw on right side.	Crushing claw on left side.	Claws similar and of toothed type.
Males	562	628	1
Females	602	638	2
Total	1,164	1,266	3

I have shown that in *Synalpheus brevicarpus*,[b] of the Bahama Islands, where the large hammer claw can be recognized even before the animal is hatched, the members of a brood are either right-handed or left-handed, that is, have the hammer on the same side of the body. This seems to be a case of direct inheritance from the parents, though not enough data were collected to settle this point.

Since the issue of that work my early observations have been extended by Coutière and our combined results are tabulated below.[c]

TABLE 8.—SHOWING POSITION OF BIG CLAWS IN BROODS OF SYNALPHEUS.

No.	Great claws of mother.	Number in brood.	Right-handed larvæ.	Left-handed larvæ.
1	....	40	....	40
2	....	[d] 30	....	30
3	Left	16	....	16
4	....	3	2	1
5	Right	6	4	2
6	Left	4	1	3
7	...do	44	....	44
8	...do	22	1	21
		165	8	157

[a] "In the Carabi and in the Carcini the right claw is invariably the larger and stronger. For it is natural to every animal to use its right side in preference to its left. In the Astaci alone it is a matter of chance which claw is the larger, and this in either sex." Aristotle: The parts of animals; translated by W. Ogle, London, 1882.

[b] Herrick, F. H.: Alpheus: A study in the development of Crustacea. Memoirs of National Academy of Sciences, vol. v. ch. v, 4th mem., p. 370–463+, pl. 1–38). Washington, 1892.

[c] Coutière, H.: Les "Alpheidae", Morphologie externe et interne; Formes larvaires; Bionomie. Annales des Sciences naturelles, 8e sér., Zoologie, t. LX, p. i–iv, 1–560, pl. 1–6, text fig. Paris, 1899.

[d] The exact number in this brood was uncertain, but all that were preserved were left-handed. No. 1–4 were observed by the writer, no. 5–8 by Coutière. No. 1–3 refer to *Synalpheus brevicarpus*, no. 4–8 to the small *Synalpheus longicarpus* which abounds in the big black *Hircinia* sponges along shore.

Out of a total of 165 larvæ all but 8 were left-handed and 4 of these last are known to have had a left-handed mother. Four "families" in which every one of the 130 members were left-handed are known in two cases at least to have had left-handed mothers, the position of the crushing claw not having been observed in the others. Where the children of the same family vary in this character, it is probable that the parents or grandparents varied also. However, as I pointed out in 1892, the position of the toothed or crushing claw is not haphazard in its primary condition, but is predetermined in the egg.

In the next section, however, we shall see that in *Alpheus* as well as in other genera a remarkable reversal of the position of the big claw may take place, as a result of loss, so that in the course of life the crusher may shift back and forth, being now on the right and now on the left side of the body. The question therefore arises whether the left-handed female (no. 7 of the table), whose 44 children were all left-handed, was herself left-handed at birth, and secondly, whether, as in the right-handed *Alpheus* (no. 5), two-thirds of whose young were right-handed and the other third left-handed, the shifting of the big hammer claw would influence the inheritance of the children. These questions can not be answered, but it is suggested that in *Homarus* as in *Alpheus*, where no loss of limbs or other serious disturbance to the processes of growth have occurred, the right or left handed condition is due to inheritance.

Emmel has recently shown that up to the fourth molt the large crusher claw may be made to develop upon either side of the body at the will of the experimenter by the amputation of one claw, thereby, as it were, throwing the greater quantity of energy into the other for the purposes of growth. This power of control, however, ceases during the fifth stage, as at all later periods when asymmetry has become established and when the amputation of either chela does not normally reverse the conditions present. Emmel concludes that the factors which control asymmetry are correlated with the conditions of growth from the time of hatching up to the fifth stage. His experiments show that the asymmetry of the big claws of any given animal is not necessarily due to inheritance, but it would appear that in the normal course of development heredity played a part, although its initial course may be subsequently changed.

SYMMETRY IN THE BIG CLAWS.

In 1895 (*149*, p. 143 and pl. 14) I described and figured a variation in the adult American lobster in which both big claws were similar and of the toothed type. This variation was exceedingly rare, as shown by table 7. Only three cases of this abnormal symmetry were found in this collection of 2,433 lobsters made in the Woods Hole region by Mr. Vinal L. Edwards, the veteran naturalist and collector of the United States Fisheries Laboratory.

Since that time several papers have appeared upon this subject by Stahr (*258*), Przibram (*220*), Calman (*45*), Emmel (*91*, *92*, and *93–96*), and myself.[a] The first of

[a] The account which follows is partly taken from an article on 'Symmetry in big claws of the lobster" (no. *155* of bibliography).

these writers seems to have found this variation of similar toothed claws much more common in the European lobster.[a] The history of development proves, as Stahr maintained upon theoretical grounds, that the toothed claw represents the more primitive and the crushing claw the more modified type. Therefore it seemed natural to infer, as he did, that the anomalous symmetry in these weapons had been brought about by loss of a crushing claw and a subsequent reversion to the primitive toothed condition in the regenerated member which took its place. This would give us a lobster with symmetrical toothed claws like the variation described.

The converse of this, or the production of a new crushing claw in place of a toothed "forceps," could not occur upon Stahr's theory of regeneration, and hence he inferred that my report of a case of similar crushing claws in a lobster was an error. It was later at first rejected on similar grounds also by Przibram, who regarded the report as incredible and "worthy of being consigned to the realm of fishermen's myths." It should be added, however, that this objection was withdrawn in a later contribution (*223*), and neither Stahr nor Przibram are to be blamed, for my report was based upon the statement of a fisherman. Still, however great the inaccuracy of fishermen in biological matters, I have yet to find a lobsterman who could not tell a "club" from a "quick" claw. It now seems that the maligned fisherman, for once at least, was right, and he should get his dues even if earlier theories have to be revised, for Dr. W. T. Calman, of the British Museum, has described a case of symmetrical crushing claws in the European lobster (*45*), and his account is accompanied by an excellent photograph, which he has kindly permitted me to use (pl. XXIX). In all other respects this animal was a perfectly normal male. It was caught near Stromness, Orkney, and its living weight was 4 pounds 10 ounces.

In a letter, under date of December 3, 1906, regarding this unique specimen, Doctor Calman says:

> The correspondence between the two chelæ as regards arrangement and size of the crushing tubercles is even closer than appears on the photograph, where slight differences of color have a little obscured the shape in one or two points. The differences are no greater than one would expect to find between the two sides of a normally symmetrical animal. In other respects the chelipeds are practically alike in size and shape, except that, as seen on the figure, the dactylus of the left is shorter than that of the right. The basal segments of the limbs show no trace of asymmetry, which is often associated with regeneration.

To return to Emmel's paper (*93*), we find that in two recorded cases, an $8\frac{3}{16}$-inch female and an 8-inch male, "crusher claws" were regenerated after amputation by autotomy of normal asymmetrical chelæ. Emmel further records the capture at the Rhode Island experiment station in 1895 of a single adult lobster with similar "nipping" claws. When these were removed by autotomy two similar claws were also reproduced, but in this instance of the "nipping" type, like those cast off.

While in the usual course of events regeneration of a large cheliped restores the normal asymmetry of an adult lobster, Emmel has clearly established the fact that it

[a] Przibram (*223*) has reported a case of similar toothed claws in a specimen of the Norwegian lobster (*Nephrops norvegicus*) preserved in the Hofmuseum of Vienna.

Male lobster (*Homarus gammarus*) with symmetrical claws, and both of crusher type. The first specimen of the kind, living under natural conditions, to be definitely recorded. For figure of lobster with both claws of toothed type see no. *140* of bibliography, pl. 14. Stromness, Orkney Islands; weight, 4 pounds 10 ounces. Reproduced from photograph by Dr. W. T. Calman.

can both produce and restore a condition of symmetry. Both Przibram (*221*) and Morgan (*203*), as well as Emmel, have called attention to the fact that when the crushing claw is thrown off the regenerated member at first suggests a transitional stage between the more primitive toothed and the more modern crushing type, but this is not always the case, for two of Emmel's lobsters developed similar crushing claws at a single molt. Emmel's experiments show that a change in the type of big claw may occur in the adult lobster, but whether this is to be regarded as a step in the process of complete reversal of asymmetry met with in the younger stages of *Alpheus* and other forms described by Przibram remains to be seen. As Wilson has already remarked, the removal of both forceps from the prawn, unlike the case of the lobsters referred to, led to no disturbance in the normal asymmetry of those appendages. In 1901–2 Przibram (*221*) showed that in the crabs similar claws could be experimentally produced through regeneration.

To follow the reversal phenomena of *Alpheus* more closely for comparisons: We have seen that this shrimp carries a huge "hammer" or snapping claw, which in some species is as large as the entire body of the animal, and a diminutive claw of more primitive form on the opposite side. Moreover, in the common *Alpheus heterochelis* of the southern coast the small chela presents an interesting sexual variation, and resembles the "hammer" more closely than does the corresponding simpler and more primitive claw of the female.

A striking example of heteromorphic regeneration or reversal of asymmetry is seen when the *Alpheus* "shoots" its "hammer," or for any cause loses its big claw, as was discovered by Przibram in 1891. The big claw seems to hold the little one in check, for no sooner is it lost than the smaller one grows apace and becomes differentiated into a "hammer" or "snapper," while, as if in compensation for this change, a diminutive chela of the primitive type replaces the great claw lost from the opposite side. Wilson (*284*) found that in both sexes the small claw, which was regenerated from the stump of the large one, was always of the simpler female type, and, moreover, that the small chela of the male was more rapidly changed into the big "pistol" or hammer claw because it was already further advanced on this line of development than that of the female. When the smaller claw is amputated, or when the "forceps" are removed from both sides of the body at once, there is no reversal, a new slender chela or hammer claw taking the place of the corresponding member lost. Many additional facts have been brought to light through the experimental studies of Wilson, Brues, and Zeleny.

Przibram (*223*) has also found by experiment that reversal of the claws takes place not only in *Alpheus*, but also in *Athanas*, *Carcinus*, *Callianassa*, *Portunus*, and *Trypton;* that the tendency is most marked in the younger stages, and that it decreases with age. His results are therefore similar to those obtained by Emmel (*92*) in the lobster, where the experimental control of asymmetry ceases, as we have seen, at the fourth stage.

In the lobster no reversal or compensatory regulation normally or usually attends the regeneration of any of its appendages. The crushing or the toothed forceps, when severed at the "breaking plane," are as a rule replaced by their like in due time after one or more molts. How, then, are we to explain the anomaly of similar claws? It seems highly probable that the reversal, which regularly takes place in *Alpheus* when its great "hammer" claw is cut off, does actually occur, though but rarely, in the lobster, or, rather, that a step in the process takes place, there being no immediate compensatory change to restore equilibrium of the system of which the great claws form a part. Thus, when a "club" claw is "shot" or amputated by the experimenter, a chela of similar crushing type is usually regenerated in its stead, but rarely a toothed claw may appear. There is a change in the appendage, bringing about an abnormal condition of symmetry, but the process stops here, and we have as the result lobsters with similar toothed claws, like the specimen illustrated in an earlier work (*149*).

In like fashion the toothed claw of the lobster is usually replaced in regeneration by a limb of similar type, as is the rule with *Alpheus*, but in rare cases a "club" claw is substituted, and we get a lobster with symmetrical crushing chelæ, like the specimen described by Doctor Calman. This case is certainly much rarer than reversal from crushing to toothed claws. There is the possibility that these abnormal conditions of symmetry may be upset by a compensatory change in the appendage of the opposite side, but there is no evidence at present that this ever takes place.

When most of the preceding paragraphs on this subject were written I had not seen Emmel's valuable paper on the regeneration of crusher claws following the amputation of the normal asymmetrical chelæ in the lobster. Accordingly, the statement that the case reported by Doctor Calman was "for the present essentially unique in the literature of the subject" applied only to the fact of its occurrence in a state of nature or freedom, the two other lobsters reported by Emmel and referred to above being regeneration products resulting from amputations.

In discussing the significance of the substitution of the "crusher" for the primitive "toothed" type of claw, Emmel does not consider that any explanation is at present possible, either on the basis of "reversal" phenomena or of "compensatory regulation," and he thinks that we must be content at present with a record of the fact that substitution by regeneration takes place. I have endeavored merely to point out the probability that in such forms as *Alpheus* and *Homarus* we are dealing with processes which are essentially similar.

CHANGES IN THE TOOTHED CLAW AT MOLTING.

The adjustment of the blood supply in the big claws and the adaptation of their tissues to the process of molting, in the course of which their great bulk of muscles is pulled through the narrow ring at the base of the cheliped, are described in chapter IV. We shall now consider the interesting changes in the armature of the toothed claw or lock forceps, which are expressed at a given molt.

The behavior of the spines of this weapon suggests the movements of a company of soldiers at drill, and offers a striking illustration of that power of regulative control which distinguishes living things. The peculiar alignment of the spines of the forceps, by means of which its serrated jaws overlap, apparently effected by concerted but reversed movements of the teeth, and the behavior of the large "lock" spine, which gradually shifts to a position far out of line with its fellows, have already been described.

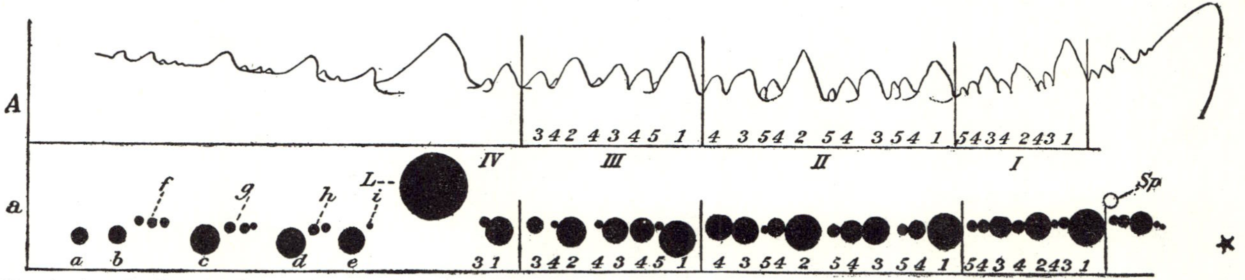

FIG. 26 and 27.—Profile and horizontal projection of larger division of right toothed forceps of male lobster immediately before the molt. Length of lobster before molting, 11¼ inches; length after, 12½ inches. Three periods and part of a fourth are present between spur (*Sp*) and lock spine (*L*). Enlarged nearly three times.

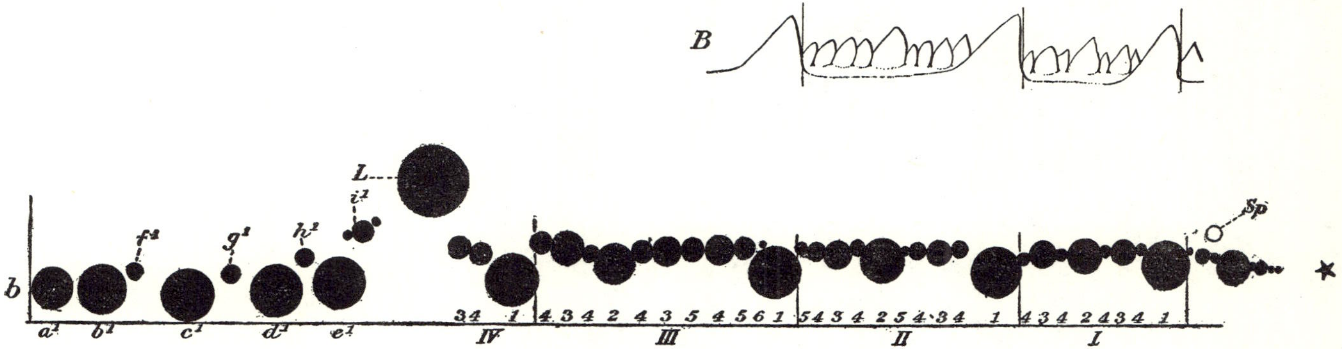

FIG. 28 and 29.—Partial profile, and projection of armature of same claw shown in figures 26 and 27, but immediately after molting; drawn to the same scale, and showing additions and losses in the periods, period 1 having acquired the typical formula in 8. Projections after camera drawings from clay impressions; periods in Roman, and orders in Arabi numerals, as in all figures, and reading from right to left. Claw-tip indicated by star.

The armature of the toothed claw of a seventh-stage lobster and that of the eighth stage from the same individual are given in figures 4 and 5, plate XLII. The formulæ for three typical periods in similar stages of another individual are also tabulated (table 5, no. 11*a* and 11*b*). It will be seen that five new spines have been gained in the course of this molt, and that one of them (the second in series III) belongs to the sixth order, while three have dropped out.

Similar changes were effected in the course of the molt of an adult lobster (lengths before and after molting, 11¼ and 12½ inches, respectively), and are illustrated in figures 26–29, where the spines are represented in profile and in horizontal projection. The "dental formulæ" are also given (table 6, no. 12*a* and 12*b*), from which it appears that five spines have been gained without corresponding loss in the three periods considered. More interesting changes have occurred at the proximal end of the jaw, where five characteristic large spines (*a–e*, fig. 27 and 29) have been retained, but the intermediate smaller groups (*f–i*) have lost from one to two members in three instances and in one case have gained two. Spine *i* has moved toward the lock spine, and bears two satellites, which seem to be thrown off as buds. The large tooth of the first order in the proximal period (IV, 1) has also received new recruits upon either hand (III, 4, and IV, 4).

Looking at the jaw as a whole, it has lost 6 teeth and gained 9, the first period alone having suffered no change in numbers. At the beginning of the molt the jaw was provided with 49 teeth, while at its close it possessed 52.

This suppression of old and emergence of new teeth probably goes on all the time in the life of this crustacean, but the changes must be compensatory, for no substantial losses or gains in the complete armature are finally registered in animals of great age. It will be observed that new spines often occur in the most crowded places, and it seems probable that such intercalated members arise as buds from their larger neighbor, as suggested above. In the earlier stages, however, there is no evidence of budding growth or division at the surface. As to why in certain parts (groups *f–h*, fig. 27 and 29) teeth are summarily suppressed, we can only hope that at some future time light may be thrown on such obscure questions.

Chapter VIII.—DEFENSIVE MUTILATION AND REGENERATION.

AUTOTOMY OR REFLEX AMPUTATION.

The casting of the big claws and of some of the smaller legs described as defensive mutilation, autotomy, or "self-amputation," is highly characteristic of the lobster. It is closely associated with the remarkable power of regeneration or replacement of lost parts, and less directly with the periodical renewal of the shell. These subjects have opened up wide fields for research, the borders of which we can only touch at a few points.

The power of reflex amputation is most perfectly developed in the large chelipeds of the lobster. When this animal is seized by the claws, and struggles to escape, amputation is likely to occur in both limbs. The animal surrenders its principal weapons, but may escape with its life. The powers of regeneration are at once enlisted in the complete renewal of the lost members. Every stage in the process can be found in animals kept alive in floating cars or in those sent to the markets. Out of 725 lobsters caught at Woods Hole, Mass., in December and January, 1893–94, 54, or 7 per cent, had thrown off one or both claws. The leg is broken off, as we have already seen, at a definite place, called the "breaking plane" or joint near its base, through reflex muscular contraction; there is but little bleeding from the old stump, and a new limb soon sprouts and is regenerated. The slender walking legs are sometimes lost and replaced in a similar way. Many, if not all, of the appendages, when mutilated or removed, are capable of regeneration, the time required for the process depending upon the proximity of the succeeding molt, the vigor of the animal, and the temperature of the water.

In autotomy the five distal segments of the limb are cast off, fracture taking place in the walking legs at the free third joint, between second and third podomeres, and in the great chelipeds at the corresponding breaking plane. On the second compound podomere of the first pereiopod of the adult the suture of basis and ischium is marked by a fine hairline or encircling groove, free from setæ, and it is always in this plane that disjunction occurs. If the terminal parts of the limb are amputated autotomy of the remaining stump usually occurs before the work of regeneration is begun. Mutilation of the claw alone, however, is not necessarily followed by the casting and renewal of the limb. Parts regenerated in any of the appendages are as a rule similar to those thrown off, except in the case of the eyes and big claws under certain conditions. The stalked eye can sometimes be made to produce an antenna-like structure, and while big crusher claw usually reproduces crusher, and lock forceps lock forceps, this is not invariably the case, and we occasionally find lobsters with both claws similar, and of either toothed or crushing type, as described in chapter VII.

Autotomy can be experimentally produced by seizing the animal by its claw or slender legs, or by stimulating the nerve of the limb directly, the reflex nerve center having been found to lie in the corresponding ganglion of the cord, but if the animal is anæsthetized it will "forget" to shoot its claw. We have seen that the basis has lost its muscles, and that the ischium possesses two extensors only; in order that autotumy should normally occur it would seem to be necessary that the part of the limb distal to the breaking plane should offer a greater resistance than the traction of the small extensors of the ischium is able to overcome; ordinarily the clutch of an enemy furnishes the opposing force required, but since the action is purely reflex, "accidental" disjunction of a limb which happens to be suddenly opposed in its movements may occasionally happen. The probable relations of autotomy to the interlocking mechanism of the coxa and ischium are described in chapter VII.

While no tendons cross the breaking joint in the adult lobster, Emmel (*97*) has shown that this is not the case in the larvæ, in which he has discovered a transitory muscle of considerable interest; this muscle originates on the inner wall of the basis, crosses what is then a free joint, and is inserted upon the inner side of the ischium. It acts as a flexor during the first four stages of life, begins to dwindle in the fifth stage, and is reduced to a mass of degenerate tissue in the sixth. It has been maintained that in the lobster the breaking plane does not represent a lost joint (see no. *235*), but that a fusion has taken place between the third and fourth segments, a statement which is not easily understood. Thanks to the peculiar interlock of spurs on the first three podomeres, it is easy to follow the changes which these segments undergo from the fourth stage onward without difficulty (see ch. VII, p. 259), and if any further evidence were needed to show that the breaking joint, which is functional up to the fourth stage, corresponds to the articulation of the second and third segments, it would seem to be furnished by Emmel's discovery of a missing flexor muscle at this point.

While autotomy does not normally occur before the fourth stage, the limbs are often snapped off at the joint destined to become the breaking plane. Lobsterlings occasionally cast a claw at the articulation between the second and third segments which has the appearance of a free joint; fusion is not completed until the fifth stage, from which time onward autotomy in its typical form becomes a common occurrence.

An interesting adjustment to prevent excessive loss of blood in the stump of the reflexly amputated limb has been described by Emmel (*97*). We have seen, in referring to his account in another place (ch. VI, p 245), that as the venous sinus crosses the breaking plane it is divided into two channels by a septum in which are lodged the two arteries and two nerves of the limb; on the proximal side of the joint the septum gives off two folds, which are swung out by blood pressure after the break occurs and acting as valves to the small openings exposed, check the bleeding at once. It would appear from Emmel's work that the severed arteries must immediately contract so that their blood is discharged proximally to the folds or valves which he describes. Whether a similar adjustment to prevent excessive loss of blood is found in the other appendages, so far as I am aware, has not been determined. To continue this account further, when a claw is shot, a short jet of blood is thrown from the stump, but the bleeding soon ceases, followed by a slight swelling of the tissues over the fresh surface; if the valves are pressed open the bleeding is resumed.

Reflex amputation in crustaceans, whether considered in relation to shock or to fear, or as an independent mechanism, must be regarded as one of the most remarkable phenomena of invertebrate life. The loss of a considerable amount of tissue is always a shock to a higher vertebrate, while a lobster in autotomy of both its chelipeds may give up with impunity one-half the weight, or even more, of its entire body. In the higher animals fear may be due to inheritance or it may directly arise through association, by experience. The lobster, indeed, shows fear by hiding or by its hasty retreat from an enemy, but reflex amputation does not appear to have any necessary relation to fear. The reflex center of the cord is aroused to activity by a stimulus coming direct through the nerves of the limb, and not from the brain. We may be sure that the same center does not at one moment give the order to flee, and at the very next compel the animal to drop any of its legs. The lobster or crab does itself a grievous injury automatically in order to escape a worse fate. This kind of reflex surgery thus seems to be an afterthought of nature, as if an attempt had been made to repair an earlier mistake, or a compensation, as it were, for having originally endowed the crustacean with a frame too vulnerable to attack, or with a mind too feeble to successfully cope with its environment.

RESTORATION OF LOST PARTS.

The power of restoring lost or injured parts through the process of regeneration is very general throughout the body and appendages of the lobster. It is exercised very perfectly and promptly in the big chelipeds when thrown off by autotomy at the breaking plane, where the process has evidently been favored by natural selection or some other factor of evolution. Regeneration is also very active in the fragile antennæ and the walking legs. All of these organs are, at the same time, very liable to injury, and are essential to the maintenance of life by directing the animal to its food and enabling it to secure it. In conveying this food to the mouth and preparing it for the stomach the mandibles and other mouth parts are quite as important; the swimmerets also serve a variety of necessary functions, but all of these structures are far less liable to injury. Whether there is a causal relation between liability to injury and facility to restore the injured parts is another question. Morgan has reached a negative conclusion in his experimental studies on the hermit crab, and concludes that "regeneration is a fundamental attribute of living beings." The question, however, does not depend upon a single relation; the relations are undoubtedly very complex, and it can not be denied that in such animals as the lobster the external organs which are most exposed to injuries of every kind and which are of immediate necessity for the maintenance of life possess the most active power of regeneration.

Emmel has shown (*89*) that the power of regeneration varies at different levels in the limbs and that even the swimmerets may regenerate more rapidly than the legs if the latter are cut off but a short distance below the breaking plane. Therefore the rate of regeneration depends upon the place of injury as well as upon the amount of surplus energy available at that point.

The regeneration of a large cheliped in the fourth and fifth stages is essentially the same as in the adult. At the moment the limb is broken off there is but little loss of blood, which coagulates and forms a protective crust over the stump. In a short time a small white papilla, which represents the rudiment of the new limb, appears in the midst of the brown, hardened clot. This papilla continues to grow independently of the molting process, though covered with a cuticular membrane, until a miniature appendage is formed. The papilla lengthens, and gradually the constrictions which mark the future joints of the new limb make their appearance. At first colorless, the new appendage becomes bright, transparent red, with bluish pigment at the constrictions of the joints. In this stage the limb is surrounded by a thickening cuticle and soon ceases to increase in size until after the next molt. If autotomy occurs just after a molt, the appendage will reach a much greater size than if it happens a short time before, but within the limiting period referred to below. When the molt finally takes place the new stump becomes very much larger, and now resembles the normal appendage in all respects except size. With each succeeding molt the normal size is gradually attained.

The large cheliped of the young lobster in the fifth stage may be regenerated in from 15 to 18 days after a single ecdysis, or it may require a month's time, during which the animal may pass two molts. The normal size, however, according to Emmel, is not attained until after the third molt. He also found that by the repeated removal of the same appendage in sixth to eighth stage lobsters the rate of growth in the mutilated limb was repeatedly reduced, but the experiment was not carried very far. This observer has also found that the thoracic legs will not begin to regenerate if removed immediately before a molt. The limit varies from 2 to 4 days in sixth to seventh stage lobsters. In more mature animals the limiting period is 16 days at its shortest duration. Accordingly, if accidents happen shortly before the molt, the animal must wait until this crisis is over before nature can give any attention to the restoration of the parts lost. Apparently in this case the energy required to renew the entire cuticular covering does not leave any surplus immediately available for the growth of new limbs and tissues. If the tips of the large chelipeds are clipped off, autotomy does not always or usually occur, and the limb is completely repaired after one molt. If the limb is injured below the propodus, it is usually cast off at the plane of fracture.

The antennæ are very liable to injury, particularly the delicate, sensitive flagella. Autotomy does not occur in these appendages, but regeneration may take place at any articulation in the flagellum or stalk.

In the young the whip of the second antenna may be completely restored without a molt taking place, while in the adult one molt at least appears to be necessary for complete restoration. In the fifth stage lobster, already mentioned, the antennary flagellum was restored in about 15 days. This appears first as a papilla or bud, which becomes sickle-shaped and finally coiled so as to resemble a small spirally twisted red wax taper.

The cuticle of the limbs in process of restoration must be elastic and capable of considerable distension, although the limit of this distensibility is, in most cases, soon reached.

According to the studies of Miss Reed upon the process of regeneration in the crayfish (*235*), the membrane or the inner half of the double fold which remains after autotomy, and the blood cells beneath it serve to protect the end of the stump, but take no part in the regeneration of the new limb. To summarize briefly her account, the process of actual regeneration begins in about 5 days after the loss of the original member by an extension of ectoderm over the opening, which thus replaces the blood plug formed at the time of injury. Later these same cells secrete chitin and form a thickened disk over the broken end of the nerve. The ectoderm pushes out into a growing, expanding tip; its cells become elongated, join the cells of the old nerve, and reconstruct those of the new one. As the bud grows out, muscles and nerve are regenerated from ectoderm cells and folds in this layer appear, thus marking out the future podomeres of the new limb. The folds, which arise as ingrowths of ectoderm, also secrete chitin; they split to form the folds of the joints and, finally, at their ingrowing ends give rise to the tendons of the muscles and to the muscle fibers which are attached to them.

Emmel (*97*) has obtained similar results in working upon the lobster, wherein the wound caused by autotomy is soon covered by a plate of migrating epidermic cells. The wall of the limb and possibly its core were found to be epidermic, the old muscle and connective tissue cells of the stump appearing to contribute little to the new appendage. Both new nerve and new connective tissue elements seemed to owe their origin to the epiblast of the regenerative bud.

MONSTROSITIES.

The curious monstrosities that occur in the appendages, particularly in the large claws of the lobster, have attracted the attention of naturalists from early times. They were noticed by Von Berniz over 200 years ago, and some good figures of the deformed claws of the crayfish were published by Rösel in 1755.[a] Among the later students of variation Bateson (*19*) has shown that in most of the cases of supposed duplication of limbs in both insects and crustaceans the extra parts are double instead of single, as where two dactyls are formed at the extremity of the claw instead of a complete claw consisting of dactyl and propodus. He has also formulated certain principles according to which supernumerary appendages make their appearance in secondary symmetry. If the normal appendage which bears the extra ones is a right leg, "the nearer of the extra legs is a left and the remoter a right."

The monstrosities noticed in the chelipeds of the lobster are mainly the result of a secondary outgrowth from one of the two terminal segments. Rarely the appendage is duplicated or triplicated. In some cases the extra appendages are perfectly formed, while in others deformation has been carried to excess, resulting in irregular branching processes or grotesque contortions. Injuries to the claws are excessively common, while duplication of the parts is rare. Defective or deformed claws, the result of injuries in different stages of repair, are met with every day by dealers, while thousands of lobsters may be examined without meeting a single case of repetition or duplication of parts.

If the tips of the claws are snipped off near the articulation of the dactyl, the lost parts are restored at the next molt without autotomy taking place. This is called simple regeneration by Przibram (*221*). This restoration is often perfect, but not

[a] Insecten-Belustigung, dritter Theil. Nuernberg, 1755.

always so. Distortions arise which may have been caused by a pinch and arrest of growth while the claw was soft or by injury to the stump. In the latter case the member might be only partially restored, and unequal growth would account for the defect.

A small budlike swelling is sometimes seen near the apex of either division of the claw, and it formerly seemed to me improbable that this could be due to a simple injury since such appearances are rare, while injuries to the big claws must be excessively common. I further assumed that, given such an outgrowth, a progressive series of changes might take place with successive molts, the swollen part becoming bifid and eventually completely divided. To continue the account upon this basis: With the growth of the animal, the superadded part, whether it be upon dactyl or propodus, seems to be shifted at each molt farther and farther back upon the claw, and meantime, in most cases, to undergo fission in a vertical or somewhat oblique plane. This fission apparently proceeds until one or both of the supernumerary dactyls are entirely separated. The opposing edges of these become gradually toothed, so that each is almost an exact copy of the original. According to the principles laid down by Bateson, the part which is nearer the original joint corresponds with the appendages on the opposite side, that which is farthest away with those on the same side of the body. Many cases occur, however, which do not conform to this and apparently to no other rule (see *149*, p. 144–148).

Since the appearance of my earlier work referred to above, the excellent researches of Przibram (*220–223*) and Emmel have added greatly to our knowledge of this subject. The former has shown that in all probability monstrous growths of every kind result from a regenerative process following upon injury. However, such growths are comparatively rare and follow only upon injury of a certain kind, or upon an injury inflicted at a certain time with respect to the molting period, or under certain conditions of the animal which are not fully understood.

Przibram found that when an injured leg was retained duplication of the part might arise through a division of the regeneration rudiment, as in vertebrates, and it was further shown by Miss Reed that when a leg of the hermit crab is thrown off, if the base is split lengthwise so as to divide the nerve, there often appear two new legs, each connected with one end of the nerve. It would thus appear that duplication of a limb is subject to the will of the experimenter, and that duplicated parts may often arise in nature through an accidental injury to the nerve rudiment. Further, in 1905 Zeleny (*290*) obtained by experimental means the regeneration of a double chela in the fiddler crab. Two cases where duplication of parts of the big claw followed directly upon injury to the claw itself or to a regeneration bud have been recorded by Przibram (*223*); the first concerned a specimen of *Portunus hastatus*, which suffered in an aquarium the loss of both points of its big right claw in an irregular manner, and regenerated within three months; after molting, the dactyl became doubled, while the propodus was unchanged. The second case arose through an artificial division of a normal regeneration bud of the last walking leg of a *Carcinus mœnas*. The operation was performed with fine scissors on May 14, 1901, and after the molt, which occurred on June 2, the protopodite showed two separated dactylopodite buds. Since this animal died on the day after the molt, it was not possible to test the hypothesis outlined above, of progressive changes following each molt. Przibram further expressed the belief that similar claws in the lobster were due to regeneration, since in crabs individuals with similar claws could be experimentally produced, a view confirmed later by the experiments of Emmel, already referred to.

Emmel (*92*) has described three additional cases in which abnormalities have been artificially produced through the process of regeneration. In two instances similar crushing claws resulted, and in a third case a triplication of the claw occurred in one of the walking legs. This adds greater weight to the conclusion that all deformities in the limbs of these crustaceans, as well as the condition of abnormal symmetry rarely met with, may arise in nature through the process of regeneration, directed by some injury or abnormal condition in the nerve end, the regeneration bud, or the growing or developing limb.

Monstrosities occur in the early and late embryos, and are therefore regarded as congenital in their origin (see *149*, p. 216). It is well known that embryonic or larval monstrosities can be produced by subjecting the eggs of many animals to unnatural and unfavorable conditions, and it is possible that the causes which produce a double-headed larval lobster are similar to those which bring about the duplication of a big claw in the adult. Perfect twins are occasionally produced from the same egg (see p. 321).

Emmel has also recorded a striking case of the triplication of a big (right) crushing claw in a 10-inch male lobster taken alive on the coast of Maine. The normal claw was the smaller and transitional in type, while the two supernumerary claws were considerably larger and typical crushers. Of these the outermost was an inverted right, with lighter colored surface uppermost, and the other a normally disposed left. The abnormal chela was removed by autotomy, in anticipation by the experimenter of some interesting results at the next regeneration, but to the regret of all students interested in the problems of regeneration this animal died in September, 1906.

Emmel remarks that if the duplication of the big claws and other similar deformities which appear in the lobster were congenital in origin, we should expect to meet with cases in the larvæ and the later stages of growth, but after an examination of over two thousand fourth and fifth stage lobsters not a single abnormal case was observed. Examination of thousands of larvæ have everywhere given the same result.

What was described in the newspaper press as a "lobster pearl" was taken from a claw of a cooked lobster by Mr. F. W. Denton, of Hollis, Long Island. Through the courtesy of Mr. Alfred Eno, of Jamaica, N. Y., the writer was able to examine this interesting specimen, an account of which, with illustrations, has been published (see *157*). The "pearl" is a globular body 11 millimeters in diameter and of the same creamy tint as the inside of a lobster's shell, with which it agrees in every physical and biological character.[a] It probably represents a freak of the regeneration process following injury to the claw, and a more or less permanent invagination of the skin at a certain point. It is safe to say that no true pearl can be formed in any arthropod.

---

[a] Dr. W. T. Calman, of the British Museum, writes under date of January 14, 1911, that a similar specimen was received from a fishmonger in London several years ago, but in this case the body was "embedded in the abdominal muscles of *Palinurus valgaris*," and is now preserved in the Museum of the College of Surgeons.

Chapter IX.—REPRODUCTION.

Since every attempt at the artificial propagation or rearing of animals must be made in imitation of nature, the more exact our knowledge of the reproductive life and habits of old and young the more likely are we to succeed. Apart from their economic bearings, however, the problems suggested are the most interesting with which the zoologist has to deal. In the case of many animals the facts which lie at the surface can be gathered and utilized with comparative ease, while in others, as with the common eel, whose breeding habits baffled naturalists for centuries, opportunities for making the essential observations are seldom, if ever, presented. In some respects the lobster belongs in the latter class; its life is spent at the bottom of the sea, and when confined in aquaria, where alone continuous observation is possible, the normal play of its reproductive functions is apt to be disturbed. While much attention has been given to the subject, and many important facts have been learned, there are certain questions to which a confession of ignorance is the best answer that can be given. In reviewing the matter in hand we shall endeavor to make it clear whenever a plausible conjecture is offered in place of well-attested facts.

SEXUAL DISTINCTIONS.

In general form and color the sexes agree so perfectly as to be indistinguishable to an inexperienced eye when examined from above. The female abdomen is relatively broader than that of the male in adaptation to the protection and safe carriage of the eggs, while length for length the male is heavier, this advantage in weight being seen in his slightly larger claws. Above the 8-inch size, as we have already observed, males are usually heavier than females of the same length, even when the latter carry eggs.

Upon turning the animal over, the sex is readily determined by a glance at the swimmerets, the first pair of which is rudimentary in the female, and bears but a single hairy blade, the endopodite (fig. 1, pl. XXXIX). This may be considered as an adaptation for the benefit of the eggs, for were these appendages of normal size they would catch so many ova at the time of spawning as to make it impossible for a large animal comfortably to fold her tail, a difficulty actually experienced by egg-bearing lobsters over 16 inches long. The seminal receptacle appears as a bright blue shield wedged between the bases of the last two pairs of thoracic legs on the underside of the body. (Pl. XXXIII, and fig. 4 and 6, pl. XLIII.) Its function is to receive and hold the sperm until the eggs leave the body and are ready for fertilization. Just in front of this organ the oviducts open close together on the basal segments of the second pair of small claw feet. Each duct is closed by a valve and faces its fellow with an inclination backward. When the eggs are emitted from the mouths of the ducts their natural course in the case of an animal lying on its back would be downward and backward over the seminal receptacle.

Turning to the male and confining our attention for the moment to external anatomy, we find correlative structures of great interest. The seminal ducts open to the outside much as do the oviducts, but on the basal segments of the last pair of walking legs. The openings face the middle line obliquely and are directed backward and downward. The underside of the tail is armed with a median row of four sharp spurs, which project downward and backward from the sternal bars of the second to fifth abdominal somites; in the mature female these protective spines are rudimentary, a condition which certainly favors the safe storage and carriage of eggs.

In place of the seminal receptacle we find in the male small corresponding wing-like processes diverging to form a deep V-shaped groove in which rest the tips of the stylets or modified first pair of swimmerets (fig. 5, pl. XLIII and fig. 1, *a*, pl. XXXIX). The inner branch of the second pair of pleopods bears a peculiar short spur, and it is to be noticed that when the swimmerets of the male are directed forward the stylets meet on midline between the wings of the sterna just mentioned to form an imperfect archway or covered passage, while in the divergent angle behind rest the short hairy spurs. That these parts are concerned in the passage of the spermatophores to the seminal receptacle of the female can hardly be doubted. Their structure and function will be more fully considered after the several organs themselves have been examined.

THE RIPE OVARY.

The ovaries, or "coral" as they are sometimes called, are immediately exposed upon opening the dorsal body wall. They consist of two cylindrical rods of tissue united by a transverse bridge, behind which each lobe gives off a short, straight duct (fig. 1, pl. XLIV). The ovarian lobes traverse about two-thirds the length of the body, extending from the forward end of the stomach to the third, fourth, or fifth segments of the tail, and when approaching maturity are of a rich dark-green color. The ripe ovaries are so much swollen that they fill all the available space in the upper parts of the body. The bead-like eggs are clearly seen through the thin ovarian wall, and when this is cut they flow out, if perfectly ripe, in an uninterrupted stream. When the congested ovary is not mature the loosened eggs stick together and can not be easily disengaged without injury. A female with eggs approaching maturity can be readily distinguished by extending the translucent membrane between the tail and carapace, through which the color of the ovary is at once apparent, but since the eggs can not be pressed from the unyielding body of the animal, there is no way of telling when these are ripe short of actual dissection.

During the long period of growth, which leads up to the production of the first generation of eggs, various changes ensue, which are essentially uniform except for variations in color imparted by the yolk to the immature ova. After the first generation of eggs is expelled a normal reproductive rhythm is established, and during each cycle which follows, from egg laying to egg laying, the ovary undergoes a definite series of changes, unless the normal rhythms are disturbed by unusual and unfavorable conditions. A complete change in environment may necessitate a change in reproductive habits, and it is remarkable how quickly this crustacean can on occasion adapt itself to new conditions, as seen in the successful transportation of the lobster 12,000 miles through the Tropics to New Zealand in 1906–8 (see p. 176).

The history of the ovary will now be considered on the basis of the periodic events noticed above and as they have been found to occur on the coast of Massachusetts.

DEVELOPMENT OF THE OVARY TO THE FIRST SEXUAL PERIOD.

The ovaries (pl. XLV) are first recognized in well-advanced embryos as minute paired ovoidal masses of mesoblastic cells below the forward end of the heart and close to the pericardial wall. Later they appear as solid rods composed of a wall or capsule and a lining epithelium. The ovaries do not originate as hollow tubes, but virtually possess a tubular structure at the time the ripe eggs are expelled by contractions of their muscular walls. Egg laying is followed by a collapse of these walls and the immediate return of the ovary to a solid condition. It will, however, be easier to understand the structure eventually attained by conceiving the organ as possessed of a tubular form, the entire wall of which is composed of two parts, namely, (*a*) a capsular layer consisting of involuntary muscle, connective tissue, blood vessels and sinuses, and (*b*) a lining epithelium. Between these parts the blood finds ready access in irregular channels after leaving its definitive vessels. The ovarian epithelium consists of a basement membrane and epithelial cells from which the eggs and egg follicles are differentiated (fig. 1, pl. XLVI). The superficial area of this epithelium becomes greatly increased by irregular inwardly directed folds or invaginations. Through the reentrant sinuses thus formed blood penetrates to every part of the organ. The egg follicles are eventually composed of a thin sheet of tissue, the cells of which, as we have seen, are homologous with the ova. These follicles separate each egg from its fellows, form a medium for the transfer of nourishment to it from the blood, and soon begin to secrete about it the transparent egg shell or chorion. Owing to the manner in which the invaginations of the ovarian epithelium arise, the ova at a certain stage are arranged in irregular, radial and longitudinally directed tiers; each tier is embedded in opposing sheets of follicular tissue, while each ovum is completely inclosed, and the largest and oldest eggs are peripheral.

Along the central ridges of the epithelial folds the primitive ovarian cells mulitply and become differentiated into the future ova and follicular elements which are crowded or discharged into what corresponds to the lumen of the ovary, or into its central parts. (Fig. 5, pl. XLV.)

The process of early differentiation and growth of the eggs seems to proceed in the following manner (fig. 1, pl. XLVI): Along the crests of the central folds referred to above the ovarian cells become columnar and often greatly elongated; each narrow cell appears to be attached to a corresponding thickening of the basement membrane, which forms the lining of a blood sinus. To this is due the "pitted appearance" mentioned by Bumpus (*41*). The nucleus of a cell destined to become an egg, which lies close to the basement membrane, swells into a large spherical vesicle, about which a thin layer of cytoplasm, without boundary wall, may be discerned. Granules of yolk appear almost immediately in the cytoplasm, and henceforth the growth of the egg is determined by additions to the store of yolk, the materials for the manufacture of which are supplied by the blood. At an early stage the eggs probably multiply by division, and where they do not immediately break away from the parent epithelium they become elongated by mutual pressure, so that their long axes are parallel to each other and perpendicular to the basement membrane. Irruptions of ova, however, always occur at certain points, so that the young eggs appear in bunches along the crests of the original folds.

The nuclei of those cells destined to form a part of the follicle are easily distinguished by their smaller size, rod-like form, and by the relation to the young eggs which they promptly assume. The nucleus or germinal vesicle grows apace and continues to expand until, at the close of the first year after a given ovulation, it attains a diameter of one-eleventh millimeter. Rarely two or more nucleoli are developed in the young eggs; there is usually but one nucleolus and this of large size.

When sections of the ovary are examined, after treatment with the usual killing, fixing, and staining fluids, we find the nucleoli of all the eggs lying against the nuclear wall in the same relative positions; that is, at the "bottom" of the nuclei or on the side which was lowest at the time of fixation. The nucleolus is apparently released from its suspension in the nuclear reticulum by the action of the fixative employed, and responding promptly to the influence of gravity, drops like a shot in a bag. The ultimate position of the nucleolus is thus solely determined by the direction of gravity, and in reference to the egg itself by the position of the tissue at the time of fixation.

The growth of the first generation of eggs is exceedingly slow, occupying from four to five or more years, during which the ova must derive their nourishment indirectly from the blood. Swarms of new cells which continue to arise along the axial folds tend to drive the largest and oldest eggs toward the outer walls, a condition which is maintained until these ova approach maturity. When the limit of growth is reached the eggs are dehisced from their capsules, fill the lumen of the ovarian tube, and crowd the germinal folds and younger eggs of the next generation farther and farther toward the periphery.

We have already referred to the variable color of the organs during this period of their growth. Bright yellow, flesh and salmon color, light olive green, with many intermediate tints, are commonly noticed, while after the first eggs are produced, uniformity in the color of the organs prevails. With rare exceptions, after the first egg laying the ovary in due time assumes a characteristic light pea-green color and becomes progressively darker with age until maturity is reached.

CYCLICAL CHANGES IN OVARY AFTER THE FIRST SEXUAL PERIOD.

We have finally to consider the changes which the ovary normally undergoes during each successive reproductive period. After the eggs are laid the collapsed organs assume a grayish-white tint and appear flecked with green spots—the residual ova which fail of emission and stick fast in the lobes and ducts. In the course of 36 hours or less the ovaries are again solid masses with central germogenal folds, the larger eggs lying nearer the periphery, where the epithelium has become decidedly glandular in appearance. (Fig. 4, pl. XLV.) These gland-like organs apparently contribute to the growth of peripheral eggs for a short period and subsequently disappear. Amœboid cells pass from them into the eggs, where their nuclei degenerate, giving rise to swarms of fine chromatin-like granules, which persist for a considerable time. In 5 weeks from the date of oviposition the gland-like bodies are reduced to shrivelled remnants, of which later no vestige can be recognized.

While the massive yolk of the eggs is mainly derived from materials drawn from the blood and laid down at first in the cytoplasmic reticulum, the migratory cells just described contribute in a minor degree toward the supply, and the glandular follicles possibly manufacture yolk directly, although the evidence which seems to support this idea may be wholly deceptive, owing to the presence of degenerative elements.

In the course of 5 or 6 weeks the ovary, flecked with degenerating eggs which failed of passage and now of a bright orange color, begins to assume a light-green tint. Examination of the larger ova shows that the pigment, a green lipochromogen, is first formed in the yolk spheres immediately around the nucleus and thence spreads centrifugally until it involves the entire yolk mass. In a year's time, or at the beginning of the summer following ovulation, the peripheral eggs, while but little larger, are more uniform in size and color, and the whole organ presents a characteristic pea-green tint. A second period of active growth ensues, followed by a second interval of quiescence during the winter. At the beginning of the third summer after the last ovulation these eggs enter upon their third and last period of active growth and are soon ready for extrusion. (Fig. 5, pl. XLV.)

Owing partly to the presence of the egg membrane or chorion, absorption of the residual eggs at each period of laying is exceedingly slow. After the lapse of 2 years traces of them can be detected, and the presence of these orange flecks in the ovary of any lobster tells us conclusively and at a glance that it has already spawned once at least.

The ripe eggs, as spawning time approaches, lie free in the lumen of the ovary, which they distend to an unusual size, its elastic walls becoming very thin in consequence. Maturation may be completed in the ovary itself, but fertilization is possible only after the eggs have been expelled from the body. The massive yolk is inclosed in a flexible and transparent shell or chorion, secreted, as we have seen, by the egg follicle or sac, and by the time the ovum has reached the ducts its nucleus (female pronucleus) has migrated to the surface. The ripe egg possesses a single membrane only.

DISTURBANCES IN CYCLICAL CHANGES OF THE OVARY.

It is convenient to notice here what the fishmonger in England sometimes calls "black lobsters." During the summer months the English lobster dealer is said to examine his stock daily and to cull for immediate sale such animals as show a tendency to blacken. It seems that whenever females with ripe ovaries happen to be caught and are either sent to market or kept in floating cars, the normal reflexes which attend the reproductive functions are apt to be disturbed. The eggs, instead of being expelled in the natural way, perish in the ovary, possibly by having their requisite supply of oxygen from the blood curtailed, and absorption of this inert mass begins, in part at least, through the agency of the blood. By taking up the green pigment from the eggs the blood becomes very dark in color, thus giving all the tissues an unpalatable greenish-black appearance, very noticeable at the articular membranes.

The green color of the eggs, like that of all parts of the integument of this animal, is due, as we have seen, to the presence of dissolved pigments of a very unstable character. In consequence of partial absorption and coincident changes in the pigment which remains, the degenerating eggs gradually assume a yellowish-orange color. Whether the animal survives these conditions and succeeds in producing another batch of fertile eggs in due course has not been determined, but the chances would seem to be wholly in its favor.

While physiological disturbances of this kind are commonly induced by unnatural conditions, a single case has been observed in which the eggs of an animal recently taken from the sea were partially absorbed. Degeneration had spread irregularly throughout the entire organ, which at this stage of the process presented a remarkable appearance, being dark green, marbled with light lemon yellow. All the tissues pervaded by the blood seemed to be steeped in a green dye, which the organism was trying to throw off.

The structure of the ovaries, as outlined, suggest certain questions of considerable economic interest, such as the age at which sexual maturity is reached, the limits of the breeding season, and the length of the reproductive cycle or the frequency of spawning. We shall endeavor to show what light direct observation and anatomy have shed upon these matters.

PERIOD OF ADULT LIFE OR SEXUAL MATURITY.

The age of sexual maturity varies greatly in individuals, extending over an interval in which lobsters vary in length from 7 to 11½ or 12 inches. Out of thousands we should expect to find here and there one of possibly less than 7 and more than 12 inches in length coming to maturity for the first time. We may safely conclude that the majority of these animals are mature when 10½ inches long. Very few are with spawn before attaining a length of 8½ or 9 inches. In order to test this question traps must be put down at a certain point, kept there for a long period, and the catch noted day by day and month after month. This was done in the harbor at Woods Hole, Mass., where traps were laid by Mr. Vinal Edwards December 1, 1893, and the daily catches recorded until July 1, 1894, the conditions as to molting and the presence of eggs being noted in each individual. A summary of the catch showing the proportion of each sex and the presence of external eggs is recorded in table 9. During a period of 6 months 1,344 female lobsters were captured, and of these 168 carried eggs; of 249 females measuring from 6 to 8 inches but 3 bore eggs, while of those under the 9-inch length but 11 were berried.

TABLE 9.—RECORD OF THE TOTAL CATCH OF LOBSTERS IN THE HARBOR OF WOODS HOLE, MASS., FROM DECEMBER 1, 1893, TO JUNE 30, 1894, SHOWING THE NUMBER AND SIZE OF EGG-BEARING FEMALES.

Length in inches.	Number of males.	Number of females.	Females with eggs.	Total.	Length in inches.	Number of males.	Number of females.	Females with eggs.	Total.
6	3	4	...	7	10⅛	...	1	1	1
6¼	1	...	...	1	10¼	62	71	17	133
6½	3	4	...	7	10½	79	103	28	182
6¾	5	...	...	5	10⅝	1	...	...	1
7	45	47	1	93	10¾	18	18	2	36
7⅛	...	1	...	1	11	31	62	20	93
7¼	10	4	...	14	11¼	10	11	...	21
7½	66	47	...	113	11½	11	30	4	41
7¾	20	9	...	29	11¾	2	2	...	4
8	168	140	2	308	12	9	14	3	23
8⅛	...	1	...	1	12¼	1	...	...	1
8¼	44	29	...	73	12½	4	7	...	11
8½	143	115	7	258	12¾	...	1	1	1
8¾	26	27	1	53	13	4	4	...	8
9	170	166	13	336	13½	1	...	...	1
9⅛	...	1	1	1	14	1	...	...	1
9¼	32	38	4	70	14½	1	2	...	3
9½	148	169	24	317	15	...	3	...	3
9¾	27	29	3	56					
10	167	184	36	351		1,313	1,344	168	2,657

The reproductive curve, based upon body length, is seen to begin with the 7-inch lobster and to rise very slowly between this and the 9-inch size.

We do not assume that lobsters are always uniformly distributed, or that had the experiment been conducted elsewhere the results would not have been somewhat different. Where thousands of lobsters are captured at any point a considerable number measuring 8 inches or less may be found to have eggs outside of the body, but the proportion of this number to the total number of animals of the same length captured in the same place for the entire period will undoubtedly be very small.

LIMITS OF THE BREEDING SEASON.

Much confusion formerly existed concerning the time when the lobsters laid their eggs. This arose mainly from the fact that the eggs are carried by the females for a period of 10 months before they are hatched, and because of occasional departures from the common rule to which the majority conform. The following conclusion was reached in 1895: "About 80 per cent of spawning females lay their eggs at a definite season in the summer months, chiefly in July and August. The remainder, about 20 per cent of the whole number, extrude eggs at other seasons, in the fall and winter certainly, and possibly also in the spring." While this statement seems to me now to be in the main correct, I consider it very probable that considerably less than 20 per cent of the whole number of spawners lay eggs out of season, as was then suggested. It is not necessary to review the data by which it was definitely proved that eggs are at least occasionally deposited in winter and fall. The only way to check these results is to determine the retarding influence of a temperature varying from 67.1° to 32.1° F. (September to February, Woods Hole, Mass.) upon different batches of eggs laid out of the usual season. When normal eggs in the egg-nauplius stage, which in summer is outlined on the tenth day and well formed on the fourteenth, are found in winter; when segmented eggs are taken in November, and unsegmented eggs in February, it is evident that the production of fall and winter eggs is not a unique occurrence in this animal.

At the western end of Vineyard Sound and in the region about Woods Hole the greater number of spawners lay eggs during the latter part of July and the first half of August. The summer spawning for each year lasts about 6 weeks, and fluctuates from year to year backward and forward through an interval of about a fortnight. This variation in the time of egg laying is not remarkable, since the period of growth of the ovarian ova extends over 2 years. Any disturbance of the vital conditions of an adult female during this period would be likely to affect the time of spawning. The spawning season in the middle and eastern districts of Maine is about 2 weeks later than in Vineyard Sound. In 1893, 71 per cent of eggs examined from the coast of Maine were extruded during the first half of August.

According to the testimony of various observers, the eggs of the European lobster are generally laid and hatched from July 15 to August 31, in the northerly parts of its range, including Scotland, the west coast of Norway, and Helgoland. Larvæ may exceptionally appear, however, at the end of June, or even as late as the first part of October. In the Skager Rack and Cattegat, at the straits of the Baltic, the hatching period, at least, is about two weeks earlier (see no. *305*), while in the English Channel, at Plymouth, Allen found that the old eggs were hatched chiefly in May and June, and the new ones laid chiefly in August.

FREQUENCY OF SPAWNING.

The conclusion reached in 1895 that the American lobster as a rule lays her eggs but once in 2 years having been questioned, the subject was again taken up in 1902, and more conclusive evidence of the truth of this general statement was given.

It was suggested that "the best way to test the question by experiment would be to take a female which had recently hatched a brood and keep her alive until the following summer, when the next batch of eggs would be due, in case the spawning period is a biennial one." I attempted to try this experiment when, on June 19, 1900, Mr. Vinal Edwards, acting under my direction, through the U. S. Fish Commission, placed in a floating car at Woods Hole 36 lobsters from which the old light eggs, when close to the hatching point, were removed to the propagating boxes. I wished to ascertain three things: (1) Whether any eggs were extruded in the fall, which, according to the idea of an annual breeding season, ought to occur; (2) what changes took place in the ovary during the entire period from summer to summer; and (3) how many lobsters among those which might survive would lay eggs in the following season, one year from date.

In order to follow the behavior of the ovary I directed that at the beginning of each month one of the lobsters should be killed and its ovaries preserved, a proceeding which Scott (*248*), in a paper on the spawning of the European lobster, quoted in another place, criticises as follows: "There is nothing to show that the eggs carried by the lobsters at the beginning of the experiment hatched out naturally and were therefore extruded during the previous year." On the contrary, all were of the class which we call "old egg" or "light egg" lobsters, which taken in June means that these eggs were laid the previous summer, and can mean nothing else, unless the rarely occurring "fall" and "winter" eggs which I have described can reach the hatching point in June, a supposition still awaiting proof. There is, further, no evidence that the removal of the mechanically attached eggs from a lobster in June alters its physiological condition. Mr. Scott says further: "There was no obvious need to kill one lobster each month to discover whether it was going to extrude eggs or not." This would seem to be an obvious conclusion, but it should have been equally clear that this step was taken for another purpose, namely, to follow the changes which were taking place in the ovary itself. The condition of the ovary tells us at once whether growth of the ova is active or slow, or whether an absorption of the eggs already formed is going on. The step was far from needless, for after July it proved that there was no preparation for the production of fall or winter eggs. In other words, it showed that in these animals there was no tendency to produce eggs in each of two consecutive years, the chief point in the experiment. It was impossible to foresee how many of these animals would die in the course of their confinement or because of it, but had all of them lived two-thirds of the total number at the start, or 24, would have had a chance to spawn in 13 months from the time the experiment began.[a]

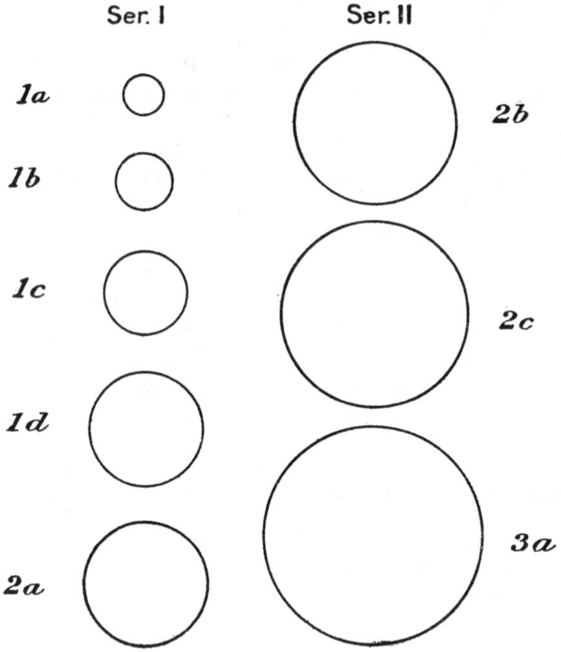

FIG. 30.—Diagram to illustrate growth in a single generation of lobster's eggs during a period of nearly 3 years, from an initial stage in ovary to time of hatching. Ser. I, internal or ovarian eggs; Ser. II, external or attached eggs. *1 a*, ovarian egg immediately after egg-laying; *1 b*, the same, 15 days after; *1 c*, the same 42 days after; *1 d*, the same 1 year after; *2 a*, the same in second growth period, 1 year and 10 months after egg-laying; *2 b*, fresh laid egg; *2 c*, "strictly" fresh, but removed from ovary or duct; *3 a*, last period of growth in shell, or egg-embryo about to hatch. Sizes deduced from averages of 10 eggs in nearly every case. Enlarged about 20 diameters.

[a] The experiment would have been more satisfactory if the directions, which were as follows, had been carried out: "Preserve the ovary of one lobster the first day of each month from July to December. If the number of lobsters should warrant it, continue to preserve the ovaries of one animal from January 1 until July. If, however, the remaining lobsters are few in number, and do not stand the confinement well, keep all as long as possible, preserving the ovary of each one that dies. * * * In case the lobsters die rapidly in late summer or early autumn, preserve ovaries of those only which die, giving the date."

By means of the animals killed it was shown that from June 19, 1900, to May 1, 1901, during a period of 10 months and 12 days, the ovaries had undergone a slow and gradual growth, a very important fact, which, if the conditions of growth were normal, is strong evidence that in the American lobster annual spawning is not a usual occurrence.

It was further demonstrated that the ratio of growth of the ovarian eggs for stated periods implied a reproductive cycle of 2 years. (Compare fig. 30.)

In conclusion we found that the theory of biennial spawning is supported: (1) By the statistics of the fishery; (2) by the anatomy of the ovary of the adult female taken at different seasons; (3) by the ratio of growth of a given generation of ovarian ova for stated periods; (4) by observation on animals kept alive for long periods; and (5) by the evidence of the rapid growth of ovarian eggs of spawners for any given year during the height of the breeding season.

Any rule to which the majority conforms may be expected to have exceptions. A lobster may exceptionally lay eggs in two consecutive seasons, and it is possible that in some cases the normal biennial period may be even prolonged.

When the preceding paragraphs were written I had not seen a paper of Appelôf (*6*) in which he confirms the theory of biennial spawning in the European lobster by an experiment conducted on a larger scale at the fisheries station at Stavangar, Norway. His statement is as follows:

> Since the matter (the question of spawning) had not been decided by experiment, I selected 100 lobsters, which were kept in a natural basin in the neighborhood for this purpose. It can now be maintained with complete assurance that in fact 2 years elapse between each egg laying.[a]

As already seen, a number of spawners, probably a very small proportion, lay out of season, in fall and winter. How can we account for these exceptional cases? An experiment tried by Mr. Cunningham (*63*) in the summer of 1897, on the European lobster, suggests an answer to the question. At Falmouth, England, five female lobsters, bearing external eggs which were nearly ripe, were placed in a floating box during the summer. After their ova were hatched these females were kept confined with two males until after October 14, when one was found to have newly spawned. This proves that it is possible for the European lobster to produce eggs in two successive years, but it does not prove that this is the common habit of the species in European waters. It also strongly suggests that these October eggs may correspond to the fall and winter eggs occasionally produced in the American form. By accelerated growth of the ovary the ova might be laid in fall or winter when not normally due until the summer following. Under such circumstances the ovarian eggs would come to maturity in 15 instead of 23 months. It would be interesting to know when these autumnal eggs hatch. The suggestion which we formerly made that they do not give rise to regular summer broods should be withdrawn, for it seems to us now that more confirmatory evidence is required before we can accept the statement that the young of the American lobster are ever hatched in the sea outside the period embracing the months of May, June, and July.

---

[a] In referring to later experiments conducted at the lobster park, at Kvitingsö, Appellôf remarks: "The conclusion that the female lobster on the west coast of Norway normally lays its eggs only once in two years, I later found year after year to be completely confirmed." (See *305*, p. 23).

A later notice of the annual spawning of the European lobster after transplantation to artificial ponds in New Zealand has been given by Anderton (5), whose observations on the molting and breeding habits of this animal under a complete change of environment are most interesting and are referred to in various parts of this work (see p. 302). At the time of writing, when his observations had extended over 3 years, several of the lobsters had laid two batches of eggs, and one, which bore attached eggs at the time of shipment, was known to have spawned three times in 3 years and 7 months. The record for the latter lobster is as follows:

Arrived with a few eggs still attached, January, 1906.
First molt, in absence of male, January, 1907.
Second molt, followed by copulation, November 21, 1907.
First spawning under new conditions, January 24, 1908.
Hatching of first batch of eggs, November 23 to December 28, 1908.
Second spawning; date not determined, but before March 12, 1909.

These animals were confined in small ponds with concreted bottom, and regulated tidal flow, and were regularly fed and skillfully cared for. It is interesting to notice that while the seasons are reversed in the southern hemisphere, the local range of temperature in New Zealand is similar to that at bottom of Vineyard Sound, Massachusetts, the lowest average temperature of 3° C. ($37\frac{2}{5}$° F.) being recorded for July (compare p. 182), and the highest average of 13° C. ($55\frac{2}{5}$° F.) from December to February.

An interval of 65 days ensued between copulation and spawning, and the fosterage period from egg laying to the hatching of the first young was 10 months to within a day. While it can not be maintained that these novel conditions give the usual spawning habits for *Homarus gammarus* until similar results are obtained within its natural range (compare Appellöf's experiments, given above), they show that the lobster is remarkably plastic and able to withstand considerable change when directed by skillful hands.

NUMBER OF EGGS PRODUCED.

The freshly laid eggs are of a dark green, almost black hue, when seen in mass, and somewhat irregular in shape, but they soon plump out and become nearly spherical or ovoidal in form. As the eggs develop they increase in size, become elongated, and, owing to the gradual assimilation of the dark yolk, lighter in color. (Compare fig. 33, *a b*.) This is most noticeable toward the close of the period of development, when the phrase "old" or "light" egg lobster is commonly used by fishermen to distinguish them from the "black" egg lobsters, which have more recently spawned.

The fresh egg measures approximately $\frac{1}{16}$ inch in diameter (1.5 to 1.7 mm.) and weighs $\frac{1}{6000}$ ounce or $\frac{1}{321}$ gram. A fluid ounce of eggs weighs about 1 ounce avoirdupois. The number of eggs laid is proportionate to the volume of the ovary and of the body, and varies from about 3,000 to nearly 100,000 in animals from 8 to 19 inches long.

TABLE 10.—PRODUCTION OF EGGS.

Length of lobster.	Smallest number of eggs.	Largest number of eggs.	Average number of eggs.	Number of lobsters examined.	Length of lobster.	Smallest number of eggs.	Largest number of eggs.	Average number of eggs.	Number of lobsters examined.
8 inches	3,045	9,135	4,822	6	13 inches	6,090	48,720	28,610	321
8¼ inches	6,090	7,612	6,851	2	13¼ inches	24,360	48,720	33,495	5
8½ inches	3,045	12,180	6,935	9	13½ inches	6,090	54,810	32,858	146
8¾ inches	6,090	9,135	7,105	3	13¾ inches	42,630	42,630	42,630	2
9 inches	3,045	18,270	7,902	143	14 inches	6,090	85,260	36,960	426
9¼ inches	6,090	12,180	9,083	35	14½ inches	21,315	60,900	42,968	90
9½ inches	3,045	20,792	9,297	241	15 inches	12,180	97,440	46,524	280
9¾ inches	3,045	15,225	9,947	55	15¼ inches	....	....	54,810	1
10 inches	3,045	24,360	10,555	514	15½ inches	24,360	97,440	53,795	45
10¼ inches	6,090	22,838	11,622	61	15¾ inches	48,720	54,810	50,750	3
10½ inches	3,045	36,540	12,905	532	16 inches	24,360	97,440	57,146	103
10¾ inches	6,090	24,360	14,067	45	16¼ inches	....	....	66,990	1
11 inches	3,045	48,720	15,410	568	16½ inches	36,540	85,260	66,053	13
11¼ inches	6,090	25,882	17,102	43	17 inches	12,180	85,260	63,336	30
11½ inches	3,045	42,630	18,668	307	17½ inches	60,900	73,080	64,960	3
11¾ inches	12,180	24,360	17,993	11	18 inches	60,900	91,350	77,430	7
12 inches	3,045	54,810	21,351	414	19 inches	54,810	91,350	77,647	4
12¼ inches	18,270	27,405	23,396	8					
12½ inches	9,135	42,630	24,812	156	Total number examined				4,645
12¾ inches	18,270	42,630	26,390	12					

In table 10 (reproduced from *149*) we have given the smallest, largest, and average number of eggs removed from the bodies of 4,645 individuals. These animals were "old" egg lobsters and were caught in Vineyard Sound and vicinity from April to June. The numbers were determined as a basis of 6,440 eggs to the fluid ounce. These tabulated results show great variability in the number of eggs borne by individuals of the same length, which may be attributed in part to loss of ova, but more to variation in the period of sexual maturity. Thus in 514 lobsters of the 10-inch length the number of external eggs varied from 3,045 to 24,360, with an average of 10,555. For the 12-inch size the corresponding numbers were 3,045, 54,810, and 21,351. We have seen that the period of sexual maturity is exceedingly variable in different individuals and that one animal may lay its first batch of eggs when 7 inches long, while another may not rear a brood until its body is 5 inches longer and has increased greatly in volume. The phenomenon is not remarkable in view of the slow growth of the ova, but it is important to recognize the fact.

Consideration of the average number of eggs produced suggested a general tendency which was expressed as follows: The number of eggs produced at each reproductive period tends to vary in a geometrical ratio, while the lengths of the animals producing these eggs vary in an arithmetical ratio. The average production in lobsters 8 inches long being 5,00c eggs, the average product for lobsters 10 inches long would be 10,000; for the 12-inch length, 20,000. This high rate of production is not maintained beyond the length of 14–16 inches. The lobsters with the largest number of eggs measured from 15 to 16 inches in length and carried upward of 97,000 eggs, which measured 16 fluid ounces and weighed nearly a pound.

Lataste (*177*) in a critical paper on that section of my earlier work dealing with the fecundity of this animal observes that the number of eggs carried by the lobster at any given time should be proportional to the volume of the body or to the cube of its length. If N represents the number of eggs carried, l the length of the animal, and k denotes a constant, according to Lataste, the relation of these quantities would be expressed by the following equation:

$$N = kl^3;$$

$$\text{Whence } k = \frac{N}{l^3}.$$

He has drawn up a table (based on table 15 of *149*), from the data of which he deduces the cubes of lengths, the ratios of the average number of eggs to cubes of length (k), and the means of these ratios.

In the lobster the reproductive powers are manifested suddenly at a certain age, after which they increase steadily, reach a maximum, and then presumably slowly decline. Accordingly during the first period only does the fertility increase proportionately to the increasing volume of the body, as expressed in the equation given above.

We have no definite information upon the duration of life, or decline of rate of growth in these animals. It is certain, however, that the renewal of the shell is quite as necessary for the continuance of life as of growth, since in the course of time death would result were not the injured and abraded shell restored. In higher animals the skin and at least some of the tissue cells are being continually renewed throughout life, while size limit of the body is early attained, and it is not likely that a dense and heavy shell like that of the lobster could be sloughed without increase in the size or volume of the body. The decline in sexual vigor may therefore result from the tax which molting continues to levy upon the capital stock of energy at every period of life. According to Lataste: $k = f(t)$, k being a function of age which has no real value, except as it is confined within certain limits.

In conclusion, we wish to observe that upon the principle of correlation of parts the ratio of the number of eggs to body length should correspond in a general way to the ratio of the volume of eggs to the total volume of the body were the latter a constant quantity, but owing to the frequent loss of the great claws this is not accurately represented by the cube of the length. All that we can say is that in the long run there is a tendency to produce in such a ratio, but the physiological condition of the animal is an inconstant and indeterminable factor. The high birth rate of the lobster teaches us to expect a correspondingly high death rate, a subject which will be later considered.

BREEDING HABITS AND BEHAVIOR IN CRAYFISH.

The breeding habits of lobsters, so far as they were then known, were described in 1895. Since that time a number of important facts have been ascertained, but our knowledge of the subject is still defective at many points. The behavior of the American lobster at the time of pairing and extrusion of the eggs has probably never been witnessed in a state of nature, and certainly but seldom in any of the higher Crustacea. We have had more or less circumstantial accounts from Chantran, Ishikawa, and Cano, regarding the time and process of egg laying in the crayfish, shrimp, and crab. The pairing habits and process of laying the eggs in the European lobster have been described by Anderton and Scott, as will be noticed later, while a remarkably full and accurate account of the habits of the American crayfish during the breeding period has been given by Andrews.[a]

Since the activities of the breeding crayfish are without doubt similar in some degree to those of the lobsters, and since they are at present far better known, I shall now give a summary of the instinctive acts and events in *Cambarus* for the period in question, drawn entirely from the work of Andrews referred to above.

Pairing in *Cambarus affinis* takes place in spring (February–April) and fall (October–November). The male catches the female by the antennæ or about the head, rolls her on her back, seizes her by the claws, stands over her body, and holds her in this position from 1 to 10 hours, during which time the sperm is transferred to the annulus or sperm receptacle on the ventral side of her thorax. This process may be repeated by "either male or female," both of which are in hard shell.

The male holds with his big forceps all the claw feet of the female in a bunch on either side, her abdomen being coiled under his, which closely presses it, he meantime supporting with his left or right fifth leg the abdominal appendages which are to transfer the sperm to the annulus. The first two pairs of abdominal legs or modified pleopods of the male are directed downward and forward against the ventral surface of the thorax of the female. Since the pleopods tend to lie flat against the body, they thus fold or close upon the the fifth leg, which stops them, forming a rigid support, and at the same time giving them the necessary elevation. The male then presses close upon the female so that his pleopods are directed toward the annulus and are forced into it, where the sperm is deposited. Spines on the legs of the male further tend to hold the pair firmly interlocked. *Cambarus affinis* has a prominent spine on the third joint (ischium) of the third pair of chelipeds, which fits into the base of the fourth pair of legs of the female. Spines or hooks of this character are wanting in the lobster. Thus rigidly interlocked, the transfer of sperm goes on slowly and may last for hours.

The vas deferens of the male is protruded or evaginated, as may be readily observed in all copulating males, forming a soft translucent double-walled tube, the lips of the opening being tightly closed. This evaginated duct fits in the groove which passes down the outer side of the first pleopod, and serves to conduct the sperm towards its tips. The appendages are rigid, sharp-pointed tools which are inserted into the annulus, and against which the modified second pair of pleopods are closely pressed. Sperm issues from the ducts as in the lobsters (compare fig. 2, pl. XLIV) in long vermicelli-like packets, or gelatinous capsules known as spermatophores, and guided possibly by the second pair of pleopods, passes slowly down the groove of the first pair to the receptacle or chamber of the annulus. The female is remarkably passive and appears as if dead, while the excitement of the male is marked.

While the spermatic receptacle of the lobster (pl. XXXIII and fig. 4 and 6, pl. XLIII) corresponds in function to the annulus of *Cambarus*, the latter appears to represent only the unpaired wedge-like middle piece of the former. The development of the seminal receptacle in the lobster proves that the middle piece in this animal is the anterior

---

[a] Andrews, E. A. Breeding habits of the crayfish. American Naturalist, vol. XXXVIII, p. 165–206, fig. 1–10. Boston, 1904.

section of the sternum of the eighth thoracic segment. The divergent wing-like processes in front of the annulus in the crayfish evidently correspond to the convergent wings, which are the modified sternum of the seventh thoracic somite, and which, united with the middle piece, form the elastic lips of the shield-shaped receptacle in *Homarus* (*st.* XIII, fig. 4).

The laying of eggs in the crayfish may not occur for some weeks after sexual union, and as Andrews remarks, some protection such as the annulus affords is necessary, since sperm can not long survive exposure to water.

PAIRING HABITS IN THE LOBSTER.

Both Boeck (*24*) and Fraiche (*109*) have referred to the union of the sexes in the European lobsters as if they had witnessed the act, but the errors which they exhibit tend to discredit their statements, however brief. Fraiche remarked that copulation in both the common and Norwegian lobsters took place in fall (October and November), and in the case of the former that it was extended into winter. "As with the crayfish, the sexual act is accomplished belly to belly, and so closely and firmly do they clasp each other, that, if taken from the water at this period, it is with difficulty that they can be separated."

But the only reliable observations under this head have been made by Anderton (*5*), of the Marine Department of New Zealand. The sexual act was noticed on a number of occasions among the European lobsters kept under observation in small artificial ponds. The general succession of events was as follows: Molting in early summer (November and December), followed in the course of a few hours by coition between a soft female and a hard male, and by the laying of eggs about two months after this event.

One of the female lobsters kept under observation by Anderton molted on November 21, at 3 p. m., and lay for some time beside her cast shell. "Two hours afterwards," to continue his account, "it was seen roaming round the pond and frequently approaching the various shelters, returning regularly and fearlessly to a shelter containing a large male. On approaching the entrance to this shelter the large claws were extended in a direct line with the body and the antennæ were thrust within the shelter. After a few moments the rostrum of the male appeared, the female meanwhile rapidly whipping her antennæ across the now projecting rostrum of the male, which in turn showed increasing signs of excitement, the antennæ being whipped very rapidly over the female in the same manner. After an interval of perhaps a minute the male gradually withdrew from his shelter, the female at the same time turning over on its back. Coition took place at once, the act occupying only a few seconds, the male retiring at once to its own shelter and the female into another. The following day both were observed to be living in one shelter, and they continued to do so, on and off, for several weeks."[a]

[a] In reply to certain specific questions regarding the pairing of lobsters, Mr. Anderton has kindly written under date of August 21, 1910, as follows: "The female lobster after casting does appear to seek out a male as soon as the distressing effects of molting have somewhat worn off. Male and female have frequently been observed living in one shelter for some days and even weeks after coition. The act of coition is very brief, and will not occupy more than half to a whole minute. They copulate, as you express it, "belly to belly," and head to head. The large chelæ do not come into use during the act so far as I have observed. The female voluntarily turns over almost completely onto her back, the excited male completing the process for her."

Three other cases of copulation were witnessed, and in every instance between a soft-shelled female and a hard male and always within a few hours after the female had cast. In one instance when the water in the pond was run off the body of the male was left partly exposed. I have already noticed two cases in which the American female lobster was impregnated when in the soft condition and when she also bore eggs; but there are other facts which show that molting is not necessary for the impregnation of the female. In the case of the American species we have found females of all sizes from 8 inches and upward in length impregnated at all times of the year, and the adult female lobster when taken from the sea, in whatever condition of shell, is likely to have her receptacle well supplied with sperm, even when preparing to molt. On the coast of Massachusetts in June and July I have found lobsters with newly laid eggs and a lobster with brood just hatched and about to shed, with receptacles full of sperm, which was in the first instance certainly, and in the last probably, newly acquired, and when the shell was hard. We know that the sperm is endowed with great vitality; that it can endure for months, and possibly for years. It is further probable that copulation is more or less indiscriminate, and more than one union is sometimes necessary to secure the fertilization of a given hatch of eggs.

Pearce [a] has presented strong evidence to show that crayfishes have no power of discriminating sex, his conclusions being based upon *Cambarus blandingi acutus* Girard, *C. diogenes* Girard, and *C. virilis*, observed in confinement. "The male," says Pearce, "tries" every crayfish which it meets, whatever the sex, a female instinctively remaining passive, while a male attempts to escape. The sexes meet by accident in the course of their random movements in the search for food. Males were found to even copulate with dead females, and in one instance with a female of another species, when the male stylets were inserted in the usual way in the copulatory pouch or annulus.

After taking into account all the facts at present known it seems highly probable that the lobsters are actuated by similar instincts when breeding and that they possess no greater powers of discrimination.

The probable method of transfer of the spermatophores is considered in a later section.

PREPARATION FOR EGG LAYING—CLEANING BRUSHES IN THE LOBSTER.

Preparatory to laying, the female *Cambarus*, as Andrews points out, retires for a number of days to the dark corners of her abode and is busily engaged in cleaning the under side of her abdomen for the reception of the fresh cargo of eggs. Her attitude and behavior in this instinctive act are peculiar. Standing as upon a tripod on the tail fan and the tips of the great claws, with her body raised high above the ground, she picks, brushes and scrapes every particle of dirt from the swimmerets and under surface of the tail, using chiefly the last pair of walking legs, the modifications of which, especially in the last two joints, render them very effective, combining as they do in one instrument the advantages of pick, comb, and brush.

---

[a] Pearce, A. S. Observations on copulation among crayfishes, with special reference to sex recognition. The American Naturalist, vol. XLIII, p. 746–753. New York, 1909.

The brush-picks of the lobster, especially those on the last two pairs of ambulatory legs, resemble similar instruments in the crayfish, as described by Andrews, and there can be no doubt that they serve a like purpose. That they are used as cleaning brushes has been often observed, but no one has yet studied the behavior of the lobster in the critical period before egg laying is accomplished.

Nevertheless I have recorded an observation (*149*, p. 47) which, read in the light of the foregoing account, suggests that the lobster has the cleaning instinct also and carefully prepares her abdomen for the reception of the ova. In two cases which I had been watching the lobsters laid their eggs in aquaria, and then industriously picked and scratched off nearly every one of them in the course of a few days. Now, these eggs were all of small size and the ovaries did not give up more than a third or a half of their contents. Under these conditions it would not be surprising to find the attunement of the instincts at fault. Interpreted in this way, the lobster by cleaning off her eggs was only preparing herself for the reception of the ova which still clogged the ovary.

In the lobster the terminal joint or dactyl of the last pair of legs (*cl. br.*, fig. 4, pl. XXXVIII) is developed as brush and pick, there being no comb on the under side. It is cone-shaped and traversed from apex to base by three nearly equidistant rows of hairs or setæ, those of the upper row being long, dense, and serrated. The subterminal joint bears three conspicuous tufts of saw-tooth hairs, quite similar to the "scouring brushes" described for the crayfish. In place of the strong spines or picks on this segment of the *Cambarus* is a single blunt spur almost concealed by the brush of hairs in the lobster. Just above it, near the base of the line of long dense setæ is a rudimentary comb or short linear series of spines.

If the short process which bears two spurs or picks in *Cambarus* were extended, it would form, as Andrews suggests, a double claw or forceps similar to those of the smaller chelate legs. In this case, however, the chelæ would all have the same relative positions or work in parallel planes. In the second and third chelipeds the claws work up and down, or in a nearly vertical plane, on the hinge joints. The great claws, however, have undergone a twist or torsion, in consequence of which their inner or anterior surfaces have become their lower sides. (See p. 257.) The dactyls consequently face and open inward, working in a horizontal plane. Now, the terminal segments of the last pair of legs have suffered a backward rotation or twist, in consequence of which their anterior surfaces are directed obliquely outward. If this limb were chelate, the dactyl would move obliquely outward and backward instead of upward, as in the smaller chelipeds, or inward, as in the great forceps.

In the lobster the torsion of the two terminal segments of the fifth pair of walking legs has gone a step further, so that the comb and spur of the dactyl, instead of being on the lower and anterior side of the limb, as in *Cambarus*, are upper and hindermost in *Homarus*, and, further, they no longer lie midway between the hinges of the joint, as in the crayfish. The torsion and other adjustments in the fifth pair of legs in the lobster evidently fit them for reaching and brushing the swimmerets and under side of the tail.

EGG LAYING.

On two different occasions, as already related, lobsters which I had under observation laid eggs in aquaria, in the night or early morning. These eggs were fertile and normally fixed in each case, but the extrusion was not complete, and the instincts of the female did not run their normal course. In the absence of any direct observations on the laying of eggs in the American species, the following account of the spawning of the European lobster, given by Scott (*248*), has a special interest:

> The lobster turns onto its back and by the aid of the two large claws and ridge of the abdomen makes a tripod of itself, the head being considerably higher than the posterior portion. The abdomen is then strongly flexed, forming a pocket, and the setæ on the edge of the abdominal segments make the space along the sides perfectly tight. A Λ-shaped opening into the pocket is formed by the telson and the sixth abdominal segment. This opening, when the abdomen is flexed, is slightly posterior to the first pair of swimmerets. The eggs then flow from the two genital openings in a continuous stream, one at a time, and pass along at the bases of the last walking legs and into the opening of the "pocket." The course of the eggs into the "pocket" is further assisted by a constant pulsation of the first pair of swimmerets, causing an indraft, which carries them rapidly inside. None of the eggs are lost on the passage from the genital openings to the "pocket" unless the lobster is disturbed. As the eggs leave the oviducts they become covered with an adhesive substance which causes them to stick together and to the swimmerets. The period of oviposition in the lobster under observation was just over four hours. Half an hour after the eggs had ceased to flow the lobster righted itself and walked into a corner of the tank, eventually getting into a nearly perpendicular position, with the head downward. It remained in this position for the rest of the day. Next day it was walking about the bottom of the tank in the usual way of a berried lobster. That the adhesive power of the eggs was imparted to them before leaving the oviducts was proved by collecting some just as they emerged from the genital openings. When these samples were placed in a glass of sea water and collected into a heap, they all became attached to one another and also to the glass. Moreover, the adhesive material only remains soft for a short time, as when the individual eggs were isolated and prevented from adhering upon the glass it was found that at the end of half an hour the adhesive property had entirely disappeared.

ARRANGEMENT AND DISTRIBUTION OF EGGS AND THEIR ATTACHMENT TO THE BODY.

Ishikawa, who watched the prawn *Atyephyra* lay her eggs in an aquarium, says that the act is performed in the early morning, and that it is preceded by a molt the night before, an order of events which has been often noticed in the higher Crustacea. The eggs were "almost rod-like" when they came from the ducts, and were laid down in an orderly manner, the anterior swimmerets receiving the first, while those deposited later were driven backward by the last pair of thoracic legs. The abdomen was incurved to form a pouch during the process, and the thoracic legs as well as the swimmerets and their corresponding segments were in constant movement.

In the lobster the ova adhere principally to certain setæ of the appendages of the five anterior segments of the abdomen (pl. XXXIX), and since hairs are absent only from the articular membranes of this region, they become bunched about the stalk of each appendage, and extend over the sternal bar and inner (epimeral) wall of the corresponding somites. In a full-berried female the swimmerets are embedded in a solid mass of eggs up to their branches, comparatively few being fixed to the free blades, and these only to their inner or proximal ends. The eggs, however, are so completely adherent to one another that if every hair were severed the entire cargo would float off in a single mass. It should be noticed that the stalks of the swimmerets are inclined inward toward the median plane of the body, and not away from it as in the thoracic region, and also that three tufts of long setæ are borne on the inner margin of each, two on the lower part of the inner blade or endopodite, and one on the adjoining end of the stalk or protopodite (fig. 3 and 4, *a*, *c*, *d*); further, that upon these setæ a vast number of eggs find anchorage, and that glands are very abundant beneath the skin of these parts. Four smaller tufts (*e*, *f*, *b*, *g*, fig. 3) also carry eggs, and like the former are non-plumose. Assuming that the cement is derived in part at least from the tegumental glands, and that the eggs are engulfed in it when they reach the abdominal pouch, it is difficult to understand how in the lobster the true swimming hairs catch so few eggs and in the prawn *Alpheus* none at all, unless it be due to gravity or the ability of the animal to direct the course of the egg stream while lying on her back and gradually changing her position. The difficulty of explaining this simple fact is not lessened by assuming that the cement originates in the oviducts.

ORIGIN OF THE EGG GLUE AND FIXATION OF THE EGGS.

Upon reaching the sea water in the abdominal pouch the eggs are fertilized by the sperm with which the seminal receptacle is charged, and, as seems probable, all are mixed in a secretion coming from the tegumental glands as well as from the oviducts by the beating movements of the swimmerets; the cement gradually becomes viscous, hardens, and eventually incloses each egg in a thin capsule; the individual eggs of the entire mass are eventually fastened to one another and to certain hairs of the abdominal appendages by the spun sheets and threads of the glue. The latter is an ectodermic product and resembles chitin in its appearance and behavior. A knowledge of its chemical and physical properties when combined with sea water, at the time of its secretion, would probably include the answer to a number of puzzling questions.

There are three subjects, apart from the more special problems of cytology, concerning the pairing of the higher Crustacea about which exact knowledge is particularly needed. These are: (1) The exact rôle played by the cement-producing organ; (2) the kind of stimulus or stimuli needed to arouse the sleeping sperm in its receptacle, set it in motion, and direct its course to the eggs; and (3) more light on the action of the rays, and the "explosive capsule," by means of which recent students have endeavored to explain the forced entrance of the head of the sperm into the egg. Direct observations are too limited at present to afford a basis for the final settlement of any of these matters.

The origin of the cement has been attributed, on the one hand, to the sexual organs and especially to the epithelial lining of the oviducts, and on the other to the tegumental glands of the swimmerets and lower side of the abdomen and to the egg itself.

The older writers, among whom were Cavolini (1787), Rathke (1840), and Erdl (1843), generally favored the first hypothesis. Lereboullet (1860) was the first to attribute the cement to the abdomen, and Braun (1875) the first to describe "cement glands" in the crayfish. Tegumental glands are found in practically every part of the body covered by the skin or invested by its folds, occurring even in the alimentary tract, the gills, seminal receptacle, and the "ear sacs." Feeding experiments with carmine seem to have shown that they have an excretory function in some degree at least, but it is equally certain that in some parts of the body they give rise to definite secretions. At the time of oviposition the pleopods of the female are swollen with what appears as an opaque whitish substance, which is seen upon microscopic examination to be composed of thousands of these organs. Each gland is hardly an eighth of a millimeter in diameter, and each opens to the exterior by a capillary duct, the entire length of which, not including the part which traverses the cuticle, is scarcely more than $\frac{1}{10}$ millimeter and its diameter only $\frac{1}{125}$ millimeter. Such organs are absent or found but sparingly in the pleopods of the male. After ovulation these glands appear to be for the most part in an exhausted condition, zymogen-like granules filling the central ends of their clustered cells. In one case examined, in which the animal had recently hatched eggs and was about to molt, the glands were shrunken and transparent.

While these facts may be entirely misleading, an observation of Prentiss (*217*) seems to show that this is not the case, inasmuch as glands of this type occur in the sensory cushion of the otocyst of the crayfish and probably in that of all crustaceans in which sand particles are adherent to the sensory hairs. Until some more probable source of the secretion is discovered, it is reasonable to infer with Prentiss that these glands furnish the glue by which the otoliths are fastened to the pinnules of the sensory setæ.

THE OVIDUCT AND ITS PERIODIC CHANGES.

The evidence regarding the part played by the epithelium of the oviducts will not be perfectly satisfactory until much more is known concerning the nature of the secretions of these organs during the period of egg laying. Our studies of the histological changes which the oviduct undergoes are limited to two significant stages, one in which the ovary was nearly ripe and the other from a female with external attached eggs in yolk segmentation.

It is evident from a comparison of the critical stages that cyclical changes occur in the oviduct, no less marked in character than those which arise in the ovary itself, and to which they are evidently related.

By the time the eggs are ready to be laid the oviducal epithelium is distinctly glandular in type (fig. 3 and 4, pl. XLVII). Its cells become greatly elongated and distended, while after egg laying they are shrunken to less than one-fourth their former size. When treated with the common hardening and staining reagents before egg laying, the cytoplasm is clear; the nuclei are also clear, elongated by the pressure exerted in the direction of the short axes of the cells, and lie well down toward the basement membrane. After ovulation the cytoplasm of the shrunken cells is more vesiculated; the nuclei are more granular, more deeply stained, oval in form, and are farther removed from the basement membrane. Furthermore, large vesicular cavities occur within or between the cells next the lumen of the glands, where products of nuclear degeneration are not wanting.

It thus seems evident that the glandular epithelium of the oviducts pour an abundant secretion over the eggs when these are delivered into the abdominal pouch. According to the account of Scott quoted above, the eggs are viscous when they leave the ducts, become adherent in sea water, but soon lose this property. So far as I have been able to ascertain, eggs to all appearance ripe, which were taken directly from the ducts shortly after egg laying, were nonadherent and showed no trace of cement or a secondary egg membrane, but at this time the action of the glands had ceased.

In the lobster with external eggs in segmentation, referred to above, the oviducts were beaded with ripe eggs, or as Duvernoy expressed it, stuffed like sausages, with eggs which failed of passage arranged in line, but they were not viscous at the time of examination, and were surrounded by the chorion only. Assuming that the oviduct contributes to the formation of the cement, some other chemical products would seem to be needed to render this effective. These are possibly supplied by the secretions of the tegumental or "cement" glands of the swimmerets in the presence of sea water. At all events it would seem that there is poured into the pouch at the time the eggs pass into it an abundant milky or turbid secretion from these glands, which under the microscope is seen to be swarming with minute floating particles or spherules. A similar secretion occurs in the crayfish, which after the setting of the cement is found to cover her eggs in a sort of protective "apron," as Andrews calls it, a sheet of grayish mucus or glair. When this is removed the eggs appear bright and fresh beneath it. This "apron" seems to be a residue of unused material, the presence of which may be needed not only to hold the eggs and sperm in the pouch but to take part in the production of the liquid hydraulic cement.

COMPARISONS WITH THE OTHER CRUSTACEA, AND THEORIES OF FIXATION.

In the lobster the glue forms a thin transparent sac about each egg (fig. 5, pl. XLIV mb^2), and the capsules of adjoining ova are united by short solid ribbons, or flattened strands of the same material. Similar bands adherent to the hairs and often coiled spirally about them hold the entire egg mass to the body. The cement is thus a continuous sponge work, which is imitated in the manufacture of certain kinds of nut candy, where the kernels are stirred in the thick sirup and held immersed when it hardens.

Coutière[a] describes a slightly different mode of fixation in the Alpheidæ (*Alpheus* and *Synalpheus*), where the eggs or egg-groups adhere only to the stalk of the pleopods, and never to the fifth pair of swimmerets, nor to the abdomen directly. The supporting hairs are bunched at the two extremities of the basal stalk and are nonplumose, as in the lobster.

Where the eggs are few in number, as in *Synalpheus longicarpus*, they are glued direct to the hairs, but where more numerous several hairs are cemented into a cable

[a] Coutière, H.: Les "Alpheidae," Morphologie externe et interne; Formes larvaires; Bionomie. Annales des Sciences naturelles, 8e sér., Zoologie, t. LX, p. I—IV, p. 428.

by a flattened membrane with double walls, representing the expanded capsule which surrounds the egg. In most cases the hairs furnish support to but a small part of the egg mass, the individual eggs being freely united with their neighbors. Thus in the prawn *Eucyphotes*, according to Coutière, the capsular cement gives off three or four flattened bands, each of which is soldered at its apex to similar strands from other eggs. The point of union is marked in each band by a lozenge-shaped or circular thickening. This would indicate that the eggs are surrounded by a layer of the viscous cement, and separated by sea water until they come together. Each lozenge-shaped thickening would then represent the original points of contact of egg with egg, the strands being spun from the sheath by a mutual pulling strain, due to the weight of the moving eggs.

This condition is especially interesting since it seems to prove that such eggs must have received their coat of cement before leaving the body. Unless it should appear, however, that the marks of contact may be completely effaced by fusion of the united strands, it offers no basis for a general conclusion regarding the origin of the cement substance in other decapod crustacea, like the lobster and crayfish. It is probable that in this as in many other particulars there is no absolute uniformity.

A much more anomalous method of fixation of the egg to the swimmeret is described by Williamson (*281*) for the crab, *Cancer pagurus*, and in Brachyura generally. According to this observer, the eggs lie thick upon the hairs of the inner branches of the swimmerets, and are attached by independent and often intertwined stalks, but there is no union of egg to egg, as in *Synalpheus*, *Homarus*, and other Macrura. The eggs are attached to single hairs, which garnish the endopodites, and usually to hairs only. There are said to be two membranes in either ovarian or attached egg, namely, a delicate vitelline membrane and a chitinous chorion. Between these a slight perivitelline space is formed upon contact with sea water. How does it happen that the eggs escape the hairs of the exopodite, and how are they suspended to the silken hairs of the endopodite without a single case of adhesion of egg to egg, and with little sticking of hair to hair?

Williamson in brief offers this explanation: "The intimate relationship between the egg and the hair is due to the hair acting as a skewer, upon which the eggs are impaled and strung." Further, the hairs are supposed to penetrate the chorion and pass through a perivitelline space without injury to the vitelline membrane. The chorion thus pierced collapses, and a little albuminous perivitelline fluid is pressed out, which becomes adhesive in sea water and serves to glue the chorion to the vitelline membrane and the egg to the hair; later the glue and chorion is pulled out into the sheets or cords by which the egg is anchored to the hair.

The solution of the problem of fixation in the eggs of the blue crab appears to carry us into deeper water than before. In order to make comparisons I have examined the eggs and abdominal appendages of the blue crab, *Callinectes hastatus*. *Callinectes* lays upward of 4,500,000 eggs,[a] and the endopodites of the swimmerets are buried out of sight by the mass. As in *Carcinus* these myriads of

[a] Smith, S. I.: Report on the decapod Crustacea of the Albatross dredgings. Report of the Commissioner of Fish and Fisheries for 1885, p. 618–619. Washington, 1886.

eggs are attached exclusively to the long silken tufted hairs of the inner branches of the second, third, fourth, and fifth pleopods. They are distributed, therefore, in 8 bunches, with over half a million eggs to a bunch. The appendages are flattened, and excepting the anterior face near the tip and a portion of the posterior face near its base, the endopodite is studded with remarkably long silken setæ. Each hair carries from 150 to 200 eggs, and each egg is glued by an independent stalk to the hair. Each egg is, moreover, extremely minute, measuring about $\frac{1}{108}$ of an inch ($\frac{23}{100}$ mm.) in diameter, or smaller than the dot of the letter *i* of this type. The hairs are extremely slender, varying in diameter from $\frac{1}{200}$ of an inch at base to $\frac{1}{700}$ of an inch at middle, beside which a human hair is very coarse and a silken thread a veritable cable. These attenuated hairs taper gradually to a sharp point.

The exopodite of the swimmeret is fringed with a dense row of plumose setæ, which are not more than one-fifth as long as the egg-bearing hairs of the inner branch and which, according to Williamson, serve in *Cancer* as a barrier to prevent the escape of the ova from the brood-chamber before they become attached. Strange to say, they do not catch a single egg.

Upon the theory of Williamson, and the assumption of an average cargo of 4,500,000 eggs, we can appreciate the nice work in fencing which would have to be performed by the silken hairs of *Callinectes* and indirectly by the appendages of the crab. Some 22,496 hairs would be required to spear and string 200 eggs each, and the feat would have to be done in the dark, as it were, and upon an egg so small as to be hardly visible upon the point of a fine cambric needle. But this is not all; the thrusts of the hairs must pierce a perivitelline space, that is, penetrate a tough chitinous membrane and be deflected from a semiliquid envelope. If this really happens, it is certainly a most wonderful performance.

Our objection to such a theory of attachment is based upon general principles, and before accepting it we should wish to have answers to the following questions. How is it possible for these delicate hairs to spear anything, and least of all solid spheres like an egg, suspended in water, and therefore in unstable equilibrium? The hairs have no more rigidity than a silken thread; they can hardly stand alone; and when loaded with eggs at their tips the spearing of additional eggs would seem to be impossible. (2) How is it possible for a spear or needle to penetrate the tough outer coat and avoid piercing the egg, for the suppositional inner membrane really does not exist at the time the egg is laid? (3) Are the almost microscopic eggs pushed along like beads on a string or birds on a spit, 200 or more crowded in line, and each leaving a viscous trail, without clogging the line, sticking together, or crowding one another off? (4) How is it possible for drops of an albuminous liquid to ooze from a hole in an egg without spreading over that egg, for a hair in contact with the egg would certainly not conduct this liquid against the force of gravity, and myriads of eggs must occupy every position with respect to the hair? Perhaps we can get a better idea of the physical difficulties involved by imagining a fly-fishing rod reduced to great tenuity and used as a spear for apples. How many apples of whatever size could its tip hold?

Before the question of egg attachment in the crab can be settled we must have very full and exact observations of the behavior of these animals during the period of egg laying. Now in *Callinectes* the endopodites are packed full of "cement" or tegumental glands; the exopodites contain fewer glands but an excess of cell disks or concretions (see *149*, p. 108). In fact, Braun called attention to the presence of glands in these Brachyura over thirty years ago.

If the secretion of the receptaculum seminis of the crabs is limited only to the uses of the sperm, as seems probable, we are inclined for the present to accept the older theory, namely, that eggs are glued to the hairs by a cement which is secreted by glands which lie at their base.

Why the eggs of the *Callinectes* are not stuck together or why neighboring hairs do not more frequently adhere is not apparent, and can not be explained until we know more about the physical properties of the glue itself. The hairs of *Callinectes* are covered with a continuous sheet of glue, but are not often adherent. Possibly the eggs stick to them before they have a chance to become entangled. Each egg is tethered by a thin spun sheet of glue, which is continuous with a narrow band or sheet, in which the entire hair is embraced up to the tip or very close to it.

As was pointed out by the writer in 1892 and as had already been demonstrated by Mayer in 1877, the crustacean egg does not possess a yolk-membrane. The ovarian ovum and the mature egg when it issues from the ovary, in crustaceans as well as in insects, is provided with a single membrane, the chorion, which is secreted by the "ovi-sac" or egg follicle. The great mass of the egg is made up of inert yolk; the protoplasm, which alone has formative power, is practically restricted to the center of the egg. When in the course of segmentation or later the protoplasm has reached the surface, a delicate membrane is secreted by the blastoderm. This often glues the egg fast to the chorion and gives much trouble to the embryologist. No doubt it was this membrane which gave rise to the mythical "Dotterhaut," or vitelline membrane of Erdl, Rathke, and the older school of embryologists.

A single membrane only, the chorion, is apparent in the eggs of *Callinectes*, but since the cord of attachment spreads out over its surface without any apparent break, the egg is probably covered with a thin layer of cement which has the same index of refraction as the chorion to which it is inseparably glued.

Williamson endeavors to extend his ingenious theory of fixation by "spearing" and the liberation of the cement from the egg itself to the lobster and other Macrura. Thus he says that the secretion "is not a true cement" capable of forming an outer envelope, but an albuminous substance, and that "the weight of the egg tends to stretch out the ductile chorion into long thin stalks." It is quite certain that the egg of the lobster, as in all the higher Crustacea, possesses a single membrane when it leaves the ovary, but the egg attached to the body has acquired a second and distinct membrane which is continuous with the stalk of attachment. The two are easily separable in picro-sulphuric acid; the second or outer layer is the "cement membrane" (fig. 5, pl. XLIV).

As we have already seen (p. 305) the eggs of the lobster are attached to the non-plumose hairs of the swimmerets as well as to the abdomen and to each other. Here at least it is impossible to apply any theory of fixation which does not involve a fluid cement substance, engulfing both hairs and eggs, capable of setting under sea water and possibly in chemical relation to it into a firm "hydraulic" cement which is non-ductile under ordinary pressures when it is once set.

I have spoken of the chorion as a tough membrane. That this is true is proved by the vicissitudes through which it passes unharmed. In egg laying the egg is compressed, being rod shaped in some forms when it passes the duct; it is therefore elastic, but it is only slightly ductile and then only when under great pressure. The freshly laid lobster eggs are spherical and as we have seen measure $\frac{1}{16}$ inch (1.5–1.7 mm.) in diameter; the egg embryo when ready to hatch is oblong, and measures about $\frac{1}{12}$ inch (2.1 mm.) on the average of the short and long diameters (fig. 33). This swelling in size, due to embryonic growth, stretches the chorion to great tenuity, until the limit of elasticity and ductility is reached, and the membrane bursts under the pressure, aided to some extent by the exertions of the larva.

THE MALE SEXUAL ORGANS.

The paired testes of the male are either distinct or united by a transverse bridge, and each give off a coiled duct or vas deferens, which opens by a valvular orifice on the inner side of the first segment of the last thoracic leg (fig. 2, pl. XLIV). The duct consists of a proximal division which conducts the sperm from the testes, an enlarged glandular part, and a terminal muscular or ejaculatory segment. A linear milk-white mass marks the course of the sperm through the transparent tube. In the coiled glandular division it is embedded in gelatinous envelopes or spermatophore-sacs (*sph.*) secreted by the lining epithelium. A sphincter muscle (*sp. mu*) produces an abrupt swelling at the beginning of the ductus ejaculatorius, the function of which is to eject the spermatophores. The latter have the appearance of semitransparent rods of vermicelli about an inch long, and consist only of opaque masses of sperm and the gelatinous medium described. When pressed out artificially, they imbibe water and swell perceptibly.

SPERM CELLS, THEIR ORIGIN AND STRUCTURE.

The sperm cells of the lobster (fig. 31) were apparently seen for the first time by Valentin in September, 1837, and he gave a brief account of his discovery in the following year. A more accurate account by Kölliker, who also remarked on the apparent immobility of the "rayed cells," appeared in 1843, with figures, and notice of the "seminal sacs" or spermatophores.

The structure and genesis of the spermatozoa of the lobster have been studied with much detail by Grobben, Gilson, Hermann, Sabatier, and more recently by Brandes (*33*), Labbé (*175*) and Koltzoff (*172*). Probably few structures in the animal kingdom have been more puzzling than the rayed cells of the decapod Crustacea. The puzzle consisted in harmonizing the following conditions as generally found in these animals. The large eggs of crustaceaus are surrounded by a tough chitinous membrane in which neither pore nor micropyle has been discovered. The sperm cells may be rounded or columnar, but whether devoid of processes of every kind, or provided only with three stiff rays as in the lobster, under ordinary conditions of observation these cells are absolutely immobile. Still every thoughtful observer who has pronounced the decapod sperm to be immovable must eventually recant, and like Galileo declare, "E pur si muove." How then could such sedentary bodies seek, find, bore through the tough shell and fertilize the egg? Brandes, Labbé, and Koltzoff have offered or worked out fertile suggestions, which afford a satisfactory solution to the general problem, subject to a course of verification and extension in different species of crustaceans.

That the "immobile" sperm cells really did move, has been maintained for thirty years or more by Owsjannikow, Hermann, and Cano. Thus apropos to this subject Grobben (*122*) remarks: "The stiffness of the rays does not prove that these cells are completely immobile. Moreover, the observation of Owsjannikow that the rays sometimes draw themselves in, and certain structures which I have examined, enables me to conclude definitely that these rays are living protoplasm and that they represent amœboid processes, remaining almost in a state of rest." (Compare the observation of Cano quoted below.)

In 1890 Hermann (*138*) had described movements of the processes of sperm cells, and in 1893 that excellent observer, Cano (*46*), spoke of seeing "certain of the sperm cells, especially the rayed ones, in amœboid movements in the sperm receptacle of the crab *Maia*."

In 1896 a remarkable statement regarding independent movement in the sperm cells of the lobster was made by Bumpus (*42*), to the effect that he had "seen the spermatozoa in active movement, swimming across the field of the microscope with the same nervous contractions that are characteristic of the Hydromedusæ."

In 1897 Brandes (*33*) asked how it was possible for the decapod sperm to enter an egg where no micropyle could be found, and especially in sperm cells like those of *Astacus*, which have no pointed head, but which are spherical and of considerable size. "I suppose that the sperms at the moment of contact with an unfertilized egg undergo a change, which consists in this, that a more or less pointed part of the anterior end of the sperm, the so-called clapper, the "tigelle" of French writers, is evaginated and so the membrane of the egg which at the moment of egg laying is perhaps somewhat yielding, is perforated." This ingenious suggestion, which was elaborated at greater length, has proved very fruitful, for it was confirmed by Labbé in 1903, and especially by Koltzoff (*172*) in 1906, who has worked out the development and structure of the sperm cell in a number of Brachyura, such as *Portunus*, *Maia*, *Pagurus*, and *Eupagurus*, and of Macrura in *Homarus*, *Galathea*, and *Scyllarus*.

According to these later observers, the sperm cell is a very complicated and delicate machine, beside which a clock or watch seems like a crude affair, especially when we consider the vast difference in size. This cell may be compared to a self-propelling torpedo, designed to move in a certain direction, and to explode the moment the cap or head strikes the hull of a vessel, or any opposing object.

In the following description of the genesis and structure of the sperm cell of the lobster, I shall follow in the main the account of Koltzoff (*172*), from which I have constructed a number of diagrams (fig. 31, *1–3*, and fig. 32). This account, whether correct in all particulars or not, is at least intelligible, and we are able to understand the remarkable evolution in form which these cells undergo in consequence of changes in osmotic pressure. It is very different from that of Sabatier, who devotes 37 pages to the sperm cells of the lobster, yet leaves it difficult to understand his descriptions and impossible to construct any consistent diagrams from his figures.

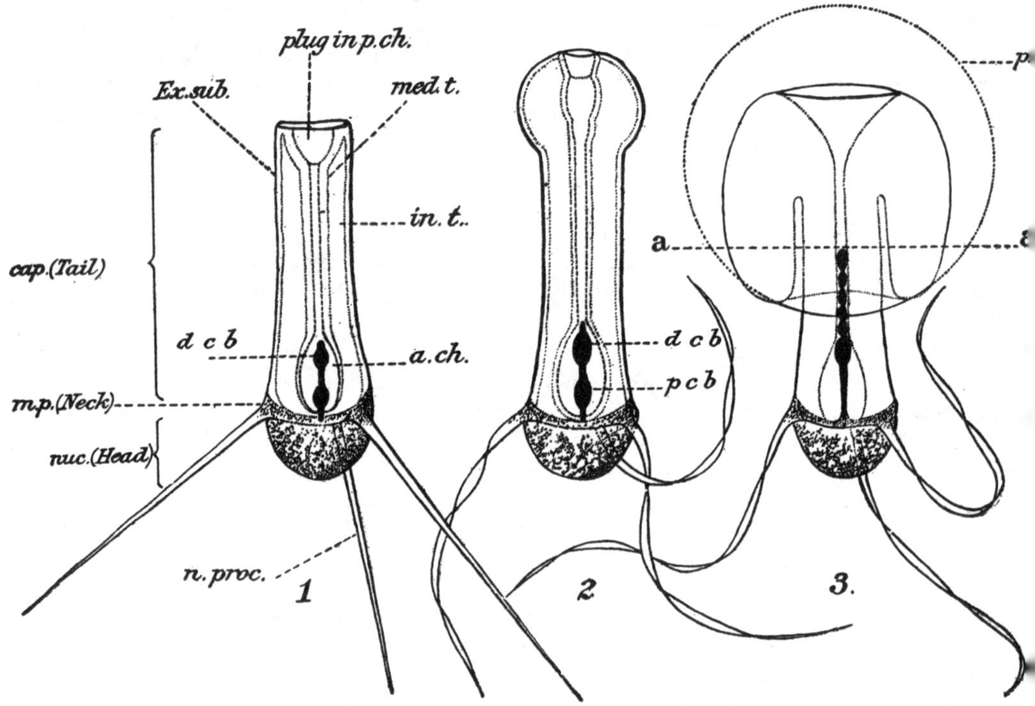

FIG. 31.—Diagrams of sperm cells of the lobster before (1), during (2), and after (3) capsular explosion, based upon Koltzoff (see *172*). *a a*, plane of section in figure 36; *a. ch.*, anterior chamber; *d c b*, distal central body; *Ex. sub.*, external layer; *in. t.*, inner tube; *med. t.*, medium tube; *n. proc*, neck process or ray; *p c b.*, proximal central body; *p. ch.*, posterior chamber; *p. mb.*, outer protoplasmic (?) membrane.

According to Koltzoff the sperm cell is derived by metamorphosis from a spermatid which in turn arises by division from a spermatocyte of the testis. The centrosome divides into two parts, and for some time remains united by fibers to the nucleus. The cell body is stuffed with granules which exhibit a difference in staining power, and in fact become differentiated into two important parts of the sperm, the mitochondral body and the capsule.

In the course of these changes the mitochondral becomes pressed against the nucleus, and molded upon it. A vesicular sperm cell is thus formed, peculiar to the decapod crustacean but comparable to the flagellate spermatozoa of other animals. The crustacean sperm becomes differentiated in three parts, namely, (1) the nucleus or head, (2) mitochondral body (a partly fibrous and partly granular structure) representing the neck or middle piece, and (3) the explosive capsule or modified tail.

The sperm cell develops processes (fig. 31, 1) which in the lobster arise from the neck. They ordinarily appear to be immobile and are distinctive of the decapod sperm. The number of these processes varies from 1 to 10, but 3 is a common number, which is found in *Homarus*, *Palinurus*, and *Galathea*, as well as in some of the crabs. The number is of physiological importance, since they are used for orienting the sperm upon the egg in fertilization. In the true crabs the processes arise from the head and are therefore nuclear in origin. It may be added that in the prawns (*Carididæ*) the usual processes are lacking, but the capsule ends in a sharp thread-like tail. In the crayfish (*Astacus*) and many crabs, as well as in *Gebia* and *Callianassa*, the neck and capsule are reduced in size and pressed against the head.

The processes are supported by a central mitochondral, skeletal fiber, or bundle of fibers. If in the course of development of the spermatid, these strong skeletal fibers project from the cell body with free ends, appearing to draw after them the more fluid constituents of the cell. The skeletal fibers can be demonstrated by plasmolyzing the cell or surrounding it with a solution of higher osmotic pressure. These skeletal fibers are really bundles of fibrils, which have a tendency to spiral winding.

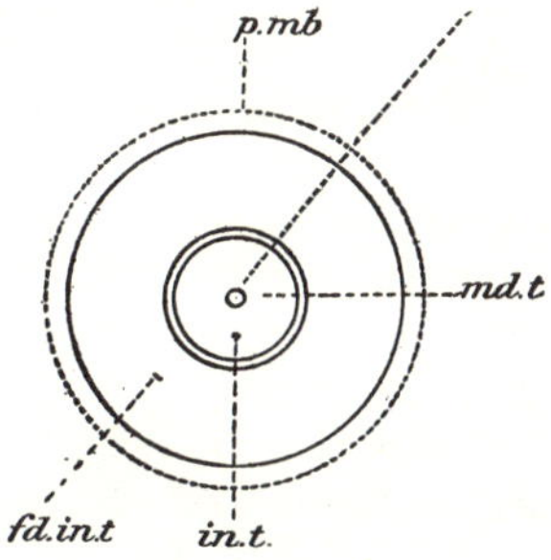

FIG. 32.—Diagrammatic section of sperm cell in capsular explosion, as seen in plane *a a*, figure 31 (3); *fd. in. t.*, folded inner tube.

The capsule (fig. 31, 1 and 2) is a double walled cylindrical body, a median tube running through it from end to end. This tube is formed by a median invagination extending from the hinder end forward to the neck, and is expanded at either extremity into a widechamber. The distal opening is closed by a plug of chitin. A peculiar rod or "Polster" of stainable substance is pressed from the central body into the anterior chamber. In the ripe sperm the outer capsular wall and the axial tube consist of chitin, and may be regarded as continuous, except at the point pierced by the "Polster." This stainable rod is often constricted into a proximal central body in the neck and a distal central body in the capsule (*p* and *d c b*).

FERTILIZATION.

In the lobster the sperm cells pass a long latent or resting period in the sperm receptacle, and may retain their vitality for from one to two years, and possibly longer. When the eggs are laid, the sperms leave their receptacle, find the eggs, and fertilize them. The spermatazoa are either pressed out by mechanical force, or else they must be aroused to activity by a definite stimulus, probably of a chemical nature.

By pressing the lips of the spermatic receptacle of a female with internal eggs nearly ripe, I have observed the sperm in a thick grayish mass which gave up its cells freely to sea water. This at all events suggests the possibility that the lobster herself is the direct agent in emptying her receptacle. In any case it is highly probable that the sperms are directed by chemotropism to the eggs after reaching the water. Nothing is known by direct observation of the phenomena of fertilization up to this point.

What are the locomotor organs by which the sperms leave the sperm receptacle or by which they seek and find the eggs in the brood chamber? In our search for an answer to this question we must remember that the lobster lies upon her back when the eggs are laid, so that the force of gravity is a bar rather than a help to the movements of the sperm at this critical period. We may assume first that in leaving the receptacle the locomotor organs of the sperm cells are the rays or processes, which I showed in 1895 to be rigid in the testes but limp in the receptaculum. This movement is probably amœboid in character, consisting in the lengthening and shortening of the protoplasmic element of the process which flows from the neck of the cell. As with the amœba a solid support is necessary for the process of locomotion to be effective, for according to a recent observer this animal probably draws itself along by the adhesion of its pseudopodia to the surface over which it creeps.

How does the cell make its way through the water to the egg? No satisfactory answer can now be given, but if Bumpus was not entirely mistaken in his report of movements of the lobster's sperm, as quoted above, we might plausibly suggest the following solution, which is of course purely hypothetical. Upon reaching the water the plug of the capsule is loosened and falls out. Water then enters and fills the inner tube. This water is subsequently ejected by contraction of the vesicle, and the cell is drawn forward by inertia. It should be added here that in some forms (*Eupagurus*) the capsule is covered by a thin protoplasmic layer, and that in this membrane contractile fibers are sometimes seen; transverse rings can be demonstrated in the lobster. The action is supposedly recurrent. The processes direct the cell, as do barbs the arrow. The eggs are big targets, and the moment one is struck orientation of the sperm upon its surface begins.

At this point speculation gives way in a measure to direct observation, and I return to the account of Koltzoff (*172*) who, like other observers, was unable to see the minute sperm enter the huge opaque egg. Disclaiming the ability to give a complete account of the movements of the sperm cells, he says: "My observations and experiments can naturally clear up only certain phases of these processes, and a whole string of hypothetical conclusions is needed to unite them into a harmonious whole."

Contact with a large and possibly moving body, or thigmotaxis, seemed to furnish the most powerful stimulus to the cell processes, which have been observed to shorten and lengthen, though not to the extent of more than one-tenth of their length. Once in touch with the egg the sperms begin to orient themselves in such a way that the cell comes to stand upon its thin elastic processes as upon a tripod, so that the head is placed in direct contact with the surface of the egg. The elastic process or processes in contact with the egg possess an adhesive power; they seem to shorten, and thus to pull the sperm cell into position.[a]

In this critical situation when the conditions for fertilization are favorable something pulls the trigger and fires the gun. That is to say the capsule explodes and shoots backward, while the head in consequence of the rebound leaps forward and is driven through the chorion and into the egg.

The space between the inner and the outer capsule is filled with a peculiar explosive substance, which according to the ideas of Koltzoff possesses the property of swelling up when it meets with water. Water must either enter through pores of the inner tube or be absorbed through the outer wall of the capsule. The extension or swelling of the explosive material is rapid and is usually attended by an evagination of the inner tube and discharge of the central body.

The sperm cell is thus deformed by the action, and since the character and degree of the evagination varies with the physical and chemical conditions present the number of these apparent artifacts is very great.

In actual conditions or in 4.2 per cent isotonic solutions of calcium chloride in sea water, it is possible to follow every step of the discharge. Labbé in 1894 described the discharge of the capsule as the final developmental stage of the sperm. The explosion of the capsule seems to liberate the elastic energy of a coiled spring represented by the central body, which may show a spiral form in *Pagurus* or a series of beads, bands, or granules.

In abnormal capsular explosion, according to Koltzoff, there is a double spring of the sperm, first forward and then backward. If the suggestion of the free movements of sperm given above, and for which I am alone responsible, should prove to be an error, these abnormal explosive movements might account for the contractile pulsations described by Bumpus.

According to Koltzoff the energy of the explosion is contained in the explosive material. When the chitin plug of the inner tube is driven out, water enters and eventually penetrates to the inner capsule and brings on the explosion. My suggestion that water might enter the inner tube and be driven out by a contraction of the protoplasmic layer surrounding the capsule, thus causing the cell to move forward, presupposes that water does not at once penetrate the capsule and reach the explosive substance. If this really happens the suggestion regarding locomotion would be untenable.

No special stimulus was found which would effect a normal capsular explosion, and it is possible that the sperms respond to a coordinated series of stimuli. Nothing is yet definitely known upon this subject.

According to Koltzoff the head and neck containing the proximal central body are driven into the egg and take part in fertilization, while the capsule, with its processes, in whole or in part, and the distal central body, are left outside and disappear.

Notwithstanding the difficulties, owing to the great size and opacity of the egg and the small size of the spermatozoa, Koltzoff observed a single case where a normal sperm

---

[a] Koltzoff also offers a different and contradictory explanation of the adhesion of the sperm cell to the egg, namely, that the egg membrane appears in many cases under the microscope to be finely porous, and that the processes are driven like so many splinters into these pores.

having oriented itself on the surface of the egg exploded and penetrated the chorion; this happened in three different species of crabs. The capsule of the normally oriented sperm exploded while in view, and the nucleus was drawn into the egg, but it was impossible to distinguish anything whatever within the opaque ovum. He inferred, but did not prove, that this series of events represented a true fertilization process.

Several attempts were made at artificial fertilization of lobster eggs at Woods Hole in 1891, but like the experiences of Koltzoff in 1906 they were unsuccessful. There are the difficulties of first obtaining perfectly ripe eggs, and, secondly, of meeting the other conditions of fertilization in which the secretion of glands from the ovaries, oviducts, or integument of the swimmerets may play a part. I made glycerine extracts from the ovaries and oviducts in the hope of finding a chemical stimulus for the sperm, but did not succeed, the primary difficulty of getting the organs in the proper state of maturity being at that time insurmountable. It was impossible, also, to get any secretions from the swimmerets by applying electrical stimulation to the ventral nerve chain, from which they are innervated.

THE SEMINAL RECEPTACLE, COPULATION, AND IMPREGNATION.

The habits of the lobsters at the time of sexual union, so far as at present known, have been already described. (See p. 302). We have now to consider how the female is actually impregnated, that is, how the spermatophores are transferred by the male to her receptacle. According to the account quoted above the transfer is quickly made while the female lies on her back, and in the three or four cases observed when her shell is soft.

While no direct observations on the further course of events are as yet available, the structure of the spermatophore, the male stylets, and the female receptacle render plausible at least the following account, which is purely conjectural. Before proceeding with this, however, it will be necessary to examine the secondary sexual structures with greater care. The seminal receptacle (fig. 6, pl. XLIII) lies on the underside of the female immediately behind the opening of the oviducts and between the bases of the last two thoracic legs. (Compare p. 301.) It presents the appearance of a light blue shield with deep median groove. When examined closely it is found to consist of a pair of wing-like processes, the enlarged sterna of the seventh thoracic somite, with a middle piece belonging to the succeeding segment wedged between their posterior extremities. The lips of the median groove are elastic, and if forcibly depressed are seen to open into a membranous pouch, in which the spermatozoa are carried. The pouch is laterally compressed and extends directly upward at right angles to the long axis of the body and is supported on the link-work of the internal skeleton. (Fig. 4, pl. XLIII, *sac.*) We should notice that this sac, far from being a delicate structure, is well adapted to receive rough treatment with impunity. Within, the middle wedge-shaped piece is continuous with a pair of calcareous rods which form a solid frame for the posterior and upper part (or bottom) of the sac, where they are firmly sutured to the endophragmal skeleton. Within the pouch this sternal bar is prolonged into a stout keel, where it is strengthened with yellowish deposits of chitin of a horny consistency. (Fig. 3, pl. XLIV, *bar.*)

The stylets or modified appendages of the first abdominal somite in the male (fig. 1, a, pl. XXXIX and fig. 5, pl. XLIII) have stout stalks and a single terminal blade. The latter is nibbled at the end, grooved along the median side, and bent in such a manner that when the stylets are opposed they form a covered way. At their hinder extremity they leave a wide open angle, but partially closed by the spurs of the second pair of swimmerets (fig. 2, a, pl. XXXIX, *sp.*) when these appendages are naturally extended forward On the anterior or upper side of the opposed stylets a deep groove on the stalk of each leads obliquely into the arched passageway. The tips of the stylets when held in this position diverge slightly, and when pressed into the seminal receptacle the elastic lips of the latter catch on the nibs and hold the appendages until they are forcibly withdrawn. The indurated tip of each stylet is interrupted by a minute oval area of soft membrane, but this does not appear to be the outlet of any peculiar organs. The tissues of the stylet itself, like those of the swimmerets, generally abound in tegumental glands and large glycogenic cells. In copulation the animals undoubtedly lie with ventral surfaces together, but apparently do not remain in this position long. After seizure of the female, the spermatophores are emitted and possibly with the aid of other appendages are conducted to the passage formed by the stylets, the tips of which are inserted nearly vertically into the spermatic receptacle and there held in the manner indicated. The spermatophores not only swell and soften in water, but possibly may be disorganized before the sperm are free to enter the receptacle, but this is not probable.

The crustacean sperm, as we have seen, is like a submarine torpedo, loaded and primed, capable of piercing the membrane and forcing a passage into the egg the moment its latent energy is set free.

While much of the preceding account is based solely upon inference derived from a study of the organs and of the changes which some of them are known to undergo, its presentation may be worth while, if only to call attention to the wide gaps still remaining in our knowledge of the whole process of fecundation in the higher Crustacea.

Chapter X.—DEVELOPMENT.

ANALYSIS OF THE COURSE OF DEVELOPMENT.

The entire course of development for each individual may be conveniently divided into embryonic, larval, and adolescent periods, which close, respectively, with hatching, the emergence into the fourth stage, and the acquisition of the secondary sexual characters and full adult power, reached in the female, according to Hadley, at the twenty-third molt. The age of sexual maturity or the entire period from larva to adult is subject to great fluctuation, owing to individual variations, changes in the environment, and to other causes. A 10-inch female lobster may be from 5 to 6 years old, or even older. There are really no sudden transitions, but only gradual progressive changes, the nature of which especially at the fourth stage is often disguised by the abrupt passage of the molt.

The embryonic life within the egg membranes is the most constant, occupying approximately ten and one half months on the coast of Massachusetts, during which the stored yolk supplies the materials and energy for growth. When this period is closed at hatching, the egg membranes burst, and together with a larval cuticle are cast off, thus leaving the animal free to enter upon an independent career. A remnant of unabsorbed yolk always remains, however, in the mid-gut region and serves to tide the little lobster over a critical interval before it is thrown entirely upon its own resources.

Pairing probably does not continue long after sexual union has been accomplished, yet when confined in ponds lobsters have been known to hold together for several weeks, and even to occupy the same shelter. (See p. 302.)

Parental instinct developed in the mother is mainly directed to the safe fosterage of her eggs. The young disperse as soon as hatched, rising to the surface, where they swim as free pelagic organisms until their larval life is over. Development proceeds through a series of metamorphoses or individual changes, externally marked by a corresponding series of molts, in the course of which the old cuticle is periodically shed in its entirety and as one piece to give place to the new covering already formed. The abrupt molts thus furnish a ready means of following the development and growth of the crustacean step by step from infancy to old age. The embryo virtually molts several times, though its cast cuticle seems to be mostly absorbed. The first of these membranes to be shed and absorbed in the egg is secreted by the blastoderm, and was mistaken for a true yolk or egg membrane by the older observers. As we have already noticed, the ripe crustacean egg possesses but a single protective envelope, the chorion or flexible shell, which at hatching time has been reduced to a layer of great tenuity.

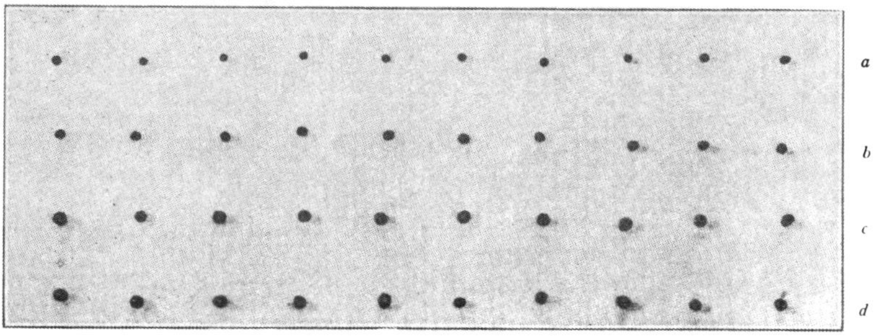

Fig. 1.

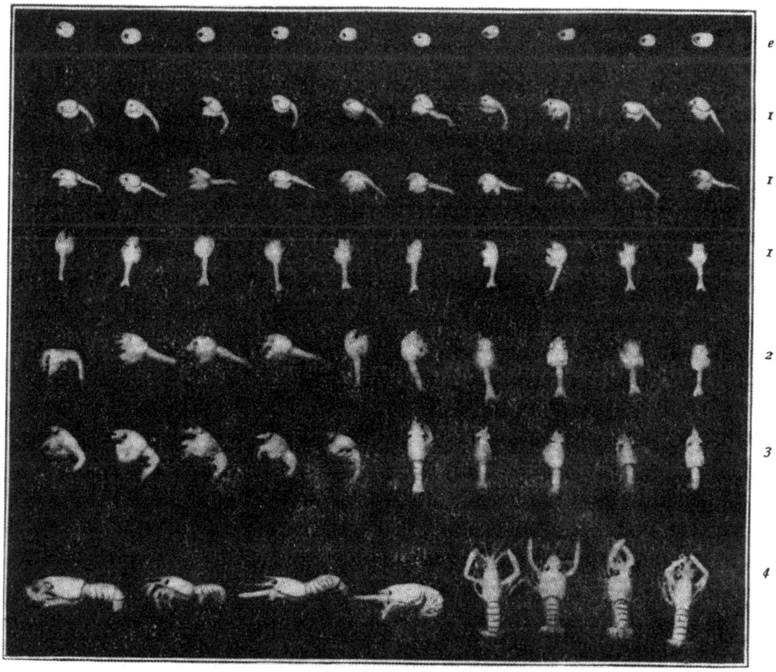

Fig. 2.

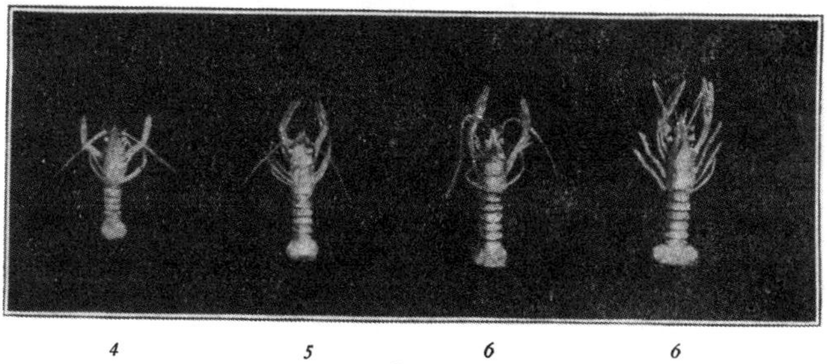

Fig. 3.

FIG. 1.—Growth stages of lobster eggs and young to illustrate relative sizes attained at Woods Hole, Massachusetts. *a*, ovarian ova in June; *b*, external egg in invagination stage, July; *c*, egg-embryo, September 1; *d*, embryo, March 1. In this and following figures, all represented in full size from alcoholic materials.

FIG. 2.—Growth stages of young lobsters continued. *e*, Embryo at hatching (July); *1* (first line), first larva, not free from first molt; *1* (second and third lines), first free larval stage; *2*, second larva; *3*, third larva; *4*, fourth stage.

FIG. 3.—Growth stages cf the lobster continued. *4*, Fourth stage; *5* and *6*, fifth and sixth stages, respectively.

When the lobster is ready to hatch, it is therefore covered from head to foot with a close-fitting chitinous tunic which must be shed before active life is possible. As explained earlier, this outer garment sticks to the egg coverings and is kicked off when these are cast aside.

Before hatching and therefore before the molt which occurs at birth, the terminal telson is forked, and in this respect recalls the more primitive protozoëa larva, which has been attributed to the lobster without any further warrant than this fact; the first larva resembles an overgrown zoëa, and the fourth corresponds in some degree to the megalops state of the crab.

Since the first larval stage is preceded by a true molt, failure to pass which is often fatal in the operations of fish hatcheries, it has seemed best to recognize this fact. The molts and stages will therefore be named and numbered uniformly; molt 1 introduces stage no. 1, and not stage 2, according to most writers on these subjects; molt no. 4 precedes stage no. 4, and so on.

The first larva (fig. 34 and pl. XXVIII) is about one-third of an inch long, and continues to swim near the surface for from 3 to 5 weeks, or until the fourth (pl. XXXI) or fifth molt, when it sinks to the bottom and passes the remainder of its life essentially like an adult animal. The life of such a crustacean is thus made up of a series of stage periods, each of which represents the time passed between successive castings of the shell. The first four periods during which growth is most rapid and change most profound are passed rapidly. After this point, and particularly after the sixth or seventh stage, except for increase in size, there is comparatively little change from molt to molt.

During the three early stages the larvæ lack the power of very precise orientation. They will move steadily for a time with nicely coordinated movements, when their equilibrium is suddenly upset and they begin to reel or turn over completely. This seems to be due to the fact that their statocysts, which are the most important balancing organs, are not well developed until the fourth stage.

Twins and monsters are occasionally born, a fact noted by Brightwell in 1835, but this seldom if ever occurs under normal conditions. (See ch. VIII, p. 287.) In two cases of twins observed by Anderton in the European species one larva was released earlier than the other, which continued to rotate in the egg until set free.

The following changes in structure and instincts take place at the fourth molt or beginning of the fourth stage, which marks the most surprising leap in the whole history of development: Loss of the primitive swimming branches of the thoracic appendages; the cuticle becomes shell-like, containing more lime; the pigments are denser, the colors brilliant, and the color pattern variable; otocysts are present and orientation is perfect; rotation of great forceps is complete; the animal, during at least a part of this stage, moves toward the light and swims steadily at the surface with great claws directed forward and held close together; the preying instinct is more marked; the fighting instinct, the instincts of fear, "feigning," and hiding are all developed at the beginning or close of this stage or in the fifth, which follows, when the animal goes to the bottom to stay.

When a bottom life is finally adopted, the instincts of burrowing, hiding, wariness, pugnacity, and preying become strongly accentuated. The animal is negatively phototactic and tends, as in all later stages upon the whole, to avoid strong light.

In the larval lobsters the big claws are prehensile organs solely, by which the food is seized and transferred to the mouth parts. At the fourth stage the great double claws are perfectly developed, similar in structure, and of the primitive toothed type. The smaller chelæ and other appendages are in perfect symmetry. At about the sixth or seventh stage a difference in the big claws begins to appear, the claw on one side developing crushing tubules and becoming larger and heavier in accordance with the greater development of its muscles. The smaller forceps, the jaws of which have developed serially arranged teeth, retains its primitive form. Whether right or left claw shall be of the toothed or crushing type is predetermined in the egg, all members of the same brood in all likelihood being either right-handed or left-handed. (See p. 274.) Injury or mutilations, however, may determine the position and character of the claw in after life.

At the seventh molt the cast shell is blue with some green and brown pigments on the tergal surfaces. Pigment is thenceforth more and more deposited in the outer calcified layer of the shell, which becomes wholly responsible for the color of the animal. The dorsal median stripe of the carapace, which marks an absorption area of distinct service in molting, is much narrower than when first observed in the fourth stage. At the time of the fourth molt this linear area is one-eighteenth of the width of the carapace at its widest part. It gradually narrows until in the adult state it is in the proportion of one-sixtieth or less.

The sex can be determined as early as the eighth stage by the openings of the sexual ducts, which in the male arise in the coxa or basal segment of the last pair of thoracic legs and in the female on the coxæ of the third pairs of pereiopods. The sex can not be determined by the modified swimmerets of the first abdominal somite until some time between the eighth and the tenth molt. At about the eighth stage also the peculiar seminal receptacle of the female begins to undergo its characteristic differentiation.

During the adolescent stages, when the lobster of either sex measures from 1½ to 4 inches in length, there are certain marked characteristics—the relatively large size of the eyes, recalling those of the shrimp *Penæus setiferus* and probably a relic of an ancestral stage, the fringe of long setæ on the tail-fan, and the tufts of hairs about the ends and along the serrate jaws of the toothed claw.

With this introductory sketch, we will examine more closely the embryo and larva, although it is not our intention to enter minutely into all the details of their structure.

EMBRYO.

The freshly laid eggs are dark green, almost black in color owing to the presence of the soluble pigment, a lipochromogen, in the yolk, and the glass-like transparency of their membranes. (Compare p. 298.) The golden yellow variation. which is often associated with dark green, as in the eggs of certain shrimps, has not been observed in the lobster, but its eggs are occasionally straw color, grayish-green, or yellow-green. When plunged in alcohol or hot water the ova respond like the shell of the animal and become light red, a more stable pigment, a lipochrome, soluble in alcohol, being formed. By adding alternately hot and cold water the eggs may be turned to red and green several times in succession.

The fresh-laid eggs, which are seldom seen, can be detected by examination with a hand lens. The transparent capsule closely invests the yolk, which presents a very fine-grained and uniform texture, quite different from that which the ova later possess. Maturation is without doubt completed by the formation of polar cells either in the ovary or during the passage of the eggs to the outside, although we have never been able to find these bodies in stained sections of the egg. External segmentation of the yolk follows in from 20 to 25 hours after oviposition, and the large yolk segments which are early formed can be detected by the naked eye. A clear perivitelline space, apparently filled in part with exudatian from the egg, soon appears between the shell and yolk. At the close of this process, or after invagination has begun, the living egg, when examined with a hand lens or low power of the microscope, is likely to be mistaken for one freshly laid. The ova, however, are not so closely adherent, are somewhat lighter in color, and the yolk has a coarser and more irregular texture. The first division of the protoplasm is central or subcentral. In the second and third segmentations, with four and eight cells, the products begin to separate and migrate outward. The greater number tend to move toward the side which marks the animal pole, where the yolk becomes distinctly flattened, and the shell correspondingly elevated. The cells which migrate toward the surface of the depressed area bring about the first segmentation of the yolk into hillocks. As they multiply by indirect division their products diffuse over the egg, and at the fifth segmentation, of 32 cells, the entire surface of the yolk is thrown into hillocks or inverted pyramids. The segmentation is rythmical, the early periods lasting about 4 hours, but the rythms of individual cells are not in harmony, and the segments are unequal. Later when about 110 cells are present the periodic divisions become more uniform over the entire egg. With each division the protoplasm approaches nearer the surface, and meantime a limited number of cells are formed by tangential divisions and migrate to the depths of the yolk. By a continuation of this process the yolk becomes surrounded by a thin mosiac of cells, or rather by a single tier of several thousand minute columnar cells or diminutive yolk pyramids of uniform size. Their "apices" blend into the central yolk mass, which harbors a few wandering and degenerating cells.

Cell division then becomes more rapid over a considerable area of the surface, which includes the animal pole, and at a certain point an invagination of superficial cells occurs. This begins by the in-wandering of a few cells, and is followed by the rapid multiplication of those thus immersed in the common food stock of the developing egg, and by the sinking of a small area of the blastoderm about this point, forming what is usually called the "egg gastrula" stage. The depression is at first shallow, and becomes a well-defined circular pit, but is never very deep. It is subject to marked individual variation, but commonly elongates transversely to the long axis of the future embryo, endures 4 or 5 days, and then completely disappears. In front of the pit a wide embryonic area is defined by rapid divisions of the surface cells. The latter, which are the direct descendants of the enormous yolk pyramids or hillocks, become distinctly separated into a single stratum of yolk-laden and columnar cells. Below the point of invagination the ingrowing plug of cells expands by rapid divisions of its elements, and like columns of smoke from a steam engine a dense cloud-like mass is spread into the yolk. Many of the cells break loose from the syncytial mass and worm their way through the yolk like independently moving amœbæ. Many of them degenerate, while others creep forward and attach themselves to the embryonic area. The cells introduced by invagination give rise, in terms of the germ-layer theory, to the hypoblast or endoderm, and to at least a part of the mesoblast. It is almost certain that the yolk-wandering cells receive many recruits from the surface of the embryonic area; the yolk cells introduced earlier for the most part degenerate before the stage of invagination is reached. By multiple divisions cell nests are formed, particularly in the embryonic region at the surface, or more commonly just beneath it in the midst of spheroidal masses or balls of yolk.

Death waits close upon the birth of new cells, and from an early stage to the later egg-nauplius period degeneration is a marked characteristic of this and many other arthropod embryos. Nebulous clouds of chromatin strew the paths of cell migration, and are carried to every part of the egg, where they remain until absorbed. In the early stages at least embryonic layers do not exist, and attempts to reconstruct them out of a mass of rapidly multiplying, degenerating, and moving elements, by the aid of theory and the imagination, have thus far proved neither successful nor profitable.

The appendages are the first of the distinctly embryonic parts to make their appearance; they are formed by paired tubular folds of the body wall. They possess solid yolk cores which are gradually absorbed and replaced by mesoblastic cells which migrate from the embryonic region. The limbs arise in pairs in the following order: (1) First antennæ, (2) mandibles, (3) second antennæ, (4) first maxillæ, and the remaining thoracic appendages in regular succession. The second antenna soon becomes bilobed, the inner branch representing the future long "whip" or flagellum of this limb. The first antennæ remain single until shortly before hatching, when the inner flagellum buds out from the inner lower surface of the primary stalk (see p. 226). The optic disks, at first paired rounded areas of rapidly dividing cells, soon become elevated into lobes and form the rudiments of the large eyestalks. The mouth appears at about the ninth day as a median pit on a line drawn through the hinder margins of the buds of the first antennæ and before the second antennæ are formed. At the tenth day the three pairs of nauplius-appendages are present as buds; a day or two later the upper lip or labrum has grown down over the mouth and a larger fold representing the abdomen and a part of the thorax has grown forward from the region of the thoracic-abdominal plate, marked by the earlier point of invagination. At 14 days of age the latter fold is divided at its extremity, which represents the forked telson-plate of the larva and touches or overlaps the lip. In 3 weeks the conical eyestalks are most prominent; 8 to 9 pairs of appendages are present, and the telson overlaps the brain. The brownish black eye pigment of the retinal cells begins to appear in the fourth week as a thin crescent at the base of each lobe, and gradually extends in area until in 3 or 4 months time it forms the large, rounded eye spots, so conspicuous a mark from this time onward. A cuticle to be later absorbed surrounds every part of the embryo, and rudimentary setæ are beginning to appear on the telson plate and antennæ.

Up to the fourth week internal changes, which we shall not attempt to describe, have led to the already complex foundations of the nervous and muscular systems, the heart, and alimentary tract. Of the latter the stomodæum or oral invagination gives rise to a distinct pouch from the epithelial lining of which the cuticular coat of the mouth opening, esophagus, and stomach sac are derived. The proctodæum, to which the anal opening and lining of most of the intestine is due, is similarly formed through a median ingrowth of ectoderm near the posterior end of what becomes the thoracic abdominal fold. The cuticular lining of the intestine when formed, like that of the stomodæum, is continuous with the outer skin and must be shed at every subsequent molt. The proctodæal invagination is at first solid or nearly so and is not sharply bounded from the yolk, which with its inclosed cells distinguished as hypoblast, represents the embryonic section of the digestive tract, called the mesenteron, and gives rise to the gastric glands and to the epithelial wall of a small section of the tract into which they open. The walls of the mesenteron become continuous with those of the proctodæum and are gradually extended forward on all sides until the entire yolk mass of the egg is inclosed within the folds of the paired gastric glands and forward division of the intestine. At a later period of embryonic life the screen which separates the stomodæum from the yolk is absorbed and its walls unite with those of the mesenteron. At the time of hatching the residue of the yolk lies in the folded walls of the lobulated gastric glands, from which it is finally absorbed. This residual yolk sometimes appears to pass to the masticatory stomach, but if this ever happens it must be due to secondary displacement, as will be readily understood from the relation of the yolk to the mesenteron just described. The functions of digestion and absorption, which the gastric glands or liver display on a large scale throughout the embryonic period, are retained in adult life as already noticed. (See p. 249.)

The intestine in the higher Crustacea, excepting only its terminal portion, is commonly described as arising from the endodermal or hypoblastic wall of the midgut, or mesenteron, but this is certainly not the case in the lobster, which sheds an intestinal cuticle during its pelagic stages. A median longitudinal section through the body of the larva at the time of hatching shows a distinct cuticle passing forward along nearly the entire length of the intestinal tube, and finally shading off and disappearing opposite the gastric glands. The epithelial lining of the intestine is therefore almost wholly of ectodermic origin and continuous with the epithelium of the skin, a conclusion which embryological study fully supports. Apparently in the adult animal the cuticular lining terminates abruptly at the forward end of the rectum, but this is not the case in early life.

During the course of development the ova increase considerably in size, and, losing their original globular form, become distinctly oblong (fig. 33, *a* and *b*). The bright red pigment cells or chromatophores, which are distributed in a characteristic manner, particularly on their basal segments and on the sides of the carapace, are prominent for a long time before hatching. These, together with the interference colors of the huge eye-spots and the rich green of the unabsorbed yolk, give the eggs of the lobster exceptionally brilliant color patterns.

EXCLUSION AND DISPERSAL OF THE BROOD.

It was found that when the eggs at the point of hatching were removed from the mother lobster and placed in jars at Woods Hole a full week elapsed before the entire brood was set free. Possibly the period is shorter when the animal is undisturbed and left to her own devices in the sea. When other conditions are favorable, the warmer the water the more rapid will the emissions occur. The individual variation in the eggs entailed by the long period of fosterage render it certain that all can not hatch simultaneously. Fullarton (*113*) found that in the European lobster the time required for the hatching of a brood varied from one to three weeks or even longer, but it is not likely that this period is extended to very great lengths under natural conditions.

FIG. 33.—Outlines to show relative sizes of lobsters' eggs when laid (*a*), and when ready to hatch (*b*). Enlarged about 3½ diameters.

The egg-bearing lobster instinctively folds its tail, thus securely inclosing the eggs in the abdominal pocket when in danger of enemies, while at other times she is seen at intervals to extend her tail and, standing upon her legs and incurved tail fan, move her swimmerets back and forth. In this way the eggs are aerated and cleaned, and such actions proceed instinctively during the 10 months of parental care which they receive. The cargo of eggs shows the effects of the treatment, for they pass the storms and stress of winter with remarkably little loss, and come to point of hatching bright and clean. It is rare to detect a single barren egg or broken embryo among the thousands of perfectly formed young. Yet when the egg-bearing lobster or crayfish are too closely confined, or the normal conditions of their environment seriously disturbed, sediment soon clogs the eggs and parasitic protozoa and other organisms attack and destroy the egg glue to such an extent that the ova fall off of their own weight and soon perish.

It might prove to be a point of some interest to determine whether the rhythm of the swimmerets is fairly uniform or not from the beginning to the end of the period of fosterage, but nothing can be said on this subject at present.

The behavior of the American lobster at the time of the emission of the young has not been studied with sufficient care under natural conditions; accordingly, I transcribe the following observations made on the European species by MM. Fabre-Domergue and Biétrix (*101*).

In order to ascertain as exactly as possible the age of our young lobsters, we determined to collect them for the space of twelve hours, a circumstance which led us first to find that hatching never takes place by day. At from six to seven o'clock in the evening not a larva was visible in the water of the float. Two hours later we could see several hundred of them swimming about. If we removed all of the latter with care, no new arrivals appeared before the evening of the following day. To what was the rapid emission of larvæ in so short a time due? The continual observation of our float during the first hours of night soon showed us the key to the enigma.

Toward seven to eight o'clock in the evening the female commenced to stir herself in her prison by presenting an attitude altogether unusual and characteristic. Her feet are stretched out almost rigid, her tail extended to the full in a horizontal direction, forming, with the rest of her body, a nearly straight line. She walks, as we might say, upon her toes, so careful is she to hold her entire body as far away as possible from the bottom of the aquarium. This feat lasts for a certain time; then quickly lowering her head and the fore part of her body until she rests upon the ground between her outspread claws, with tail on the other hand raised at an angle of 45 degrees and kept stretched, we see her violently shake her swimmerets with such rapidity that the eye cannot follow the movement, and a veritable cloud of larvæ are sent far to the rear and dispersed in all directions.[a] This phenomenon lasts from 15 to 20 seconds, and the female thereafter returns to her habitual attitude, to depart therefrom no more until the following evening. We have repeatedly verified the fact by observing always that the larval emission is produced in certain cases by two series of distinct movements, lasting some minutes, the second producing much fewer larvæ than the first.

The hatching does not therefore proceed independently of the mother and does not take place at all times of the day and night, but is confined to the hours of eight to nine o'clock in the evening.

The first molt which follows hatching is effected in the hours which precede the emission, and it is without doubt the movement of the larvæ under the abdomen of their mother which causes in her these signs of agitation and unrest already described. If, in short, one tries to draw the female out of the water when in this condition, we can see in her movements of defense the downfall of a great number of larvæ previously hatched but doubtless united to their mother by the molted membrane which her violent movements sufficed to break or to detach. Unfortunately we have been unable to assure ourselves whether, as Laguesse has observed in the crayfish, the young are found attached by the telson to the debris of the shell or of the molt (compare p. 167).

It should be noted that on occasion larvæ appear to be normally hatched in the daytime, and that a few may even resist the movements of their mother to disperse them, and remain for some little time attached to her body, though capable of swimming. In regard to the hatching of the European lobster when confined in ponds at the marine fish hatchery and biological station at Portobello, New Zealand (see p. 298), Mr. Anderton has written to me as follows: The hatching "almost always takes place at night. I say *almost* advisedly, since this last season a batch has frequently been hatched during the afternoon by a violent aeration of the tank water. I think about 1,700 has been the largest number hatched from a single individual during one night."

THE HATCHING PROCESS.

As already observed, what we shall consider the first molt of the larva is passed at the time of hatching, and in this act the larval cuticle and shell membranes are shed together. The stalked secondary egg membrane, representing the glue or fixative by

---

[a] With this specific and graphic account compare the brief statement of Coste, made nearly a half century before, that "The brood females straighten their tails, which up to now have been carried bent against the plastron, gently oscillating those appendages to which the bunched embryos are attached, as if to scatter the larvæ, and to aid them in breaking the shell, and hus free themselves in the course of a few days of their entire cargo." (*55*, p. 205).

which the eggs are attached to each other and to the body of the mother, in consequence of internal pressure, splits lengthwise of the embryo and its two halves separate like the skin of a pea. The primary eggshell or transparent "chorion," reduced by distention to a sac of great tenuity, adheres to the outer capsule at a point usually beneath its stalk and is in turn apparently adherent in some degree to the embryonic cuticle. Further, the invaginated hairs or setæ of the larva about to issue stick by their tips to the cuticular sheaths of the corresponding setæ. Consequently, successful hatching in the lobster means shedding the egg membranes with the old cuticle and the pulling out of the invaginated hairs of the new chitinous covering at the same time. Hatching and molting thus go hand in hand, and the first larval stage, like every period which follows, is preceded by a molt. The fact that hundreds of the larvæ which are hatched by artificial means get clear of the eggshells, but die through inability to cast this embryonic cuticle, illustrates the importance of these nicely adjusted relations.

It is thus evident that we can not help the little lobster out of its shell, but must let it escape in its own way, and if healthy it will cast in a few minutes. Its old covering must be shed in one piece and with the loss of as little energy as possible. The infant lobster hatches, molts, and unsheaths its swimming hairs at the same time, as was explained more fully in an earlier chapter (see ch. VI, p. 236). The eggshell, as we have also seen, sticks both to mother and child, while the cuticle of the latter is in turn glued to the swimming hairs of the new skin, so that every tug at the shell helps to free the little lobster from its hampering cloak and at the same time to perfect its swimming apparatus.

The young lobster is very compactly folded in the egg, which becomes ovoidal in consequence of growth. At the time of hatching this marked ovoidal form of the embryo is largely determined by the form of the carapace, which is longer than broad. The body is bent, but not twisted, the tail, as in all crustaceans, being folded against the thorax and head, the tips of the telson plate even reaching beyond the compound eyes and to a point overlying the masticating stomach. The mouth is thus covered by the overlap of the hinder part of the fifth somite of the abdomen, which also presses against the downwardly bent rostrum and the mouth parts. The antennæ are directed backward along the free borders of the carapace, while the thoracic appendages with their outer branches, like a double bank of oars, are directed downward over the abdomen and forward toward the middle line. Hatching thus implies not only release from the egg membranes, but casting off a complete cuticular molt and at the same time the evaginating or drawing out of every telescoped hair and spine of the body, including the rostrum; further, in addition to this and aided by it, the unfolding of the abdomen and the straightening of the telson and the various appendages.

Little difference in the size of the eggs was noted by Anderton (5) in the European lobster until the last month of development, when they increased as much as 3 millimeters in length in conformity to the shape of the embryo, and when convulsive movements of the embryo itself were often violent enough to move the egg from under the object glass.

THE FIRST LARVA.

[Pl. XXVIII and text fig. 34.]

When the lobster has successfully escaped from the egg capsule and shaken itself free from its cuticle, it emerges as a free-swimming animal and eventually rises to the surface, where it remains rising and sinking, but probably never far removed from the actual surface until its pelagic life is over.

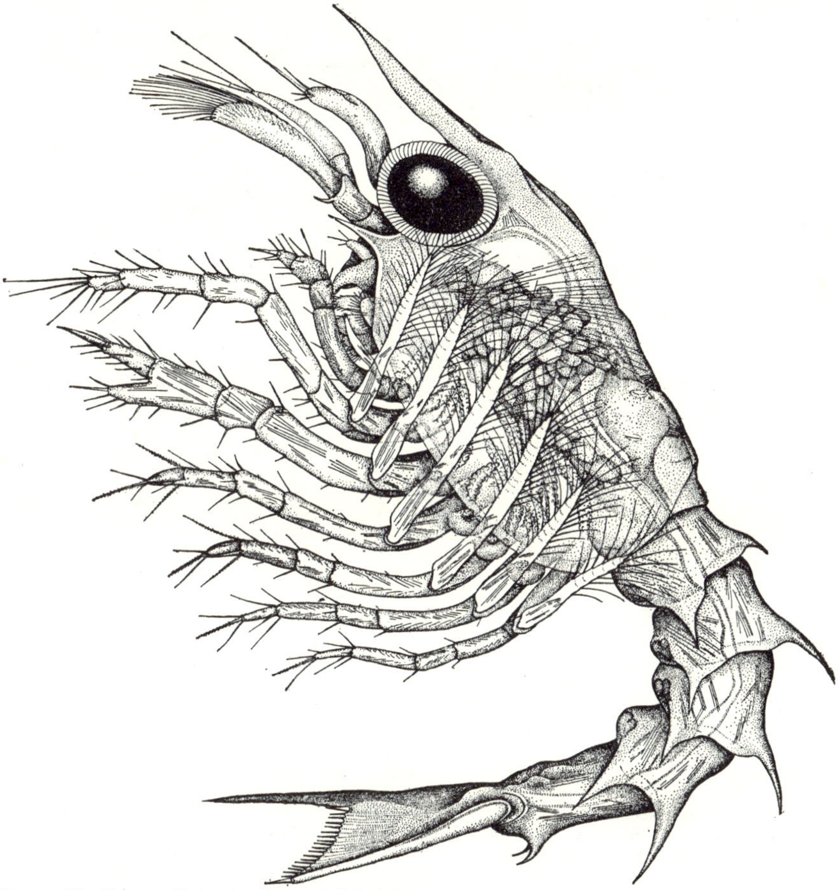

FIG. 34.—First larva, or first swimming stage of the lobster in profile. For drawing colored to life, see plate XXVIII; for natural swimming hold page sidewise with head of animal down, and compare figure 40 of text. Length about 8 mm., or a little less than ⅓ inch.

The animal is but little over a third of an inch long. The body is segmented as in the adult form, the most striking characteristics being the enormous eyes, the conspicuous rostral spine, which projects like a sharp spear in front, the triangular telson, and the biramous swimming legs, which, from their resemblance to the permanent swimming organs of the schizopods, have given to this and the two succeeding larvæ the name of the "schizopod" or "mysis stage." Functional appendages are wanting only in the abdominal segments, where, however, very small buds of the adult swimmerets can be seen beneath the cuticle in the second, third, fourth, and fifth abdominal somites.

The cuticle of the larval lobster is now as translucent as glass, and such organs as the heart and blood vessels, the alimentary tract, and the rudimentary gills are seen with

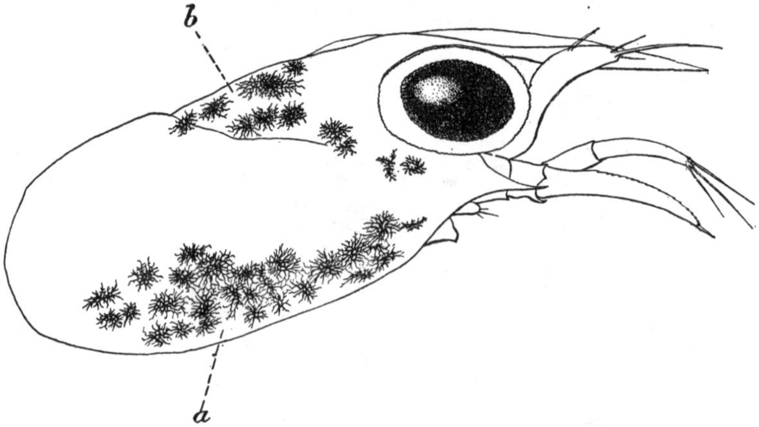

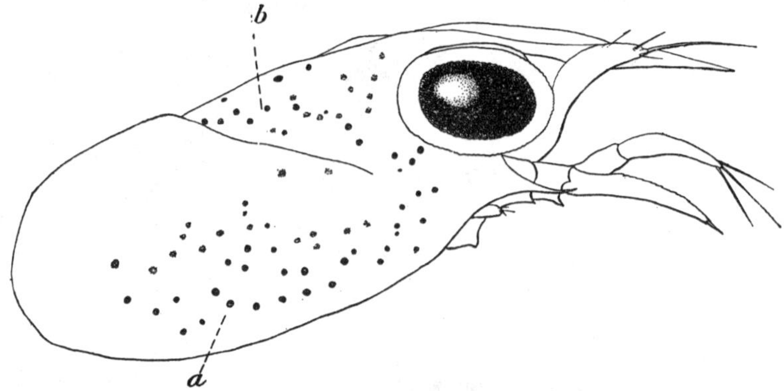

FIG. 35.—Cephalothorax of lobster in first stage when under stimulus of pressure, drawn immediately after reddening, through expansion of chromatophores. *a*, *b*, Lateral and dorsal red chromatophore groups; yellow pigment not here shown.

FIG. 36.—Cephalothorax of the same individual 10 minutes after release from pressure, and after paling from contraction of chromatophores. Both the red (solid) and yellow (dotted) pigment cells are indicated.

great clearness. The green food yolk has disappeared entirely or is reduced to a mere remnant now more yellow than green, in the masticatory stomach. Perhaps the most conspicuous internal organ is the yellowish-brown "liver," or gastric glands, the form of which on either side of the body resembles a cluster of grapes.

Color of the larva.—The gay coloring of the larval lobster, aside from that contributed by the internal organs and contents of the ailmentary tract, is produced by a blue pigment dissolved in the blood plasma and by red and yellow chromatophores which lie in the dermal layer of the skin, besides the pigment cells of the eyes. The distribution and grouping of the red chromatophores is very characteristic, and it is to these that the brilliant colors of the larvæ are largely due. The red cells are the larger and play the most prominent rôle. The expansion and contraction of the chromatophores, by which the animal becomes brightly colored or pale, ordinarily requires from 10 to 15 minutes when stimulated by pressure and released (fig. 35 and 36). The chromatophores are distributed in a number of well-defined regions, namely the carapace, in front of the cervical groove, the gill covers or sides of the carapace, the large claws and bases of the cephalo-thoracic appendages, and the dorsal surface of the abdominal segments, including the telson. These centers of color distribution are well marked from a late embryonic period to the lobsterling or fourth stage, when the change in the lobster's coloring is no less profound and abrupt than that of its structure and habits. When the chromatophores contract under the influence of a stimulus the animal becomes pale blue and very translucent; when they expand the vermilion cells give it a much more decided color. Pale blue at night, bright red by day is the rule, and among external agents sunlight seems to provide the main stimulus which causes the chromatophores to expand, but other changes, like raising the temperature or applying pressure to the body, will produce a like result. If the young lobsters are suddenly placed in darkness they tend to become paler and if returned to the light to redden more or less promptly. But the internal conditions or physiological states of the animal evidently present another and highly variable factor. All larvæ do not redden in the sun and all do not pale in darkness, while some respond more promptly to all such changes than others.

When the larvæ are seen struggling on the bottom of an aquarium, to get free from their old cuticle, when crippled in any way, or as Hadley remarks, when starved for some time, they so often turn red that this color has been regarded as a sign of weakness. On the other hand, if thousands of larvæ hatched and reared indoors are suddenly set free in more brilliantly illuminated water outside, a large proportion of them will redden, though not all. It has been asserted that the young and adult in all stages are upon the whole more active by night than by day, and that the young tend to move toward the source of light, or toward the surface where they find their suspended food. If the latter statement were true, we should expect to find the young larvæ at the surface of the ocean in the daytime and in active movement. Prof. S. I. Smith has taken the larvæ in all stages in the surface waters of Vineyard Sound in the daytime, and in several instances when using an electric light at night. These larvæ are often seen to pursue their prey by sight, and it has been shown that they can orient themselves through the medium of the eye. We thus seem to become entangled in a web of contradictory statements. The larvæ are more active in twilight or at night, but seek the light, and pursue their prey in the daytime, by the aid of sight. Red is a symptom of weakness, but they redden in the light.

The difficulty seems to lie in the fact that any given reaction is the resultant of complex conditions, which can be regularly repeated only when those conditions remain uniform. The life of the lobster during all of its free swimming life is apparently one of incessant activity, whether swimming at the surface or at whatever distance below it, and at all times of the day or night. In the account of their reactions to light, which later follows, it will be seen that their behavior is very complex and very variable. Certain responses may not only vary but even disappear altogether in consequence of changes in the organism or in the stimuli which affect it. Further, since the chromatophores as well as the muscles of locomotion are under reflex control of the nervous system, it is not more surprising to find variations in the responsive behavior of the pigment cells than in the activities of the body as a whole.

All that can be definitely said at present concerning the gay and plastic coloring of the larvæ is that it is an expression of chemical and physical changes in the body, due to stimuli, some of which are unfavorable, and that they have no protective significance. If every larva remained pale while swimming at the surface in the daytime, and took on color only at night, which is not the case, there would be no reason for supposing that there was a relation between the origin of the habit and the protection which it afforded because of the vast indiscriminate destruction which all such larvæ suffer at the hands of inanimate nature. That any such hypothetical protection would really count for nothing is further shown by the fact that the young lobster emerges at the fourth stage in a richly colored dress which renders it more conspicuous at the surface where it still swims than it would be if it remained colorless. For the continuance of the race a single lobster in the fourth stage is worth many hundreds in the first, and we should hardly expect to find nature at one moment using certain measures to protect life and at the next the same means for destroying it.

Both the blue pigment of the blood and the yellow and red pigment of the chromatophores, as already remarked, are lipochromogens, which are converted into lipochromes under a variety of conditions whether the animal is dead or alive. The stomach and liver are sometimes bright red, which recalls an observation by MacMunn, who concluded from spectroscopic evidence that in the lobster (*Homarus gammarus*) the enterochlorophyll of the liver might be carried to the hypodermis and converted into a lipochrome.

Structure and habits.—The most striking habits of the little lobsters immediately after birth are their incessant and apparently aimless activity, their preying and fighting instincts, and their voracity, which invariably results in cannibalism whenever the food supply is insufficient or unsuitable and where the young are too closely crowded in either vertical or horizontal limits; their seeking or avoidance of light under the variable sum of all the conditions which influence their behavior; their unstable, vacillating movements in the daytime or when stimulated by strong light; the total absence of the instincts of fear and concealment so clearly expressed at a later stage; their sharp vision for small floating particles at close range; their lack of precise discrimination, snapping up many inorganic particles or dead organic substances which are useless as food; their pursuit and often successful capture of copepods and other members of the plankton or floating population, showing that they can direct their movements with a certain degree of precision when necessary or when the light and other conditions are favorable.

The body of the little lobster is armed at most vulnerable points with defensive spines, and its various appendages bristle with tactile hairs or setæ, as well as with more diminutive spines, which may afford some slight degree of protection against smaller enemies when they do not assist it in seizing and tearing its prey.

The free margin of the "paddle," or forked telson plate, as commonly seen in the larvæ of the higher Crustacea, is garnished with very uniform and symmetrical spines and plumose hairs.

It is interesting to observe that certain spines and the setæ whatever their size or function, from the rostrum or tips of the claws down to the smallest microscopic hair, agree in their essential structure, and are all developed as tubular folds or outgrowths of the integument. In the course of the prenatal molt all the spines as well as the hairs are telescoped or invaginated. (Compare p. 269–270.)

In swimming the young lobsters use the outer branches or exopodites of the thoracic limbs (segments IX–XIV, table 4), by the beating movements of which they are slowly driven upward, downward, or forward (compare fig. 40), and the abdomen, by the sudden folding of which and by the aid of its broad telson plate, they dart rapidly backward. Each thoracic leg, in conformity to the type of decapod limbs, consists of a short stalk or protopodite and two diverging branches, the outer branch or exopodite which serves as a flexible "oar," being flattened and fringed with long feather-like hairs.

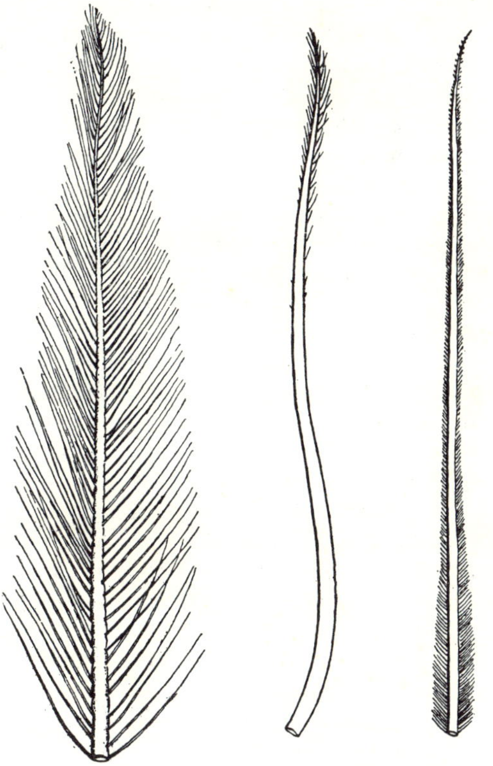

FIG. 37, 38, and 39.—Parts of setæ from cheliped of larval lobster, showing different degrees of reduction from typical plumose type. Enlarged 85 times.

The "oars" work independently of the inner branches, which in the larva are mainly prehensile organs, and which with the stalk alone give rise to the adult limbs. The concerted vibratory strokes of these minute flexible oars is so rapid and so uniform in vigorous larvæ that at a short distance from the eye it is impossible to follow their movements.

The exopodites atrophy, and are reduced to microscopic rudiments in the fourth stage, and completely disappear in the fifth. No doubt in this respect there is variation, however, as Williamson (*282*) has found to be the case in the European lobster.

In rising with head inclined, the body is usually bent into a quadrant, and according to Hadley (*131*) when the appendages are extended forward the exopodites strike somewhat forward as well as downward and thus drive the lobster upward and backward (fig. 40, *c*); when on the contrary the thoracic legs are contracted or drawn backward the larva is driven forward and upward. Whatever the direction of movement, as this observer has also pointed out, the animal always heads away from the source of light. In swimming near the surface the thorax is sometimes held horizontal with tail bent at an angle of 45°, more or less (*a*); when riding down another larva, feeding upon its carcass, or grappling with a lobster's egg the body is straightened (*b*); in the ascending currents of a hatching jar the young frequently come to the surface tail uppermost, and body vertical (*d*). By bending the body the weight is concentrated, which is especially advantageous in swimming upward. As Williamson remarks, the position of the body is correlated with the beats and direction of motion of the exopodites.

In hovering over the bottom, "standing on their heads," and as it might appear, probing the sediment with the rostrum (fig. 40 *f*), they are not trying to escape the light, as one observer has suggested, but are oriented for rising, being too weak, however, for any sustained effort. In every hatching jar or container many weakened individuals gradually settle into the sediment, a veritable trap for them, at the bottom, at first kicking away with strokes of the tail or standing erect with every oar in motion, but finally keeling over on their backs and beginning the death struggle to which there is usually but one ending.

The mutual destructiveness of the young lobsters when too closely crowded in aquaria has already been mentioned. When one lobster attacks another under these conditions the pursuer usually endeavors to get astride of his victim and with its sharp-pointed prehensile legs nip into the abdomen at its junction with the carapace. When the prey is an object too heavy to float, the lobster is frequently carried to the bottom; but if the animal is healthy it will be usually seen swimming about the aquarium dragging its prey with it and feeding upon it as it goes (fig. 40 *b*).

The beating of the heart and circulation of the blood begins at about the fifth week of egg development, or even earlier, and in the larval stages the heart and blood vessels have acquired the same general relations that we find in the adult.

The lobster at first possesses 19 pairs of filamentous gills distributed as in adult lobsters. The podobranchs are rudimentary, as are also the gill separators or epipodites, which are minute reniform plates exposed below the free border of the carapace. In the second stage these plates are taken completely into the gill chamber and the rudimentary gill of the eighth somite appears, which completes the branchial formula (see p. 246).

The nervous system of the lobster is highly developed in the larva and indeed before hatching, as shown by the admirable researches of Allen, (*2*), and brain, nerve cord, motor and sensory elements, as well as the complex stomato-gastric system, have essentially the same relations as are found in an adult animal.

Natural food of the larva.—It is not to be doubted that the incessant activity of this larva, which apparently knows no rest day or night, is needed, as Mead remarks to

FIG. 40.—Swimming attitudes of young lobsters in the first free stages; *a*, lobster swimming with body bent in the usual quadrant form, the head directed downward and often at a greater angle; the swimming branches (and the permanent limbs rather more than here shown) directed backward, in "posterior" position of Hadley; resulting movement upward and backward; *b*, young lobster playing cannibal, swimming astride the carcass of another which it has nipped at the junction of the carapace and abdomen and holds with its prehensile legs; *c*, swimming with the thoracic legs directed forward; in "anterior" position of Hadley; resulting movement upward and forward; *d*, rising position occasionally assumed; *e*, slowly moving or "floating" position sometimes observed; *f*, lobster "standing on head," apparently probing the bottom with rostrum, but really too weak to rise.

bring them into contact with the minute suspended bodies upon which they feed. All the rearing experiments that have been conducted by Mead and others with any degree of success during the past 15 or 20 years, whether in Europe or the United States, have clearlv shown that the larvæ must have their food suspended and in fine particles; the water must be gently agitated so that larvæ will not settle and become smothered in a mass of decomposing food and sediment at the bottom.

The natural food of the larval lobster consists of minute pelagic organisms, whether animals or plants, which through their own movements or their lightness remain suspended near the surface, such as diatoms and other protophytes, copepods, the larvæ of crustaceans, echinoderms, worms, and mollusks, the floating eggs of fishes, and, in fact, any member of the pelagic fauna which comes into their zone and is not too large for them to master.

The young lobster does not show, however, a very precise discrimination in its food. It will snap up almost any moving object, living or dead, which it is able to seize and swallow. Thus I have found in the stomachs of the older larvæ vegetable fibers, the scale of a moth or butterfly, and fine granules of sand.

An examination of the stomachs of a number of larvæ which were reared in aquaria to the fourth and fifth stages, when they measured 13 to 14 millimeters in length, revealed the following substances: (1) Diatoms in abundance, chiefly *Navicula* and the long tangled ribbons of *Tabellaria;* (2) remains of crustacea, probably parts of young lobsters; (3) bacteria in great numbers; (4) cotton and linen fibers and parts of algæ; (5) amorphous matter, with sand grains. The sediment of the jar contained the same species of diatoms in abundance, and amorphous débris similar to that found in the stomach and intestine.

Analysis of the stomach contents of a lobsterling captured in Vineyard Sound August 12 (length, 15 mm.) gave the following organisms: (1) Parts of crustaceans; (2) diatoms; (3) shreds of algæ. In another young lobster taken at the same time (length 17 mm.) there were (1) parts of crustaceans, (2) large numbers of diatoms, (3) filaments of green algæ and thin sheets or shreds of vegetable tissue, (4) the scale of a lepidopterous insect, (5) bacteria, and (6) amorphous matter in large masses. The diatoms and small amorphous particles of every kind may be regarded as partly or wholly incidental—that is, taken in with more important food material.

Williams (*279*) carefully examined the stomachs of one hundred larval and fourth-stage lobsters, which were being reared in the hatching bags at the Wickford (R. I.) station, and were fed with finely chopped clams. Thirty-seven contained copepods to the amount of 37 per cent of the total quantity of food present, and these favorite crustaceans were especially abundant in the stomachs of the second and third stage larvæ. Larval lobsters were almost invariably absent from their menu, from which he concludes "that a lobster in the presence of abundant food will not attack his kind."

A further discussion of food for artificially reared lobsters is given at the close of this chapter.

The length of the stage periods and the size attained by the lobster in each are subject to variations to be considered later: Length of first larva, 7.50 to 8.03 millimeters, average 7.84 millimeters (of 15 individuals); stage period, 1 to 5 days (Woods Hole, Mass.); length, 8.2 millimeters; period, 2 to 3 days, which may be extended to 25 days with the temperature at 60° F. (Mead and Hadley for Wickford, R. I.)

THE SECOND LARVA.

[Fig. 41.]

Under favorable conditions the first larval stage of the lobster lasts from 1 to 2 days. Upon molting for the first time after birth, the animal emerges into its second larval, free swimming stage.

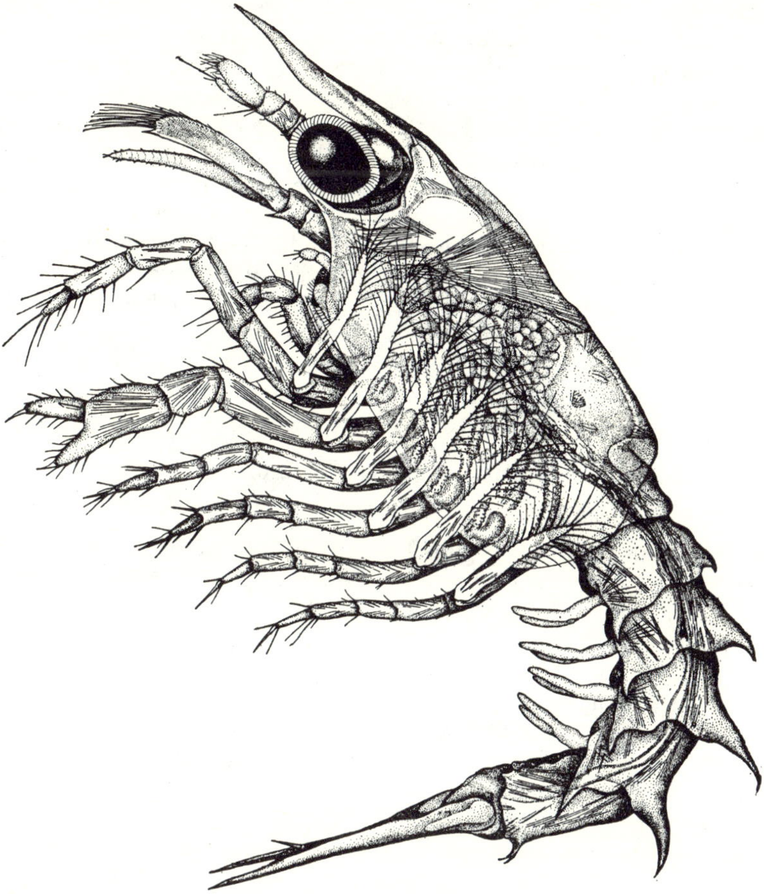

FIG. 41.—Second larva, or second swimming stage of lobster in profile. For natural swimming position hold page sidewise with head of animal down, and consult figure 40 of text. Length 9 mm., or 0.35 inch.

In habits and color the second larva resembles the first closely, but is distinguished by its slightly larger size and by the presence under the tail of four pairs of swimmerets on the second, third, fourth, and fifth abdominal segments, which appeared as minute buds beneath the cuticle of the first larva at birth. These appendages lack the swimming hairs, and do not become completely functional until the fourth stage.

Slighter structural changes which appear upon closer examination of the second larva are as follows: The rostrum is broader and its margins are serrated; the sides of the carapace completely cover the gills and separators; the sixth pair of abdominal appendages, the uropods of the tail fan, can be seen through the transparent cuticle as rudiments at the base of the telson; the stalk of the antennule is divided into three segments as in the adult, and its inner secondary flagellum, which is present in the first larva as a minute bud on the lower side of the primary flagellum, is much larger and shows traces of segmentation, while the stouter primary branch bears on its inner margin numerous clusters of sensory hairs. The long terminal spine of the outer flagellum has disappeared; the second antenna shows a reduction in its exopodite, the outer leaf-like scale with fringe of plumose hairs, which progresses with the following molts, and an extension of its segmented whip or endopodite; the chelæ or double claws borne on the first three pairs of walking legs are more perfect, and those of the first pair, which are destined to become the big claws of the adult, are perceptibly larger but otherwise similar. Both of the "great claws" gradually develop into the primitive toothed type, reached in the fourth stage, with teeth arranged in periods of eight; the primary and secondary spines only are present in the second larva. (See ch. VII.) Average length of second larva, Woods Hole, Mass., 9.3 millimeters; extremes, 8.3 to 10.2 millimeters (47 measurements); stage period, 2 to 5 days; Wickford, R. I. (Hadley), average length, 9.6 millimeters; average duration of stage period, 3 days; extremes, 2 to 7 days.

THE THIRD LARVAL STAGE.

[Fig. 42.]

Molting for the second time after hatching, the larva enters upon its third free swimming stage, in which the exopodites of the six pairs of thoracic legs (segments IX–XIV) are still functional. In habits, in color, and in general appearance the first three stages in the pelagic life of the lobster show no striking differences. The third larval stage, however, is readily distinguished from the second by the larger size of the animal, the presence of the completed tail fan, and the less rudimentary condition of the swimmerets upon the second to the fifth abdominal somites. The telson is reduced, though relatively much longer than the uropods; its terminal border is still incurved as in the first larva, but its lateral spines are longer. The inner whip in both antennæ is relatively larger and distinctly segmented, that of the second pair being considerably larger than the scale.

The "big" claws, though somewhat larger, still conform to the same type. They present a series of uniformly spaced spines, corresponding to the largest teeth of the lock-forceps or toothed claw of the adult, with rudimentary intermediate spines of the second order, or, if the latter are not present, the ducts of tegumental glands only, which mark their future position, may appear on the surface of the shell.

Like the earlier larvæ, they swim with head pointed downward, and with incurvated tail when rising, falling, or moving either forward or backward in the water, and they dart rapidly backward by sudden flexions of the tail. Yet Hadley observes that toward the close of this period they become more sluggish, as if already affected by those profound changes which at the next molt deprive them of their rowing organs and start them upon a new career. Upon the bottom, however, the third-stage lobster is nearly as helpless as at an earlier period, and while it may make the attempt to steady itself upon its legs, it can not long maintain an upright position. Its future balancing organs

FIG. 42.—Third larva, or third swimming stage of the lobster, drawn to a scale reduced from that of figures 34 and 41. See legend of figure 34. Length 11.1 mm., or 0.44 inch.

are still in an undeveloped state. The swimmerets are now fringed with short rudimentary setæ, but do not come into full play until after the next molt.

As Hadley has pointed out, at birth the larval appendages are less concentrated in the head region than in the adult state, and this is most noticeable in the maxillipeds, the exopodites of the third pair of which are used for swimming. From the first stage onward there is a gradual forward movement of the appendages—maxillæ, maxillipeds, and pereiopods—until the fourth stage, when they attain essentially their adult condition. Average length of third larva, Woods Hole, Mass., 11.1 mm.; extremes, 10–12 mm. (79 measurements); Wickford, R. I. (Hadley for 1904), average length, 11.4 mm.; stage period, 5 days.

THE FOURTH OR LOBSTERLING STAGE.

[Plate XXXI.]

The young lobster makes a surprising leap at the fourth molt, or the third after hatching, when suddenly it seems to undergo a literal metamorphosis and to become a new animal, and when for the first time it truly resembles a diminutive lobster. In form, color, habits, and instincts it differs strikingly from every preceding stage.

The oars or swimming exopodites of its twelve thoracic legs are reduced to functionless stumps, which as a rule are no longer visible to the naked eye. Yet it still swims at the surface with greater agility, precision, and speed than at any former stage. The balancing organs, formerly called the "otolith sacs," at the base of the first pair of antennæ, are fully developed, and the reeling, uncertain gait of earlier stages is no longer observed. Nor is the body bent in swimming, but is straight as an arrow, and as the lobsterling glides swiftly along by the action of its swimmerets, now for the first time in complete working order, the big claws are extended straight in front of the head and held close together. While it uses the same organs in swimming as an adult animal, unlike an adult it swims at the surface and with a relatively much higher rate of speed. As in earlier stages it darts backward by quick jerks of the abdomen, according to one observer even jumping out of the water, a feat which it is never again able to perform, and which is possibly equaled in the higher Crustacea only by certain kinds of surface-feeding shrimp. The great chelipeds are long, slender, and end in symmetrical claws of the toothed type.

The incessant and apparently aimless activity of the young in all their swimming stages has been often remarked. While this activity does not protect them from their enemies or enable them to stem a current of much strength, it is not useless, for it enables them to keep afloat and thus brings them into contact with suspended food, which has been found to be an important requisite in every hatchery. It has been further observed that when at apparent rest the motion of the swimmerets in the third and fourth stages tends to keep the little lobster from sinking.

Like the larvæ, the fourth-stage lobsters continue to feed on copepods and small pelagic organisms of various kinds, even snapping up floating insects, according to Williams (*279*), who saw a swarm of lobsterlings seize, drag under, and devour a full-grown cricket which happened to fall into their tub.

In a number of fasting fourth-stage lobsters, which Williams also examined, the stomachs were found to be empty or to contain only masses of clam cuticle, which they commonly reject, from which it appeared that such lobsters, even when very closely confined in a finger bowl and "hungry enough to eat what they ordinarily refuse, will not attack one another (unless perhaps one or more of their number is newly molted)."

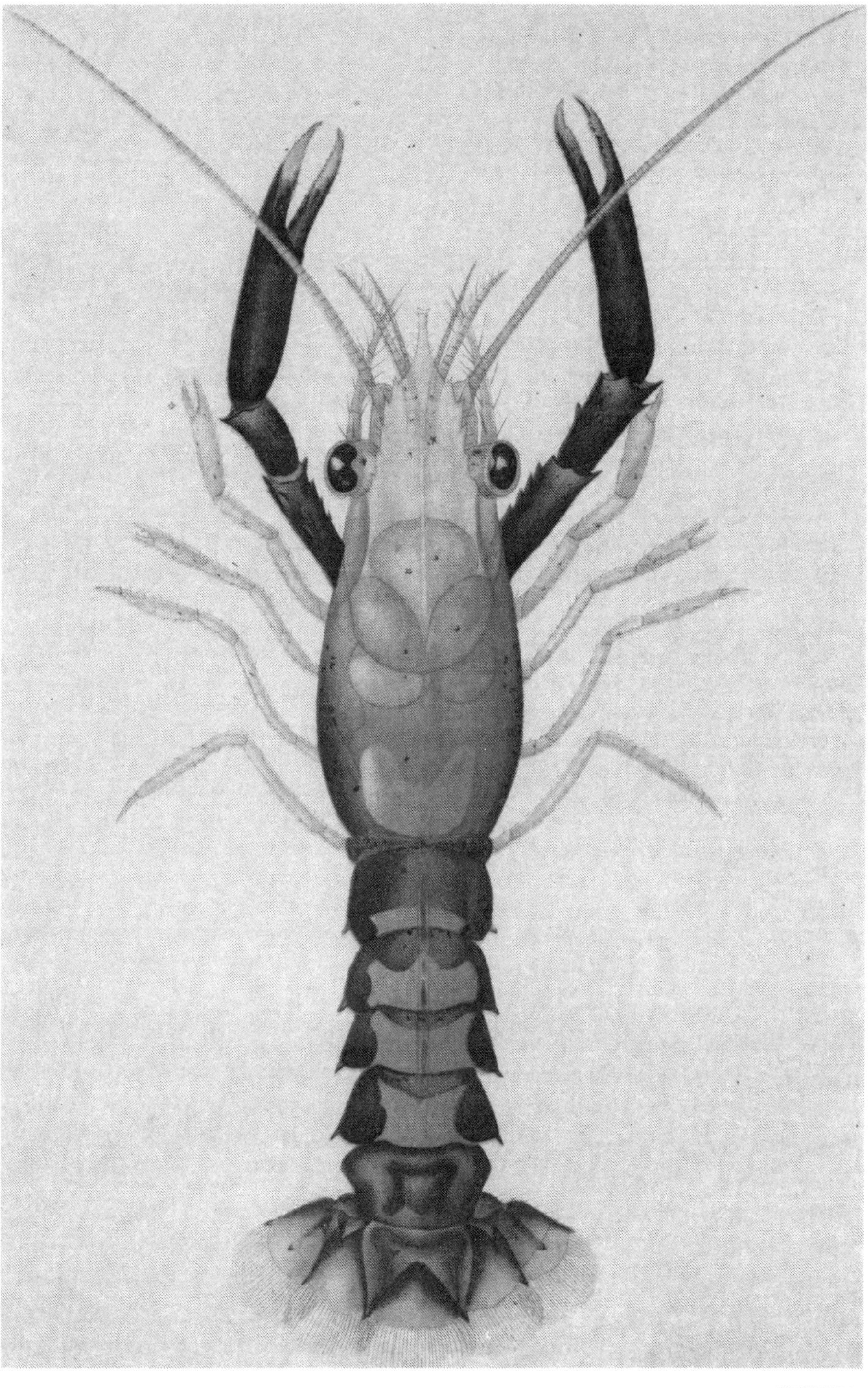

A.Hoen & Co Baltimore

FOURTH STAGE OF THE LOBSTER
LENGTH 14.6 MM.

Perhaps the most interesting morphological change which appears at the fourth stage, though by no means the most striking, is the torsion of the great chelipeds, described in chapter VII. The differentiation of the big claws, which come in time to equal one half the weight of the entire animal, is preceded by a permanent twist which has chiefly affected the fifth segment. While the lobster in the fourth stage is limber in every joint, the fusion of the second and third podomeres occurs shortly after this molt.

Lobsters after the larval period, and preeminently in the fourth and fifth stages, often exhibit the phenomenon known as "feigning death." When stroked with any object or when water is squirted on them with a pipette they will roll over and straighten out as if paralyzed. Their appearance when in this state is very different, however, from that of a dead animal. The phenomenon appears to be a somewhat sporadic reflex response, but it is interesting to find it appearing for the first time when the animal is about prepared to sink to the bottom, and to assume more fully the habits of an adult animal. (See 149, p. 184.)

Fourth-stage lobsters when approaching the end of their period frequently go to the bottom in shallow aquaria, hide under stones or any accessible objects, and even burrow in mud or sand.

The instinct of fear also appears in this stage and for the first time, associated with the hiding and burrowing tendencies. These are possibly evoked by the development of that contact-irritability which, as Hadley remarks, seems to come suddenly into play toward the close of this period. Burrowing is a kind of behavior in which the lobster frequently indulges from this time onward throughout life. The burrows serve a fourfold purpose—for concealment and therefore for protection, as a point of vantage from which to watch and seize their prey, and probably as a means of avoiding strong light, especially when adult, and particularly when confined in relatively shallow "parks" or pounds.

Digging the hole is an instinctive act; but returning to the same burrow of holding to the same crevice for the purpose of defense, for hiding, or for seizing the prey, so marked in all the later stages of both young and adults, is a distinct mark of intelligence, a habit of returning to the same spot being formed through association.

An interesting phase in the behavior of the fourth-stage lobster, as described by Hadley, is its rheotactic response or tendency to head into the current, which, with its other reactions, will be later discussed.

Color in the fourth stage.—At this period the range of color variation is much greater than at any previous stage, but color change no longer follows so promptly change in temperature, in the illumination, or in the intensity of other effective stimuli. The chromatophores or pigment cells of the skin have so multiplied as to form a continuous screen to the parts below. The former transparency of the larva is thus reduced in the same degree that the depth and brilliancy of its colors are enhanced.

The exoskeleton is now reenforced for the first time with considerable deposits of mineral salts, especially of lime. It is still quite translucent, but of a delicate light-blue tint, as appears at the molt. The body of the lobster, and the cephalo-thorax in particular, is studded with sensory hairs. The hair pores constantly increase in number up to the adult state, when the shell is finely stippled with them, while the setæ themselves have for the most part disappeared.

Microscopical examination reveals a multitude of minute, closely crowded chromatophores in the skin, containing pigments of various tints, chiefly red and yellow. The color pattern is due mainly to the distribution of these cells; the quality and degree of color which in the same individual is subject to more or less constant variation, especially before and after the molt, is determined by the expansion of the variously colored chromatophores, the contents of the alimentary tract at the moment, and the variable tints of the underlying gastric glands. The bluish tint and slightly diminished translucency of the shell, when preparing to molt, has a considerable influence on the color of the animal as a whole.

The general cast of color may be either (1) yellow and red, (2) red, (3) green, or (4) green and reddish-brown. In the first instance the carapace is light yellow, translucent, and sprinkled with red chromatophores. The abdomen and large chelæ are reddish-brown, and there is a quadrilateral yellowish-green area on the terga of the fourth and fifth abdominal segments. In the red individuals the animal is bright red, especially on the abdomen and large chelæ. The carapace is yellowish, spotted with red, and the abdomen is marked in the way just described. In the green variation the whole animal is bright green. Bright-green areas are noticeable on the abdominal terga as before, and upon the hinder portion of the carapace. There is also some brown pigment on the large chelæ and tail fan. In the fourth variety the abdomen and chelæ are rich reddish-brown, with light peacock-green on the terga of the abdominal rings, as is commonly seen, and on the carapace next to the abdomen. The rest of the carapace is greenish-brown. The characteristic tendon marks on the carapace in this and in all subsequent stages define the areas of attachment of certain tendons or muscles to the shell. They become most conspicuous after the fifth or sixth molt. Average length at fourth stage, Woods Hole, Mass., 12.6 mm., extremes, 11–14 mm. (64 measurements); stage period, 10–19 days; Wickford, R. I. (Hadley for 1904), average length, 13.5 mm.; stage period 12 days.

THE FIFTH STAGE.

The lobsterling which has not made its descent to the bottom at the close of the fourth stage continues to swim at the surface until the end of its fifth period, but whether pelagic or an inhabitant of the bottom its behavior closely tallies with that manifested in the preceding stage under similar conditions. Hadley has shown, however, that fifth-stage lobsters exhibit a stronger repugnance to light and a greater tendency to seek sanded areas and to burrow.

The structural changes which the lobster undergoes in passing from the fourth to the fifth and again from this to the sixth stage are often so slight as to be unrecognizable by anyone who has not followed each stage under the microscope molt by molt.

The salts of lime and the pigment which begin to appear in the shell at the fourth stage increase, and the carapace is in most cases fairly opaque, excepting immediately after a molt, when, as often happens in crustaceans, the body for a time becomes quite translucent. From this period onward the color of the lobster is mainly due to shell pigments which are subject to change within certain limits, and are due to the direct activity of the chromatophores of the underlying soft skin. Every chromatophore at the surface of the skin stamps its image and counterpart upon the hard, unyielding shell.

The characteristic colors of the fifth stage are seal brown or maroon, or some combination of brown and green, which bring into strong relief certain snow-white or cream-colored spots on the body and chelipeds. The carapace at this stage presents four and sometimes five prominent white spots, the tendon marks already referred to, two on each side and one crossing the middle line of the back just in front of the cervical groove and in contact with it, marking in part the area of insertion of the posterior gastric muscles. Of the lateral spots the larger is a circular or oval disk-like impression below the cervical groove and in contact with it, while the smaller spot above the groove marks the tendinous insertion of a small muscle. From this time onward it is a constant character of the carapace, although it gradually pales and ceases to be prominent. Another triangular tendon mark which later becomes noticeable and remains throughout adult life lies just above the level of the last, at the intersection of the branchio-cardiac lines and the cervical groove, its angles meeting this line and the transverse and lateral divisions of the groove or fold.

The external geography of the carapace, which still remains unexplored territory to a large extent, shows other small spots destitute of hair pores and a great variety of surface marked by depressions and elevations by the varied distribution of hair pores, and by spines many of which bear the ducts of tegumental glands, not to speak of the tendon spots already described, by grooves and larger protective spines, slightly roughened areas of muscle-insertion which are prominent just behind and in front of the transverse division of the cervical fold, as well as by areas of absorption which are essential for the molting process and are developed in correlation with the gradual deposition of mineral salts in the shell, such as the median stripe and the scalloped edges of the gill-covers. (For adult conditions see chapter VI.)

Further, the pleura of the first abdominal somite are snow-white; while the tips of the big claws, the rostrum, and the blades of the propeller or tail fan are washed with dull white or cream color. A light spot is also sometimes seen on the fourth segment of the great chelipeds.

It should be clearly recognized that here, as at every other stage, the color is subject to a considerable range of variation even in the same individual, due in a large measure to periodic changes involved in molting, to the temporary effects of light, and possibly to food and to other causes. At the crisis of the molt the little lobster is capable, as we have seen, of some quite chameleon-like performances.

But slight morphological changes are noticed in the fifth stage; the antennæ are extended in length, the big claws have become somewhat shorter and thicker, and it is common to find that the dactyl is bent so that the edges of the toothed forceps do not meet. The microscopical rudiments of the swimming exopodites have been further reduced but do not, as a rule, wholly disappear until the sixth stage. Average length at fifth stage, Woods Hole, Mass., 14.2 mm.; extremes, 13.4–15 mm. (15 measurements); stage period, 11–18 days; Wickford, R. I. (Hadley for 1904), average length, 15.5; stage period, 9.5 days.

THE SIXTH STAGE.

[Pl. XXXII.]

The sixth-stage lobster resembles the preceding stage in all essential respects both in structure and behavior, barring the fact that apparently all or nearly all animals in this period are bottom inhabitants. In color the two stages are nearly identical and subject to a similar range of variation. The tendon marks, and the cream-colored or dull-white spots on the tips of some of the appendages, which begin to show as early as the fourth stage, are even more pronounced than before. There is a prominent light spot at the distal extremity of the fourth podomere of the great chelipeds, as already mentioned for the fifth stage.

The modified abdominal appendages of the first abdominal somite commonly appear in the fifth or sixth stages as minute tubercles or buds, which at first lie upon the sternal surface across the long axis of the body, thus facing each other or pointing toward the middle line. After segmenting into two divisions, which in some cases does not happen until the eighth stage, this appendage becomes bent downward until it stands at nearly right angles with the underside of the tail. I was not able to determine the sex by the abdominal appendages alone until the tenth stage, but Hadley (*124*) maintains that this distinction can be made in the eighth or ninth stages, or even as early as the sixth or seventh stages, by means of the position of the openings of the sexual ducts. My material did not enable me to fix the sex by means of these ducts earlier than the eighth stage, but this was not extensive, and it can not be doubted but that in all such matters considerable individual variation exists.

The development of the crusher type of claw or the transition from the symmetrical to the asymmetrical condition of the great chelipeds begins in the sixth or seventh stage, and is marked by a blunting to be later followed by a fusion of the teeth to form crushing tubercles, but the change proceeds very slowly and is not conspicuous for some time. The future crusher gains at first in girth or breadth rather than in length (see ch. VII, p. 271). Average length at sixth stage, Woods Hole, Mass., 16.1 mm.; extremes, 16–17 mm. (12 measurements); stage periods, 14 days. Wickford, R. I., average length, 18.6 mm. (12 measurements); stage period, 12.7 days.

THE SEVENTH STAGE.

The seventh stage is sometimes distinguished from the sixth period, as already remarked by the first noticeable differentiation of the crushing and toothed claws, but aside from this there are no characteristics in size, form, or function by which this and subsequent stages can be distinguished with certainty unless one has watched and recorded every molt.

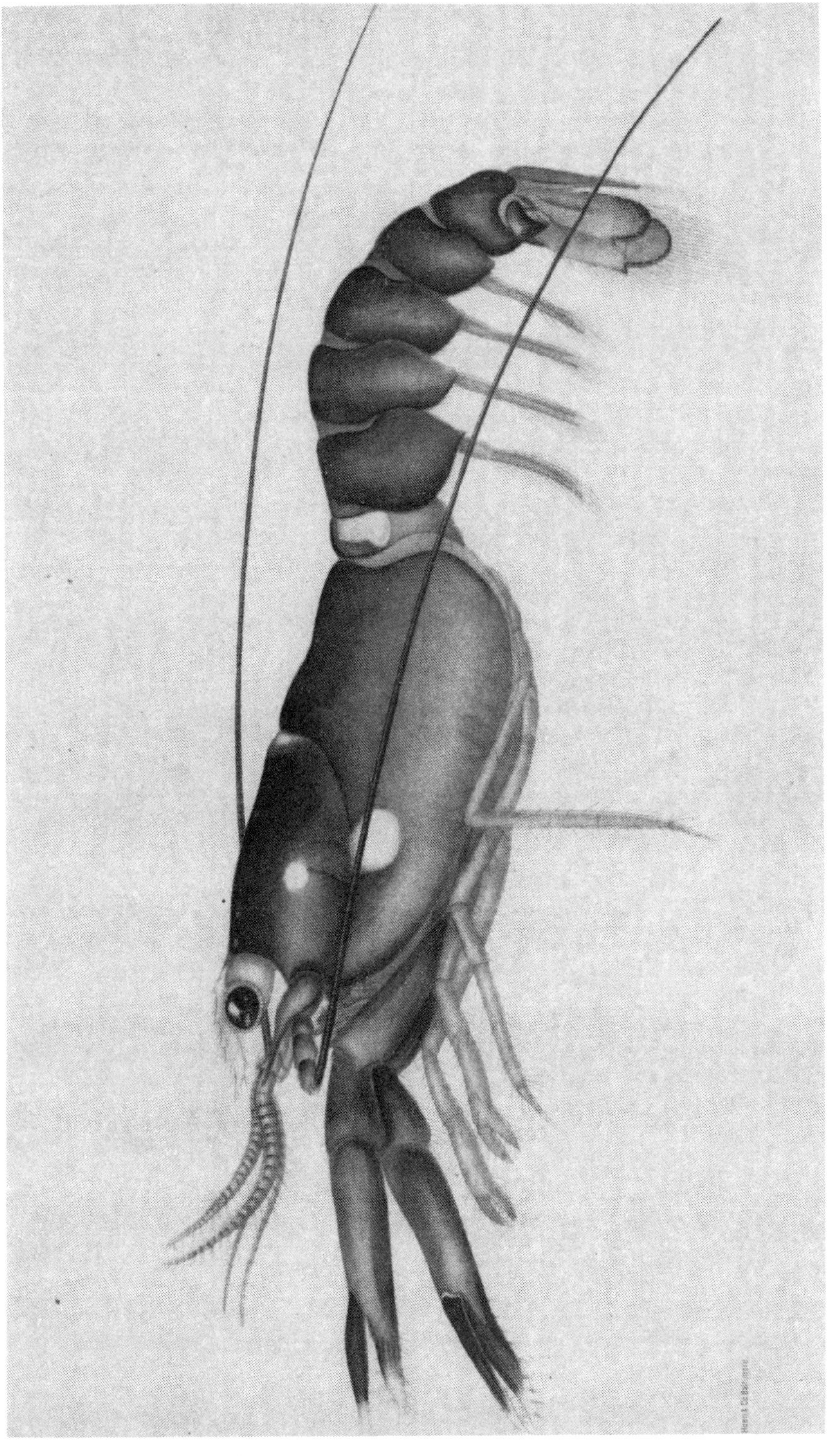

SIXTH STAGE OF THE LOBSTER
LENGTH 16 MM.

The seventh-stage lobsters keep as steadily to the bottom as the adults, and in crawling about make use chiefly of the last three or four pairs of thoracic legs. The large claws and smaller chelate legs are often extended forward in front of the head.

In the case of a lobster which was observed to molt from the sixth to the seventh stage the body was translucent, the general color being reddish brown, with a slight tinge of green on the carapace. The large claws were of a bright terra-cotta color. There was a whitish crescentic spot at the cervical groove on the back, and the characteristic tendon marks on each side of the carapace were as prominent as in the sixth stage. The pleura of the first abdominal somite were also snow white, and the uropods were tipped with cream color.

At the seventh stage pigment has been deposited below the enamel layer of the cuticle in an amount which, though at first very slight, increases with every molt and thus makes the color pattern more and more complex.

According to Hadley (*124*) the color of the seventh stage is usually and characteristically pure slate, becoming darker during the progress of the period, showing further the modifications of blue slate, green slate, and cream slate. The white spottings, as I have frequently observed, show a tendency to become creamy or buff in color in contrast to their porcelain-like whiteness in the fifth and especially in the sixth stage.

I have recorded numerous observations to show that the same animal may undergo no inconsiderable changes of color during the stage period. The color at this time is due to the pigments of the changing cuticle and to the changing pigments of the soft skin beneath it. With the advance of the stage period a new cuticle or shell is gradually formed beneath the old, which is later shed, with the tendency to become darker or more opaque. The color is also affected in some degree by any stimulus or change of the physiological state which affects the more responsive chromatophores of the soft skin.

It is therefore a difficult matter to standardize these ever-changing color effects, and not possible unless the animals are compared in the same stage period, immediately after molting, and under similar physical conditions. It is certain that the activity of the chromatophores is not dependent upon the direction or intensity of the rays of light alone, but rather more, as recent experiments seem to show, upon the physiological states, which follow upon complex and little understood changes.

Further, the act of molting by the stimulus sent into the chromatophores will sometimes bleach a brilliant animal into a pale shadow of its former self, as I have witnessed in the adult shrimp *Alpheus*, as well as in the adolescent lobster. Accordingly I consider it highly probable, if not certain, that the blue-slate or slate color is often due to the advancement of the stage period and to the peculiar opacity which always follows upon the development of a new cuticle beneath the old. It should also be observed that the cast shell, from at least the fourth stage to the present, which veils the brighter colors of the new cuticle, is blue, suffused at this time with green and brown in its pigment layer.

Hadley remarks that the adult structural type is possibly reached in the ninth stage, and the adult color pattern in the eleventh. Inasmuch as single structural characters, such as the differentiation of the big claws, are by no means regular or invariable in their appearance, we should hardly expect to find the sum of such characters expressed at a definite molt, which after all is but an incident of growth. Even at the fourth stage, as Williamson (*282*) has shown in the European lobster, the swimming organs are not shed in the same degree of completeness in all cases. Far less is it possible to fix upon any definite stage when the sexual characters and sexual maturity are reached. The data do not seem to be sufficient to make the determination of averages very precise. Average length at seventh stage, Woods Hole, Mass., 18.6 mm.; extremes, 18–19.5 mm. (4 measurements); stage period, 14–21 days; Wickford, R. I. (Hadley for 1904), average length 22.5 days; stage period, 14.3 days.

THE EIGHTH AND LATER STAGES.

The external structural changes which immediately follow the seventh stage are very slight and concern chiefly the accessory reproductive organs, such as the differentiation of the seminal receptacle of the female and the first pair of pleopods in both sexes.

The eighth stage is similar in color to the seventh, but according to Hadley there is a greater modification of the slate color, with a tendency to develop the blue slate and cream slate, or, in a less marked degree, the green slate and brown slate. According to the same observer, the blue color is more pronounced in the ninth stage, when the prominence of the white or cream colored spots is beginning to wane. It has been further noticed that in the tenth stage the olive green and olive brown combinations become more prominent; the spottings are seldom seen, and the dark mottled character of the coloring of the adult begins to assert itself. This characteristic mottled color pattern was still more pronounced in the eleventh stage, when it was apparently established.

HABITS OF ADOLESCENT LOBSTERS.

From the close of its free-swimming life until the later adolescent period the young lobster drops out of sight so completely that for a long time its habits during this interval were quite unknown (see *149*, ch. XI). After reaching the bottom we know that many of the little lobsters begin to travel toward the shore, in all probability slowly at first, but more rapidly when at the age of about 3 months they have a length of 1¼ inches, more or less.

The instinct of fear, suddenly developed in the fourth stage and present at all later periods, prompts the little animal to display great caution in all its movements, and to hide under stones or in the crevices of any protecting object whenever danger assails it.

Whenever the lobster sinks in very deep water, as must often be the case, it possibly gradually moves shoreward. At all events many adolescent and small lobsters are found along the rocky shores of bays and small inlets, where they apparently remain until driven out by ice. These small lobsters live under stones and submerged rock piles, the tops or surfaces of which are sometimes laid bare at unusually low tides in fall, when they may be found by digging and turning over the stones, at depths of but a few inches at low water, but where at the flood the sea rises to a height of 5 feet or more. The smallest, from about 1¼ to 3 inches in length, go deep down among the loose stones, where no enemy is likely to reach them. At a later period, when from 3½ to 4½ or 5 inches long, they issue from their retreats more freely and explore the bottom with greater boldness. They also dig caves under stones, from which, as at an earlier period, they stealthily crawl in search of prey, but quickly return when an enemy appears. We have seen that this characteristic burrowing instinct develops as early as the fourth stage.

As the lobster increases in size it becomes bolder and retires farther from the shore, but it never loses its instinct for digging nor abandons the common habit of concealing itself when the necessity arises.

A LOBSTER 413 DAYS OLD.

As is well known, size, whether of lobsters or of mankind, is not a certain criterion of age. In the crustacean it depends upon the number of molts successfully passed, while unfavorable conditions tend to lengthen the molting periods. Some of these conditions will be considered in a later section. This was well illustrated by the young lobster whose history follows. This animal was reared in a small glass aquarium at Woods Hole, Mass., and was fed with minced clams and the eggs of the lobster and cod. It lived from June 20, 1893, until August 6, 1894, when it had attained the length of 36 millimeters (1.44 inches).

In its final stage the colors of the animal had apparently reached the limit of their brilliancy and the mottled color pattern was as complex as in an adult animal. The body was of a light umber color freely speckled and mottled with darker tints. The appendages were reddish brown and slightly translucent. Small light spots or suffusions were found in certain parts of the body; the tendon marks, corresponding to those characteristic of the fifth and later stages, were prominent, the round spot just below the cervical groove being over a millimeter in diameter; the pleura of the first abdominal somite were snowy white, while the free edges of the segments of the body and of the appendages were bright blue; the large chelæ were white tipped. The openings of the oviduct were plainly visible, while the lips of the copulatory pouch or seminal receptacle were not yet closed. The color of the appendages on the under side was light reddish brown, and the tail-fan was of the same hue, edged with deep red; the big claws, which were tufted with setæ at their tips, showed but little differentiation. The compound eyes had acquired the large size and prominence of the later adolescent stages.

WHEN DOES THE YOUNG LOBSTER GO TO THE BOTTOM TO STAY?

Over 15 years ago I raised the question which is now placed at the head of this section, and answered it in a tentative way, but its importance seems to have been underestimated, for it has received little attention from other workers up to the present time.

It was shown that young lobsters did not uniformly make their descent to nether regions during the fourth stage or even at its end, and that the swimming period often lasted to the fifth stage, probably until its close, and possibly into the sixth stage. I have records of young lobsters captured under natural conditions at the surface of the sea (see *149*, table, p. 187), varying in length from 15 to 18 millimeters. The largest, taken 7 miles southwest of No Man's Land, near Marthas Vineyard, 18 millimeters long, was probably in the fifth stage, though possibly in the sixth, as seemed to me very likely at the time. Hadley's measurements for Wickford (R. I.) lobsters, which average much higher than those obtained by me at Woods Hole, Mass., are for the stages in question as follows: Fourth stage, average length, 13.5 millimeters (extreme, 15.5 mm.); fifth stage, average length, 16 millimeters (extreme, 18 mm., two records only); sixth stage, average length, 18.8 millimeters (extreme, 24 mm., one record). (See also later measurements quoted above.) The average length for lobsters raised in aquaria at Woods Hole in the same stages are as follows: 12.6 millimeters (extreme, 14 mm.); 14.2 millimeters (extreme, 15 mm.); 16.1 millimeters (extreme, 17 mm.). Inasmuch as size is a very unsafe criterion of either stage or age, it can not be said that at present there is any satisfactory evidence that the American lobster remains at the surface beyond the fifth stage. It is interesting, however, to notice a record by Meek (*200*) of the capture by surface net of a young specimen of the European lobster, which measured 20.5 millimeters ($\frac{13}{16}$ inch), at Alnmouth Bay, Northumberland, England, in the afternoon of September 7. Its age was estimated at 2 months. Now according to Ehrenbaum (*87*), whose work was conducted at Helgoland, such a lobster should be in either the sixth or seventh stage and upward of 61 or 87 days old, respectively (sixth stage, length, 18–20 mm.; seventh stage, length 21–22 mm.). We should therefore hesitate to affirm that in the American form the swimming life at the surface is never extended to the sixth stage.

The experiments of Hadley and others on the reactions of the larvæ show that the light-shunning, bottom-seeking, and hiding tendencies begin to assert themselves in animals artificially reared toward the close of the fourth or else in the fifth stage.

The bearing of this question upon the artificial propagation of the lobster is very evident, for, if a considerable number of fourth-stage lobsters remain suspended at the surface, the careful rearing to this stage and subsequent liberation in the sea is only feeding the fishes. A small force of predaceous tautog, or cunners, would play havoc with myriads in a short time. As we remarked in 1895, "the problem of the artificial propagation of the lobster will be solved when means are devised by which larvæ, after hatching, can be reared in inclosures until the fifth or sixth stage, when they can take care of themselves." This time limit should have been modified to read "until they go to the bottom." The lack of precision which the lobster displays in his desire to discover the bottom is very disappointing, but it seems evident that liberation of the carefully reared young at the very beginning of the fourth stage is only to court disaster, with the attendant waste of time, money, and labor.

FOOD AND CAUSES OF DEATH IN ARTIFICIALLY REARED LOBSTERS.

The yolk of hard-boiled eggs, crushed crab, boiled liver, minced fish, beef, lobster's liver, the soft parts of clams, and menhaden have all been tried as food for young lobsters by different experimenters in America and Europe with varying degrees of success.

Emmel (*95, a*) in a series of experiments upon the rate of molting of 90 selected lobsterlings which had reached the fourth stage on the same day, and which were divided into lots and were fed on different foods, obtained the following results: Beef-fed lobsters advanced to the fifth stage in an average period of 11.2 days; when fed on minced muscle of soft-shelled clams, in 11.3 days; on shredded lobster muscle, in 11.5 days; on shredded fish, in 11.7 days; on beef liver, in 12.3 days. While his tests showed a slight advantage for the beef fed over those supplied with clams, the lot which received no food other than the natural plankton of the water were twice as long in passing to the fifth stage, or 24.6 days.

In the experiments on the artificial rearing of the lobster conducted at Woods Hole, Mass., by the United States Fish Commission in 1902, the flesh of the menhaden, which is saturated with oil so that it does not readily sink, was found to answer admirably as a food until many of the larvæ began to sicken and die. The fish were shredded in a meat grinding machine, and a teacup full of this finely triturated flesh taken twice daily was found to meet the needs of about 5,000 larvæ. The voracious young can hardly be fed too much, provided the waste is not allowed to accumulate in the rearing tanks or bags, and as they grow older their ration must be increased. In June it was noticed that many of the menhaden-fed fry in the rearing bags were attacked by a fungus, which Gorham (*121*) thought was attributable to the oily fish upon which the young had fed. According to this observer, the mycelial filaments of this fungus spread from the point of infection until all the animal's tissues were destroyed and the lobster's body was reduced to a chitinous shell packed full of the mycelium.

In 1893 I described a case in which a parasitic fungus, probably belonging to the family Chrytridiaceæ, had attacked the late egg embryos of the shrimp *Alpheus*, a relative not far removed of the lobster. In this case the eggs were crammed full of the encysted parasite.[a] No internal egg parasites have yet been reported for the lobster.

The chief causes of death in the artificially reared lobsters are organic sediments, cannibalism, which is caused chiefly by overcrowding or a lack of proper food, and the exceptional fungus growths under the conditions of feeding referred to above. The sediments cling to the hairs of the appendages, interfere with the locomotion of the larva, and send it to the bottom, thus cutting off its supply of food. In this way it becomes crippled, and, being too weak to molt, it usually starves to death. Various algæ, bacteria, stalked protozoa, and diatoms occur in these sediments, but the chief offenders are diatoms.

Gorham (*121*), who has made a careful investigation of the causes of death in artificially hatched fry, names 24 species of diatoms which were found on lobsters reared at

[a] For figures and description, see appendix II, ch. V, of The embryology and metamorphosis of the Macrura, Memoirs National Academy of Sciences. Washington, 1893.

Woods Hole, of which the four most common species were *Licmophora tincta*, *Diatoma hyalinum*, *Rhabdonema arcuatum*, and *Tabellaria unipunctata*, named in the order of their relative abundance.

I have seen the fry almost buried out of sight by diatoms in neglected jars at Woods Hole, especially by *Tabellaria*, which at times was very abundant and destructive.

Other organisms found by Gorham to infest the young lobsters at Woods Hole were a filamentous green alga and a stalked protozoan, *Ephelota coronata*, which was more abundant in the waters of Wickford, R. I.

Cannibalism may be reduced by supplying the young with proper food, by agitating the water and thereby keeping both the young and their food suspended, and by avoiding overcrowding. The growth of diatoms can be checked or prevented, according to Gorham, by filtering the water; by selecting a suitable station for the rearing apparatus where diatoms do not abound, and where the temperature is high or most favorable for hastening growth and molting, by which the little animal escapes for the time being at least from all its troublesome messmates; by frequent cleaning, coating, or renewal of the rearing bags; and by reducing the light and thus hampering the diatoms by cutting down their food supply. (Compare, p. 281.)

THE SIGNIFICANT FACTS OF LARVAL AND LATER DEVELOPMENT.

Some of the most important facts concerning the larval life of the lobster may now be summarized:

(1) The young are hatched in great numbers, 5,000 to 100,000 eggs or young being produced at one time by a single animal according to its size, the number increasing rapidly in proportion to the cube of the body length or to the total volume of the body. This leads us to expect great destruction of the young in nature, an expectation which is unfortunately realized. It is a vulgar error to assume that the abundance of this animal or of any other species is proportional to the number of young born, since it neglects the equally important question of the destruction of the young or their rate of survival. The rapid rise in production beyond the 10-inch size proves that the older the animal the more valuable it becomes for reproductive purposes, barring the question of sexual decline, which is of little importance in an animal so seldom permitted to grow old.

(2) The larvæ are hatched at the bottom of the ocean in relatively shallow water at night or in early morning. A molt occurs at the time of hatching; parental instinct ceases; the larvæ are soon dispersed, and leaving the bottom lead a free-swimming, pelagic existence for a period of from 3 to 6 weeks (see p. 348), according to circumstances. Summer eggs on the coast of Massachusetts are hatched from May 15 to July 15, the majority being extruded in June.

(3) The movements of the larvæ in a natural state are not fully understood. Under certain conditions they rise toward the stronger light at the surface; under other conditions they retreat from the light, sinking to greater depths. They have been taken near the surface in the townet in both strong sunlight and at night, both with and without the aid of artificial light. At the present time they are seldom found at the surface under

any conditions. Since the young feed upon moving or suspended prey, their life can not be spent far from the surface. Their behavior at any given time is the resultant of all the conditions which affect them at that time, and therefore varies with the varying conditions of their life. The rarity of the larval lobsters at the surface in areas where the adults are known to abound may be ascribed to the following causes: (1) Wholesale destruction of the breeding animals, which has caused the present depletion of the fishery; (2) the great destruction of the young, which must take place under natural conditions; (3) the wide dispersal of the young by tides and currents which their swimming habits favor, and (4) the variable character in their reactions or movements, leading to a variable or irregular vertical distribution.

(4) The food of the larval lobster consists of minute pelagic or floating organisms, such as copepods, crustacean larvæ, algæ, and probably to some extent protozoa. The stomachs of young lobsters taken at sea have been found to contain fragments of crustaceans, diatoms, algæ, fine sand grains, and amorphous matter. They seem ready to attack and seize any small moving object, living or dead, which they are able to master. Since they follow moving objects like copepods by sight they discriminate to some extent, but their powers in this direction are slight, and would seem to be unnecessary if they early acquire the adult habit of regurgitating the indigestible residue of their food.

(5) The preying instinct, which is closely associated with that of pugnacity, is very strong in young lobsters from the time of birth. Their disposition to attack and devour one another, as seen in aquaria whenever they are too closely crowded or not supplied with the proper food, is the obvious result of an indiscriminate instinct to seize floating objects which are neither too large nor too active. Another lobster is as good a mark as a floating egg, or as a swimming copepod, which is more apt to elude them. Indeed they often give chase to crustaceans larger than themselves. The fighting instinct, if we may thus describe the tendency referred to, is closely associated with the primary instinct to seize and devour, in accordance with which the character of their activities and the structure of their bodies is distinctly correlated. It is thus evident that the organic food of the young lobster must be finely divided and floating, and that crowding in too close quarters can not be otherwise than destructive.

(6) The body of the larva is covered with a cuticle, which includes the lining of the stomach sac, and at least a part of the intestine. This is continuous with every spine, seta, or hair with which the body is protected or garnished, as well as with the internal skeleton which is produced from folds or pockets of the skin. Active growth entails the shedding of this cuticle, which is cast off in one piece, and the duration of the molting intervals or stage periods depends on the vigor and health of the individual. Each molt is a crisis in the animal's life. If the cuticle is not properly shed, the swimming hairs can not be properly evaginated, and the animal becomes helpless.

A healthy larva is always clean and transparent, while in a weakened or sickly one the hairs tend to gather sediment and parasites. Sea water of normal density in which the plankton or floating population of animal and plant life is properly balanced and an undue amount of sediment is not present, are important conditions for rearing the young, and the warmer the water, within certain limits, the more rapid the growth.

At certain points on the coast it may be possible to rear many marine animals with comparatively little difficulty, or to keep them alive in the adult state for long periods, while at other places every aquarium may become the grave of all but the hardiest species or individuals, and that in a short time. The difficulty seems to arise from the nature of the plankton, and from the tendency of certain prevalent organisms, such as diatoms, parasitic bacteria and fungi, to increase in an inordinate degree. The larvæ become weakened, and can not pass their molts.

(7) In the fourth stage the young lobster, as if in one bound, seems to justify its name, to lose its old swimming organs and acquire new ones, to lose the rolling uncertain gait of the larva and to acquire new strength with greater precision and speed. It loses in large measure its former transparency, and, together with a greater hardness and opacity of its shell, it gains a far greater brilliancy and variety of coloring. The fourth stage also marks the rise of new instincts such as fear, burrowing for concealment, not to speak of far greater pugnacity, and the dawn of intelligence or power of association, displayed in the lobsterling's holding to the same hole or retreat for hiding, to which it will return repeatedly and will defend with spirit. Perhaps more important than any of these characteristics is the fact that many of the fourth-stage lobsters probably go to the bottom and stay there. This at least is their habit when reared in confinement.

The fourth-stage lobsters seem to swim at the surface more regularly and continuously than the larvæ, and accordingly are more often taken in the net, while it is evident that the earlier stages must be thousands of times more numerous.

(8) The rate of growth is greatest during early life, and according to Hadley is 18 per cent at each molt at Wickford, R. I., up to the seventeenth stage, when it begins to slowly decrease. I found the rate to be less in the slightly colder waters at Woods Hole in the case of artificially hatched and reared young. The time interval between successive molts is indeterminate, being subject to every change which affects the physiological vigor of the animal. The advancement of the larva is to be measured by the number of its molts and not by its age. Under favorable conditions the three larval stages are passed in 10 or 12 days; the fourth stage lasts as long, so the swimming period may be over in about three weeks, or may be extended to four weeks or longer when the bottom is not sought until the fifth stage.

The approach of the molt seems to start the lobsterling on its course to the bottom; accordingly when this is delayed until after the fourth stage, it probably does not often occur until the approach of the succeeding molt. (See p. 348.)

Chapter XI.—BEHAVIOR AND RATE OF GROWTH.

BEHAVIOR OF YOUNG LOBSTERS.

Having considered the general habits of the lobster in its successive stages of development, we shall now discuss their behavior in more detail.

In the summer of 1894 I tried a number of simple experiments to test the effect of light upon the movements of the larval lobster. Twenty-five thousand young in the first stage were placed in the observation pool at the Fish Commission station, Woods Hole, Mass., in order that their behavior might be watched. The sun was intermittently obscured by clouds during the greater part of the forenoon. When set free, the larvæ soon swarmed in a large cluster near the surface, where they remained for a short time. Presently all of them went down to a distance of from 1 to 2 feet, and some of them to the bottom to a distance of 3 feet more. A lot of small cunners then appeared on the scene and snapped up the larvæ right and left. Two hours later the remnant were dispersed over the whole pool, a large number remaining close to the surface. At 1 o'clock in the afternoon the surface on the lee side still swarmed with larvæ. Occasionally one could be seen to attack and drag another down. They swam with their usual aimless activity, now rising and falling and changing their direction frequently. The majority of them had now become quite red. Later in the afternoon nearly all of the little lobsters had disappeared, having been swept out by the tide or destroyed by the cunners or other fish in the pool.

Various boxes were then constructed to admit diffuse light from above or direct light through one end, and larvæ in the first stage were found to move toward the source of the light, whatever its intensity. In similar experiments made at another time this reaction, which then seemed characteristic, was reversed, "showing possibly that under certain conditions the larvæ are negatively heliotropic." At this time the subject of animal behavior had hardly emerged as a branch of experimental biology, with its more exact analytical methods and criteria which have since been evolved.

The experimental work of Bohn (*27*) on *Homarus gammarus* and of Hadley (*131*) in particular on the American lobster have illustrated the importance of studying the behavior of such an animal throughout the entire course of its development, and at the same time have revealed the great variety and complexity of the problems involved. The following paragraphs are little more than a summary and running commentary on some of their results.

For the analysis of certain problems in behavior the lobsters are unsurpassed, since with the proper apparatus they may be hatched in unlimited numbers and maintained to any required age or stage during the summer months. The results of studies thus far made show that while the crustacean larvæ may respond promptly and in a definite manner to a certain stimulus, their behavior is complex and essentially variable, and that at any given point of time it is the result of all the influences at work.

It is evident from the preceding chapter, as Hadley has already pointed out, that the life of the lobster may be divided on the basis of behavior into three periods: (1) The three larval stages, when the animals frequently swim with head depressed, upward or downward and forward or backward, according to circumstances, by the use of their thoracic exopodites; (2) the fourth stage, when the animal is a free swimmer at the surface, the abdominal swimmerets being now functional, as in the adult; and (3) the later stages, when the swimming organs are the same, but the animal remains constantly on the bottom after its final descent in the fourth or fifth period.

REACTION TO LIGHT.

The response of the pelagic larvæ of the higher Crustacea to light, as well as the effect of light upon the growth of these animals, are questions not only of great scientific interest, but in the case of the lobster of practical importance in view of the necessity of understanding their behavior in a state of nature and of placing them as far as possible under natural conditions in the hatchery. It has been shown in general that swimming larvæ of crustaceans, in common with many other organisms, exhibit two types of response to the light stimulus, known as phototaxis or reaction to the directive influence of the rays of light and photopathy or response to changes in the intensity of light. The phototactic response is composed of two elements or components—the turning and progressive movements or, as Hadley calls them, the body and progressive orientation; the animal turns so that the long axis of its body coincides with the path of light, and it always heads away from the source; this reaction is primary, constant, and typically reflex. On the other hand, the "progressive" response which follows this stereotyped form of orientation may be positive or negative—that is, the animal may move upward or downward, backward or forward—that is, toward or away from the source of light. The photopathic response is also variable, the animal moving toward or from a more brilliantly illuminated region, according to conditions.

Thus, according to Hadley, apart from the orientation of the body there is no constant type of reaction for the larval lobster. The variable responses vary in accord with changes in the environment of the individual and changes in the individual itself or its physiological state, and are especially marked at the beginning and close of the stage periods. While the phototactic response is eminently variable, the photopathic reaction is usually positive.

In the fourth stage the conditions are somewhat reversed, since in the laboratory lobsters at this period usually give a negative phototactic reaction, while their photopathic response is at first positive and later negative. Light-avoiding reactions of whatever kind are strongly manifested in the fifth stage and may begin at the close of the fourth. So strong indeed was the tendency to shun the light that the little lobsters, as Hadley demonstrated, would even allow themselves to be stranded, with possible fatal results, rather than to approach the light, and thereby gain deeper water. It was further shown that at this time also the thigmotactic reaction, or response to contact with solid bodies, began to assert itself and thus to modify the previous sensitiveness to light, apparently leading the animal to crawl under shelter and to burrow in the sand or mud at the bottom.

Previous to the fifth stage an increased intensity of light in certain cases may reverse the response, while in others it does not. After the fifth stage no reversal of the response can be effected in this way.

We will now review some of the observations of Bohn on the movements of the larvæ of *Homarus gammarus* of Europe, reported in 1905. He believed that the newly hatched young were immediately attracted to the surface, since they are positively phototactic. At first they approached the light, while later, at the end of some days, they moved toward regions of greater obscurity. Upon the swimming movements and unstable equilibrium of these larvæ this observer remarks as follows: The back of the lobster does not remain constantly directed upward, but is alternately inclined to the right and left, sometimes as much as 90°. It can likewise tip over by turning on the long axis of its body. The displacement of the body is effected not by the position of the longitudinal axis alone, but by that of the vertical axis of the cephalothorax as well. If the carapace is elevated, the animal both advances and rises; if it is inclined to the right, the larva advances by deviating to the right, and the more considerable the rotation the more pronounced the deviation.

In their rolling gait the larvæ tend to keep the back turned upward—that is, toward the surface illuminated by the vast expanse of sky—while the head is bent downward toward the region of shadow. When this position is maintained the eyes are illuminated in a peculiar manner. At their most elevated points, opposite to the illuminated surface, there is a lighted area, while at their most anterior ends, which are directed toward the regions of obscurity, there is an area of shadow.

"All of these observed movements," says Bohn, "such as repulsion and attraction, rolling and other rotations, are made with rapidity and precision and have the character of irresistible movements, according to laws which appear very exact, but which vary with the physiological states." Bohn concludes that the larvæ are guided in their movements mainly by the stimulus of light which enters the eyes, and that the eye acts before the "otocyst" as an organ of orientation.

In regard to the question of any real distinction between the photopathic and phototactic response, or between the intensity as distinguished from the direction of light, Hadley remarks that the direction of the light is effective in determining which eye shall be stimulated most and what parts of both eyes shall be stimulated equally. In the first instance the long axis of the body is swung into line with the rays, so that both eyes are equally affected, while in the latter the body is so placed that the anterior lateral surface of the eyes receive the strongest and the posterior lateral surface the weakest illumination.

Hadley found that when blinded in one eye the larvæ rapidly rotated on its long axis in a definite direction or performed "circus" movements, moving in circles, toward or away from the position of the uninjured eye according as the animal was negatively or positively phototropic. It was also noticed that these reactions were seldom negative except in the fourth or later stages of the lobster. Each eye is thus apparently connected with a reflex mechanism which controls the movements of a definite side of the body.

If the light which strikes a larval lobster is suddenly blocked, Hadley found that a reorientation of the body was usually effected so that the animal faced the former light source.

Generally speaking the movements of the larval lobsters seemed to Hadley to support the tropism theory, and to represent simple or complex reflexes, in the latter case of serial form, and resolvable, with sufficient data, into a number of simple components.

Both Bohn and Hadley have tested the effects of "screening" upon young lobsters, or their behavior against white and dark backgrounds, brought to bear upon them from any direction, and while the results of the observers are not wholly in accord, Hadley concludes that the larvæ orient themselves to the white and black screens or backgrounds by essentially the identical reflex movements by which they respond to direct illumination and shading.

In the case of red monochromatic light on a white ground the lobster in the first stage was found by Hadley to be negatively phototropic, but on a white ground in blue light positively phototropic. In this respect, moreover, the second and third stage lobsters responded in the same way, while against black the lobsters retreat from both red and blue in all their stages.

The fourth-stage lobsters, on the other hand, were observed to rise from black backgrounds in light of any intensity or color; that is, to display positive phototropism, and the stronger the light, the more marked was the reaction. Against white also the fourth-stage lobsters rise to any light except red, from which they tend to retreat.

The older lobsters of the fourth stage did not respond so promptly in a positive manner, and when preparing to molt they showed a negative reaction; that is, they sought the bottom, a response commonly assumed in the fifth stage, whatever the character of the light or background.

The results of Hadley's experiments were in harmony with observations of the behavior of the larvæ confined in the 12-foot canvas rearing bags, where they showed "at all times a marked tendency to sink to the bottom, except perchance at night, when more active swimming is observed in all the stages. This tendency during the daytime could not be controlled in any way. At night, however, it was possible to evoke a seemingly positive phototactic reaction from any of the young larvæ in the large canvas bags. This was accomplished by means of the acetylene light so directed against a certain area of the white field of canvas that large numbers would at once group themselves thickly about the illuminated area, manifesting in the case of the third and the fourth stages, such an effort to come into the light area that they would often throw themselves out of the water, causing thereby numerous surface ripples" (*131*).

REACTION TO OTHER STIMULI.

The results of galvanic stimulation are particularly interesting, since they apparently represent a fundamental response of living matter, this particular form of energy being unknown under natural conditions. It was noticed by Hadley (*129*) that the young lobsters reacted very definitely to the galvanic current by gathering at the anode. Under the influence of the ascending current a progressive orientation to the anode took place, providing the long axis of the body came into certain relations to the current.

Hadley has also described an interesting rheotactic response in lobsters of the fourth stage, in accordance with which they head to the strong circular current which is maintained in the rearing bags or boxes at the fisheries station at Wickford, R. I. Even within a minute after molting to this stage the lobster would face about and head into the current, swimming so actively as to make some progress if the force was not too strong. "This characteristic manner of swimming, says Hadley," "was evinced in an ever-increasing number of lobsters, until the whole body of them had passed into the fourth stage, and then it was a most interesting sight to observe the young animals, with hardly an exception, heading into the current and as a great phalanx following their circular course—but, because of the force of the current, backward."

This rheotactic response is if anything stronger by night than by day. It may be modified or lost by passing from shadow to full light in the daytime or from darkness to strong light at night, the phototactic response overcoming the influence to swim against the current. Rheotaxis is due in some measure to a stimulus which, as Hadley believes, reaches the nerve centers through the eye. It is gradually lost in the fifth stage.

MOVEMENTS OF THE YOUNG LOBSTER IN A STATE OF NATURE.

We will now review the probable behavior of the young swimming lobsters in their natural state in the sea, in order to ascertain to what extent experimental work in the laboratory has enabled us to understand their complex movements. It must be admitted that comparatively little is definitely known through direct observation upon the subject.

Under natural conditions the young of the lobster, as in many of the higher Crustacea, are presumably hatched at twilight or at night at the sea bottom, their dispersal taking place in the way already described (p. 327). Possibly under some conditions they swim to the surface during the night of their birth, while as a rule they may not make the ascent until stimulated by the light of returning dawn, but remain at the higher levels for a few days only. This is confirmed by captures with the tow net by both day and night (p. 331) and by the experiments of both Bohn and Hadley, already recorded.

Then follows a period of greater fluctuation, embracing the latter part of the first and the two remaining larval stages, during which their movements are variable. Though still coming to the surface and within reach of the net, their capture in this way, at the present time at least, seldom occurs under any conditions. Presumably in shallow waters they even settle at times upon the actual bottom, but their usual beat or range of movement, especially in deeper waters, is not known. Experiment has shown that while they tend to hold the body constantly with back to the light source they may move up or down, back and forth; that is, toward or away from the source of the stimulus, as a result of a variety of contending and conflicting influences, now one winning the day, or the hour, now another. The issue may indeed vary from hour to hour, and one might almost say from moment to moment.

With the wonderful change registered at the beginning of the fourth stage, the young lobster mounts to the surface and holds more persistently to it than ever before, at times even jumping out of the water like a shrimp, though having discarded its larval swimming organs and having brought into play the permanent swimmerets under the tail. Every observer is agreed that of all the free-swimming stages the fourth is that most commonly taken at the surface of the ocean, and especially in the brightest sunshine. This surface-swimming habit has further been observed by every experimenter who has reared these young or turned them loose into the sea. At this point the experimental testimony seems to conflict with the natural behavior of the lobsterling, since during the early part of its fourth stage it has been observed to avoid the light. The explanation would seem to be that this, like most of its similar reactions, is subject to reversal, under conditions which are not as yet fully understood, but which, as Hadley suggests, may be due to an increased intensity of the light stimulus or to an impulse which leads it to seek its food at the upper levels of the water.

At the close of the fourth, or at some time probably near the end of the fifth period, the little lobster makes its complete and final descent to lower regions. (Compare p. 348.) Thereafter the bottom of the sea becomes its fixed abode, which it seldom or never leaves unless snapped up by an enemy, or in after years it is hauled to the surface in a lobster pot.

In the fifth and all later stages the light-shunning tendency becomes more and more pronounced, but it can not be said that it is never subject to change, for more than once I have seen adult lobsters exploring the bottom in shallow water on sunny days. Yet their avoidance of strong light and their impulse to hide and to burrow after the fifth stage is fairly constant. In a word, their behavior is no longer essentially variable, but is in a measure stereotyped.

VARIATION IN THE RATE OF GROWTH AND DURATION OF THE STAGE RERIODS.

The following table shows the size and age of lobsters during the first eleven stages, and is based upon data obtained at different points on the coast under different conditions of temperature upon a varying number of individuals and by different observers. New measurements of any number of individuals made under approximately similar conditions would possibly give a different result, but this difference would not be great.

I found that the fourth stage was reached at Woods Hole on the average in 14 days, while Mead has determined this period for Wickford, R. I., to be a little over 12 days, the average duration of the first three periods varying from 9 to 16 days, with an individual variation of 3 to 7 days, according to the temperature and other conditions.

Assuming that the lobster goes to the bottom to stay at the close of its fourth stage, the pelagic life of the Woods Hole lobsters would be about 30 days, while at Wickford it would last 23 days, or a little over 3 weeks. Assuming that the bottom is not definitively sought until the close of the fifth stage, the free swimming life at Woods Hole would last 46 days, or a little over 6 weeks, and at Wickford about 30 days.

TABLE 11.—AVERAGE SIZE AND DURATION OF STAGE PERIOD IN THE FIRST ELEVEN STAGES.

Stage.	Wickford, R. I. (Hadley).		Woods Hole, Mass.	
	Average length (millimeters).	Stage period (days).	Average length (millimeters).	Stage period (days).
1	8.2	2.0	7.8	1–5
2	9.6	4.0	9.2	2–5
3	11.4	5.0	11.1	2–8
4	13.5	12.0	12.6	10–19
5	15.5	9.5	14.2	11–18
6	18.6	12.7	16.1	14
7	22.5	14.3	18.6	....
8	26.5	16.0	21.03	....
9	32.1	....	24.5	....
10	37.9	....	28.23	....
11	42.9	....	....	....

Assuming, further, that under natural conditions the molts are passed more rapidly, and that the bottom is sought some time between the close of the fourth and of the fifth stages, the pelagic life will be found to cover a period of from 3 to 4 weeks.

CONDITIONS WHICH DETERMINE THE RATE OF GROWTH AND THE DURATION OF STAGES.

The length of the stage period or the period between molts from first to last depends upon (1) intrinsic and (2) extrinsic causes. Among the intrinsic causes the following must be considered: (*a*) Inherited characters or the individual constitution, which gives a certain bent or direction to activities and limits their scope, and (*b*) acquired characters, such as the loss of limbs, which is certain to retard the rate of growth of the body as a whole by diverting energy to the regeneration of the lost parts.

Thus if the fighting and preying instincts, due to inheritance, are stronger in larva A than in larva B, A will get more food, grow faster, molt sooner, and, its inherited capital being equal in all other respects, it will distance B in the race from the start and, barring mishaps, forge ahead at every step of the way. The early advantages gained by A are cumulative in their effects. The parable of the talents is applicable even to the lobsters, and the laggard in the race, though of the same age, may not attain one-half, or even one-quarter, of the strength of its more strenuous rival, and will be fortunate if it is not cut into pieces and devoured, a contingency quite likely to happen when its running mates are crowded or underfed.

Among the acquired characters are to be reckoned any weakness which may be due in the first instance to congenital defects, such as imperfect or undersized eggs, accidents like the loss of a limb, mutilations of any kind, which, as Emmel (*90*) has shown, increase the stage period and therefore diminish the rate of growth, or parasitism which may be encouraged by a lowered vitality or improper food.

Of extrinsic causes the most important are (*a*) food of the proper sorts, (*b*) changes in temperature, a powerful factor under ordinary conditions, and (*c*) changes in light, to which the lobster, whether as larva, adolescent, or adult, is very sensitive from infancy to old age.

Every stage period culminates in molting, a result and expression of growth which is subject to the causes above enumerated and therefore indeterminate. Consequently the rate of growth in lobsters is subject to wide variation. Every individual has its own rate, which may vary from that of others or from its own rate at a later period of life by 100 per cent, and which may be different at different times of the year and at different places, as well as different at corresponding times in different years at the same place. Moreover, beyond a certain stage the rate of growth varies in the sexes. Variation in the rate of growth is far from uniform in man and the higher animals, but it is not subject to such rapid changes and wide fluctuations.

Notwithstanding the drawbacks and difficulties of the problem, it is possible to determine the average rate of growth and age of maturity, provided our statistics are ample, which is not the case at present except for one or two points on the coast.

RATE OF GROWTH AND AGE AT SEXUAL MATURITY.

In 1895 I made the first systematic attempt to determine how long it takes an adult marketable lobster to grow, and remarked: "It is impossible to answer the question with certainty, since complete data for solving the problem have not been gathered. We can, however, give a tentative answer which is probably not far from the truth."

It was further pointed out that in order to ascertain the average age of a lobster 10½ inches long (weight 1¾ pounds) it would be necessary to know, first, the number of molts which the animal had passed through, and, secondly, the time interval between each molt. We showed that the number of molts could be approximately determined by certain means discussed. The time interval could only be ascertained by keeping the animals alive for a period of years and carefully recording their growth. Both of these factors, as we have already seen, are highly variable quantities. Thus, to give further examples, the length of a certain yearling lobster which was raised from the egg was only 36 millimeters, while three other lobsters measured from 35 millimeters to 51.8 millimeters when only 5 months old. Even more striking individual differences have been given by Mead (*195*) and Hadley (*126*); two of Professor Mead's lobsters each 4½ months old (June 1 to October 7) measured about 55 millimeters and 30 millimeters, respectively, the smaller being not much larger than one of the big claws of the former. Of three lobsters figured by Hadley, each having attained an age of 1 year and 4 months on October 23, 1902, the larger had reached a length of about 120 millimeters (nearly 5 inches), the smaller but 58 millimeters (about 2⅓ inches). Lobsters that live in harbors where they find abundant food undoubtedly grow much faster than those farther from shore and on poor feeding grounds. It could hardly be expected, moreover, that lobsters kept under artificial conditions would grow as rapidly as when free in the ocean.

I also gave a record of the molts of eight lobsters varying in length from 5½ to 11¼ inches, and found the average percentage of increase (ratio of increase to total length before molting) to be 12.01. Then using the records of the lengths of lobsters reared from the first to the tenth stages at the laboratory of the United States Bureau of Fisheries at Woods Hole, Mass., the percentage of increase for a total of 246 young individuals gave the percentage of increase as 15.3 for each molt. The table follows:

TABLE 12.—ACTUAL LENGTH OF LOBSTERS DURING THE FIRST TEN MOLTS.

Number of molt or stage.	Average length.	Extremes in length.		Number of lobsters examined.
	mm.	*mm.*		
1	7.84	7.50 to	8.03	15
2	9.20	8.3	10.2	47
3	11.1	10	12	79
4	12.6	11	14	64
5	14.2	13.4	15	15
6	16.1	15	17	12
7	18.6	18	19.5	4
8	21.03	19.75	22	5
9	24.5	24	25	2
10	28.03	26.6	29.5	3

It should be added that the measurements here recorded were not made with this problem definitely in view, and are therefore uneven in number, and further that the number of young considered in the last four stages are too small to give satisfactory results.

Assuming the average length of the first larva at Woods Hole to be 7.8 millimeters, a table was drawn up giving the estimated length of lobsters during the first 30 molts as follows:

TABLE 13.—ESTIMATED LENGTH OF LOBSTERS DURING THE FIRST 30 MOLTS.

Stage.	Length.	Stage.	Length.	Stage.	Length.
	mm.		*mm.*		*mm.*
1	7.84	11	32.55	21	135.17
2	9.04	12	37.54	22	155.86
3	10.42	13	43.28	23	179.70
4	12.02	14	49.90	24	207.20
5	13.86	15	57.53	25	[a] 238.90
6	15.98	16	66.34	26	[b] 275.45
7	18.42	17	76.49	27	317.59
8	21.24	18	88.19	28	366.16
9	24.49	19	101.68	29	422.21
10	28.23	20	117.24	30	[c] 486.81

[a] 9.5 inches. [b] 11 inches. [c] 19.1 inches.

We called attention to the fact, which has since been verified, that the increase is similar from period to period during the larval and early adolescent stages. According to Hadley (*126*), during the first 17 stages, when the young have reached an age of 2 years and 3 months, the increase per cent is 18.

The frequency of molting or the stage period was next considered with the following result: We concluded that during their first year, lobsters as a rule molted from 14 to 17 times, and attained a length of from 2 to 3 inches, with the probability that this limit was often extended. Examining all the data available at the time we further concluded that the 10-inch lobster was between 4½ and 5 years old, the higher degree of probability favoring the smaller number, and had molted from 25 to 26 times. "The reader is reminded," we then added, "that this is only an estimate, based, it is true, upon rather slender data, but upon the only facts which we possess. In future years some experiments will be made by which this result can be tested."

The words just quoted were written in 1894; twelve years later the problem of the rate of growth in the lobster was taken up by Hadley (*126*), who has given an excellent discussion of the question in all its bearings and has supplied many of the data which were then lacking. His work was conducted at the Wickford hatchery of the Rhode Island Commission of Inland Fisheries under conditions which the experience of many years and of many workers has brought to a high degree of perfection. His results are therefore more complete and more valuable than those of any previous students.

Hadley's final conclusions (see *126*) so far as general results are concerned do not differ greatly from those reached by me in 1895, as may be seen by the following comparisons: Thus, I estimated that a lobster in the first year of life molted from 14 to 17 times, and reached a length of from 2 to 3 inches; Hadley determines that the yearling molts 12 times and attains a length of 2⅛ inches. According to the table (here reproduced as table 13) the 10-inch lobster has molted from 24 to 25 times and was estimated to have reached the age of 4½ to 5 years; according to Hadley a male 9⅞ inches long has molted 23 times and is 5 years old, while the female of the same length is 1 year and 5 months older. Thus at this juncture the estimates are from one to two molts apart, and for the male in essential agreement as to age.

TABLE 14 (AFTER HADLEY).—AN ESTIMATE OF THE RATE OF GROWTH OF THE AMERICAN LOBSTER FROM TIME OF HATCHING TO ATTAINMENT OF A LENGTH OF 22¾ INCHES.

Stage.	Approximate age.	Length.		Increase.			Approximate time of molt.	Stage period.	Sex.
		Millimeters.	Inches.	Millimeters.	Inches.	Per cent.			
No. 1	0	8.2					June	2 days	M. F.
No. 2	3 days	9.6		1.4		18	do	4 days	M. F.
No. 3	7 days	11.4		1.8		18	do	5 days	M. F.
No. 4	12 days	13.5		2.1		18	do	12 days	M. F.
No. 5	24 days	16.0	⅝	2.5		18	July	11 days [a]	M. F.
No. 6	35 days	18.8	¾	2.8		18	do	12.5 days	M. F.
No. 7	7 weeks	22.5	⅞	3.7		18	August	14 days	M. F.
No. 8	9 weeks	26.5	1 1/16	4.0		18	do	15.5 days	M. F.
No. 9	3 months	32.0	1¼	5.5		[b] 18	September	21 days	M. F.
No. 10	5 months	37.9	1½	5.9		18	October or November	25 days	M. F.
No. 11	9 months	45.0	1¾	7.1		18	April	5 months	M. F.
No. 12	1 year	53.0	2⅛	8.0		18	June	1½ months	M. F.
No. 13	1 year 1 month	62.0	2½	9.0		18	July	33 days [c]	M. F.
No. 14	1 year 3 months	73.0	2⅞	11.0		18	August or September	51 days	M. F.
No. 15	1 year 6 months	86.0	3⅜	13.0		18	October or November		M. F.
No. 16	2 years	102.0	4 1/16	16.0		18	April or May		M. F.

[a] The fifth stage period is generally shorter than the fourth.

[b] For female lobsters bearing eggs, there can naturally be no molt during the period that the external eggs are carried; this is at least for 11 or 12 months.

[c] The midsummer stage period is usually the shortest.

TABLE 14 (AFTER HADLEY).—AN ESTIMATE OF THE RATE OF GROWTH OF THE AMERICAN LOBSTER FROM TIME OF HATCHING TO ATTAINMENT OF A LENGTH OF 22¾ INCHES—Continued.

Stage.	Approximate age.	Length.		Increase.			Approximate time of molt.	Stage period.	Sex.
		Millimeters.	Inches.	Millimeters.	Inches.	Per cent.			
No. 17	2 years 3 months.	121.0	4¾	19.0	…	18	August	…	M. F.
No. 18	2 years 6 months.	141.0	5⅝	20.0	…	16	November	…	M. F.
No. 19	3 years	162.0	6⅜	22.0	…	15	May	…	M. F.
No. 20	3 years 6 months.	180.0	7¼	18.0	…	11	Autumn [a]	…	M. F.
No. 21	4 years	200.0	8	20.0	…	11	Late spring	…	M. F.
No. 22	4 years 6 months.	222.0	8⅞	22.0	…	11	Autumn	…	M.
	4 years 6 months.	222.0	8⅞	22.0	…	11	Late summer or autumn	…	F.
No. 23	5 years	247.0	9⅞	25.0	…	11	Summer	…	M.
	6 years 5 months.[b]	247.0	9⅞	25.0	…	11	Late summer or autumn	…	F.
No. 24	6 years	275.0	11	28.0	…	11	Summer	…	M.
	8 years 4 months.	275.0	11	28.0	…	11	Autumn	…	F.
No. 25	7 years	300.0	12	25.0	…	9	Summer	…	M.
	10 years 4 months.	300.0	12	25.0	…	9	Autumn	…	F.
No. 26	8 years	327.0	13 1/12	27.0	…	9	Summer	…	M.
	12 years 4 months.	327.0	13 1/12	27.0	…	9	Autumn	…	F.
No. 27	9 years	356.0	14¼	29.0	…	9	Summer	…	M.
	14 years 4 months.	356.0	14¼	29.0	…	9	Autumn	…	F.
No. 28	10 years	380.0	15¼	24.0	…	7	Summer	…	M.
	16 years 4 months.	380.0	15¼	24.0	…	7	Autumn	…	F.
No. 29	12 years	406.0	16¼	26.0	…	7	Summer	…	M.
	18 years 4 months.	406.0	16¼	26.0	…	7	Autumn	…	F.
No. 30	14 years	431.0	17¼	25.0	…	6	Summer	…	M.
	20 years 4 months.	431.0	17¼	25.0	…	6	Autumn	…	F.
No. 31	17 years	457.0	18¼	26.0	…	6	Summer	…	M.
No. 32	20 years	480.0	19¼	23.0	…	5	do	…	M.
No. 33	23 years	505.0	20¼	25.0	…	5	do	…	M.
No. 34	26 years	525.0	21	20.0	…	4	do	…	M.
No. 35	29 years	546.0	21⅞	21.0	…	4	do	…	M.
No. 36	33 years	568.0	22¾	21.0	…	4	do	…	M.

[a] After the eighteenth stage it is very doubtful whether the lobster molts oftener than twice in a year.
[b] It is uncertain at just what time the spring or early summer molt for female lobsters not bearing external eggs is first omitted, but it is probably near this stage.

It is shown, however, by Hadley, that the rate of growth is more rapid in the young Wickford lobsters (stages 1 to 17), that it begins to fall at the age of about 2½ years (stage 18), becomes differentiated in the sexes in favor of the more rapid growth of the male at the twenty-third stage, and continues to decrease, the stage period becoming longer and longer with age, especially in the female, where the production of eggs proceeds at a very rapid rate. Thus, according to Hadley, the increase in the 12-inch lobster has dropped to 9 per cent, or about one-half that in the first 17 stages, and while both sexes have molted 25 times, the male is but 7 years old, while the female is 10 years and 4 months. Thus he thinks that the female is outstripped in the race with the other sex on account of the drain upon her vitality due to the periodic production of a rapidly increasing egg supply, and that this accounts for the fact that so far as observed giant lobsters beyond 18 or 20 inches in length are invariably of the male sex.

Following Hadley's estimate still further, for the larger lobsters, upon the age or rate of growth of which no data are yet available, a male lobster 19½ inches long is 20 years old and has passed successfully 32 molts, while a mammoth measuring 22¾ inches from beak to telson has entered upon his thirty-sixth stage, and attained to the green old age, for a lobster, of 33 years. According to my earlier estimate a lobster at the thirtieth molt had attained a length of 19.1 inches.

That the stage periods increase with age no one can deny, for this is only another way of saying that youth is the period of most active growth. There is no theoretical limit to the growth of such a crustacean, although there is a practical limit. Thus lobsters do not attain a weight of 100 pounds, but they have tipped the scales at 34 pounds. Again, there is no *a priori* reason for assuming that the percentage increase in weight in the adult lobster at each molt may not be fairly uniform up to the period of decline. But since molting is not only the prelude to expansion in size, but also of the greatest use to the animal in freeing it from troublesome parasites and messmates and at the same time keeping its cuticular glandular system in order, as well as in the repair of injuries through the restoration of appendages and other lost parts, we should surely expect to find so useful and necessary a process limited only by the duration of life itself. This is apparently the case, and since the tendency, in all the higher organisms, at least, is to lose vitality with age we might expect the percentage of increase in weight or in the expansion of the body to decrease gradually in old age until it was practically nil, or reduced to the ability of renewing the shell or exoskeleton only. This would seem to be actually the case, although we have no direct observations upon which to found the opinion, and it is possible that death from old age in the lobster, if it come at all, would follow from final failure to cast the heavy armor, rusty with age, and scarred in many a conflict.

As has already been noticed in considering the rate of growth of the ovary (p. 299) the volume of any part or of the body as a whole does not increase proportionately with the length but more nearly with the cube of the length. In other words the percentage increase in the length of the body at each molt does not accurately express the true rate of growth, which concerns the entire volume of the body. Therefore it may be found that after a period is reached corresponding to the length of from 8 to 10 inches, the lobster, and more particularly the male, may increase more rapidly in volume and become stockier, especially to be noticed in the enlargement of the big claws, while increase in total length of the body may be relatively less.

I have shown that the male, length for length, weighs more than the female, and that a female with external eggs is lighter than one of the same length without eggs (*149*, p. 118–120, table 31); it is therefore only natural to expect to find the female handicapped by the male after reaching sexual age (7¼ to 12 inches).

We will now briefly consider the rate of growth of Woods Hole lobsters, average increase per cent 15.3, and that of Wickford lobsters with average of 18 per cent for the first seventeen stages, or 18.4 per cent as given in another place. Hadley in attempting to account for this discrepancy concludes that the former figure is too low and that it does not represent the growth of young lobsters under natural conditions at Woods Hole.

I think it highly probable now, as I did in 1892–1894, that lobsters grow more rapidly in nature than when confined in glass jars in a hatchery, but that the measurements of the early stages of the lobster which were then made were correct for the place and time there can be no doubt. They were taken upon a standardized scale, and made with care under a hand lens or dissecting microscope.

The lobster in the first stage, according to our table, was found in fifteen measured individuals taken from the hatching jars to have an average length of 7.84 millimeters (extremes 7.50 to 8.03 mm.), against an average length of 8.2 millimeters as given by Hadley for Wickford, R. I. The eggs from which these young were hatched at Woods Hole were stripped from old lobsters, taken in June to July, and placed in the McDonald type of jar then in use. The mean average temperature of the sea water at the U. S. Fish Commission wharf for a period of five years from 1889 to 1893 was for June 62.1° F., and for July 69.1° F. The water in the hatching jars was found to average one degree higher than that outside. Since I could not begin operations until the latter part of June, the eggs with which I had to deal directly or indirectly had reached a late stage of development under natural conditions, and were near the hatching point when taken. Accordingly these eggs were probably not undersized and the larvæ may be regarded as normal for Woods Hole for the period in question.

What is the average length of first-stage lobsters hatched in the waters of Vineyard Sound? Although during six consecutive seasons (1889–1894) I never succeeded in taking, with the net at the surface of the sea, under natural conditions, a single larva of the first stage, and but one of the third stage, this question can be partially answered by the earlier observations of Smith (*256*) made in 1871, who says that "the lobsters in the first stage were first taken July 1, when they were seen swimming rapidly about at the surface of the water among great numbers of zoeæ, megalops, and copepods." * * * "They were frequently taken at the surface in different parts of Vineyard Sound from the 1st to the 7th of July, and several were taken off Newport, R. I., as late as July 15, and they would very likely be found also in June, judging from the stage of development to which the embryos had advanced early in May in Long Island Sound. These young lobsters with two exceptions were taken at the surface in the daytime (forenoon) from July 1 onward, but not so commonly as young in the fourth stage." Smith gives the measurement of the first stage as 7.8 to 8 millimeters. It therefore seems probable that the average length of Woods Hole lobsters in the first stage is under 8 millimeters, and not above this measure as found by Hadley for the same stage at Wickford, but probably above 7.84, the average found for the artificially hatched young. If this be the case, it is quite certain that the rate of growth up to at least the tenth stage is slower than at Wickford, as is further indicated by the longer stage periods.

Hadley concluded that a 11-inch male lobster from Wickford was 6 years old, while a female of the same length was 8 years of age, whereas upon the Massachusetts coast this length is not attained in less than 7 and 9 years, respectively. Accordingly a 10-inch Wickford male would be about 6 years old, and a female of the same length somewhat over 7 years. I am inclined to doubt whether the difference is really as great as is here implied.

While we can not make direct comparisons with confidence without knowing the number of individuals in each case concerned, figures which neither Smith nor Hadley give, I am inclined to believe that while the rate of growth for Woods Hole lobsters during their earlier stages may be greater than 15.3 it is less than 18 per cent, and that while my former estimate of the age of a 10-inch marketable lobster to be from 4½ to 5 years may need the addition of a plus mark, especially in the female, it is probably not far from the truth.

Female lobsters are found bearing eggs for the first time when measuring from 7¼ to 12 inches (18.5 to 30.5 cm.). Amid limits so wide it is impossible to say at what time the average female lobster reaches the reproductive age, but it is probably not far from the 10-inch length, which according to Hadley would represent the twenty-third molt and an age of about 6½ years. We have no data upon the time of sexual maturity in the male, but should expect that it would be reached at the same or at a slightly earlier period.

Regarding the questions of rate of growth in *Homarus gammarus* of Europe, I shall give the general conclusions of Ehrenbaum (*87*), whose studies at the Helgoland laboratory are well known:

It is possibly not superfluous at the end of these observations to state again clearly that the results which the American naturalists and we in reliance upon them have reached in regard to growth and the relations between size, age, and life-stage cannot be regarded as completely reliable.

The numerical results which are given in the works referred to and which have been partly reproduced, can in the most favourable cases be regarded as of only average value, especially when we reflect that all biological relations possess a certain variability and cannot be expressed in absolute figures.

If, moreover, we reach the result that the Helgoland lobster lays her eggs for the first time in her seventh year of life, it by no means contradicts the idea that in many individuals this may happen in the sixth year, while occasionally females of only 23 centimeters (9⅛ in.) in size have been observed with extruded eggs, and moreover it may happen that in single cases the first egg-laying is delayed until the eighth year of life.

But even disregarding this natural and anticipated variability, it cannot be denied that our figures, even as averages, possess a certain untrustworthiness, since only one element rests upon direct observation, while another is based upon combinations. This uncertainty is sufficiently reflected in my earlier contributions (see communication of 1903, p. 154), wherein I came to the conclusion that female lobsters were in their sixth year of age when for the first time they carry eggs, while now, standing upon a basis not much more extended, I have accepted the seventh year in preference.

Moreover the American authors waver between the sixth and seventh year as regards the period in question, and find a way out on the supposition that the period is six years for the southerly state of Rhode Island, and seven years for more northerly Massachusetts and Maine. Accordingly it is well to lay it down as a general rule that the first egg-laying takes place in the sixth or seventh year of life, with the higher probability favoring the longer period. This statement would then hold good for both American and European lobsters throughout their areas of distribution. Moreover, it can be accepted as fixed that this egg-laying takes place in from the twenty-third to the twenty-fourth stage of life.

Chapter XII.—THE PRESERVATION AND PROPAGATION OF THE LOBSTER.

The lobster is easily the king of the crustacean class, and though neither "fish, flesh, fowl, nor good red herring," he is excellent eating, and that his race may increase is a wish generally felt and often expressed. Unfortunately, for many years past we have watched this race decline until some have even thought that commercial extinction, and that not far remote, awaited the entire fishery. What is the matter with the lobster?

If this is primarily a scientific question, the zoological history of the animal should give us the answer. The lobster has attracted many naturalists and other observers, both in this country and in Europe, especially during the past 15 years, until it has become the focus of a wide literature, as a glance at the bibliography at the close of this work will show. Indeed, few marine animals are now so well known. The main biological facts concerning this classical type are well in hand, and excuse can no longer be offered on the ground of ignorance.

If the question is only an illustration of "many men, many minds," we may as well give it up and let the process of extermination take its usual course. However, we consider that this problem is primarily a scientific and not a social one. When the causes of the evil are definitely known, it becomes necessary to evoke the law. If ideal legislation can not be secured, we must then strive for the best within reach. It is obviously useless or even worse to enact laws which can not be enforced, and statutes which are a dead letter and have no moral effect had better be expunged.

We have already given a brief history of this valuable fishery (p. 170), and shall now consider in a little more detail the evidences of its decline and what we consider the most effective remedies for its restoration.

THE FACT AND CAUSE OF DECLINE.

It is no exaggeration to say that in practically every known natural region of the North Atlantic coast the lobster fishery is either depleted or in a state of decline. The evidences of this condition are to be found in steadily increasing prices and in the statistics of the fisheries.

The market price, or cost to the consumer, has steadily advanced in direct ratio to the steady decrease in the market supply. Thus, in 1889 the annual catch of lobsters in the United States was somewhat over 30,000,000 pounds, valued at over $800,000; in the course of a decade, or in 1899, the annual crop was reduced by one-half, while its value had more than doubled. Since 1899 the failing supply has not been sensibly checked. Statistics of the fisheries of the two New England States—Maine and Massachusetts—which are most interested in the lobster question, have the same story to tell. In Maine, which in some years has produced two-thirds of the entire output of this fishery, the catch amounted to 14,234,182 pounds, with a market value of $268,739. Twenty years later the product had fallen to 12,346,450 pounds, a decline of over 2,000,000 pounds, while its value ($1,062,206) had advanced fourfold. The product of the fishery for 1880 in Massachusetts was 4,315,416 pounds, which sold for $158,229, while the catch of 1900, though only half as great, was worth more than that of 10 years before.

The average price per pound in the shell in Canada was 9.12 cents in 1883, 14.10 cents in 1893, while in 1898 it had risen to 18.72 cents (*187*). Large lobsters which 25 or 30 years ago could often be bought at 5 cents apiece are now sold in the shell at 20 to 30 cents a pound,[a] which at the latter figure represents a cost of about 55 cents a pound for all the edible parts, and over a dollar a pound for the clear meat of the tail and claws alone. (See table 3, p. 214). Thus, from being one of the cheapest food products of the ocean, this delicious crustacean has become one of the dearest luxuries. Once the regular summer visitor to the country villages throughout the New England States, it has now practically disappeared from the markets of all but the larger centers, and is there to be had only at many times the former cost. The fame of the live broiled lobster has spread over the Eastern and Western States, but, regardless of size or quality, the consumer must pay from 60 cents to a dollar or more for a single lobster.[b]

The former abundance of these animals on the Atlantic coast of Canada and New England was incredible, and probably for many years in succession more than 100,000,000 have been marketed, representing a cost to consumers at present prices of upward of $40,000,000. The shores on certain sections of the coast have been often described as strewn with lobsters in "windrows" after a storm. (See p. 218.) The animals were so common it is not surprising that their value was not appreciated.

A fisherman at Southwest Harbor, Maine, who had trapped lobsters for half a century, gave me the following account of his experience: About the year 1875, when the annual shrinkage in the wild crop had already been felt in many places, he took at one haul from 100 traps, which had been down 2 days, 1,985 pounds of lobsters. All but 15 of his pots contained lobsters, and from one, which was filled to the spindle, 35 animals were taken. As a contrast to past conditions, few of marketable size were at this time to be caught (July 27, 1902). The day before our interview this fisherman's son pulled 60 traps, set off Bunkers ledge, between that point and the Duck Islands, once a famous fishing region for this crustacean, and took only 9 lobsters of marketable size. Illustrations of this kind could be extended indefinitely, but the fact of decline is the one subject upon which all are agreed. It is the burden of nearly every report on the fishery which has been issued for a score of years.

The causes of the decline of the fishery are plainly evident. More lobsters have been taken from the sea than nature has been able to replace by the slow process of reproduc-

[a] Thirty cents a pound at Cleveland, Ohio, April, 29, 1907. Wholesale prices at T Wharf, Boston: Large live lobsters, 24 cents per pound; boiled, at 28 cents; chicken, live, at 18 and 20 cents; boiled, 20 and 22 cents.—(The Boston Globe, August 4, 1910.) Retail prices at the same time, 30 cents per pound; earlier in season, 25 cents.

[b] Lobsters are not cheap in the restaurants of London, where boiled lobsters are sold for 8 pence to 2 shillings or more each, according to size. One and six is a common price for the half of a boiled lobster. (1903.) (Compare p. 173.)

tion and growth. In other words, man has been continually gathering in the wild crop, but has bestowed no effective care upon the seed. The demands of a continent steadily increasing in wealth and in population have stimulated the efforts of the dealers and fishermen, who must work harder each year for what they receive in order to keep up the waning supply. The natural result has followed, namely, a scarcity of numbers and a decrease in the size of the animals caught, with steadily advancing prices paid for the product. This is precisely what we should have been led to expect, had we based our judgment upon any sound principles of common sense and human economy, not to speak of a knowledge of the mode of life and general natural history of the animal in question.

THE PROBLEM.

The problem before us is how to aid nature in restoring and maintaining an equilibrium of numbers in the species, or how to increase the number of adult animals raised from the eggs. It concerns not only the fisherman who earns a livelihood through the fishery, or the dealer who has capital at stake, but the public of many lands; in fact, everyone in the Western Hemisphere at least who likes the lobster for food. When the decline of the already depleted fisheries became a serious menace protection was sought in legislation, but since the lobster supply of this country is drawn from several States and from Canada and the maritime provinces as well, no uniformity of laws or methods was to be expected. Each state enacted its own laws, which were often widely at variance, unscientific, and subject to continuous change. Up to the present time every effort to check the constant and ever-increasing drain upon this fishery has signally failed, which shows that either the laws are defective or that the means of enforcing them are insufficient.

A sound and essentially uniform code of laws for the entire fishery is plainly demanded if legal restrictions are to be of much avail.

HOW THE PROBLEM HAS BEEN MET.

What means have been adopted in this country and in other parts to check the decline of this fishery so general and so universally acknowledged? The more important restrictive measures enacted at sundry times and in divers places have been as follows:

(1) Closed seasons of various periods in different localities.

(2) A legal gauge or length limit—namely, 9 inches in New York, Rhode Island, and Connecticut; 10½ inches in Maine, New Hampshire, and also in Massachusetts, until reduced to 9 inches in 1907; 8 inches in Norway and England; and 8, 9, and 10½ inches in different districts of Canada; in all cases penalizing the capture and sale of all lobsters under these limits, and legalizing the destruction of all adults above the gauge.

(3) "Egg-lobster" laws,[a] or the prohibition of the destruction of female lobsters carrying their external eggs. In addition to such legislative enactments, efforts of a constructive character have been made as follows:

[a] The phrases "egg lobster," "berried lobster," or "lobster in berry," or "lobster with external eggs," are all synonymous, and always mean a female with her cargo of eggs, new or old, attached to the swimming feet under the tail.

(4) To increase the supply of lobsters in the sea by fry or larvæ artificially hatched and immediately liberated, and as practiced chiefly in Canada, by holding the berried lobsters in large inclosures, called lobster pounds, ponds, preserves, or parks, and later setting them free when the young are ready to hatch.

(5) By the rearing method later introduced of holding the fry artificially hatched and rearing them until the fourth or fifth stages, when they go to the bottom and are able to take care of themselves. We need not enter here into other legislative channels, such as laws prohibiting the sale of broken or picked-out lobster meat, the operation of canneries, and the construction of gear, however necessary they may be for this fishery. We shall devote our attention mainly to those questions of most vital concern to the fishery as a whole.

CLOSED SEASONS.

A closed season for any animal, during which it is made illegal to hunt or fish for it, can only be completely justified and placed upon a scientific basis when it is made to correspond to the breeding season of the species as a whole, and when this season is limited to a relatively small part of the year. Neither of these things is possible in the lobster, since the question is complicated by the fact that this animal spawns but once in two years, so that not more than one-half of the adult females reproduce annually, and the eggs when laid are carried about by the lobsters through nearly an entire year. Closed seasons of this character are therefore not to be recommended, since they serve merely to restrict the total amount of fishing done in the year, and do not touch the root of the difficulty.

There is a closed season in the maritime provinces from June 30 to January 14, and in 1889 the Norwegian fisheries laws prohibited the taking and sale of lobsters from July to November. The apparent aim in these cases is to protect the lobsters during the spawning season and for a longer or shorter period after it, but the females only can receive much benefit, and then only provided the law against the destruction of their eggs is observed. Closed seasons set a limit to the period of destruction and may help to preserve the females by taking them into the protected class, after they have emitted their eggs.

As we have already shown, the lobster is a very sedentary animal, so far as any extended coastwise migration is concerned, and many which escape the traps in the fall will undoubtedly enter them again in the spring and upon the very same grounds.

PROTECTION OF BERRIED LOBSTERS.

A certain percentage of lobsters captured at all times of the year bear spawn, and how best to save these animals and their eggs is a serious question. The Maine laws impose a fine of $10 for every berried lobster destroyed or offered for sale. It is an easy matter to brush or comb off the eggs, however, and thus evade the law, which it is impossible to enforce completely; but however difficult of enforcement it is not wise to invite the destruction of the seed, upon which we depend for every future crop.

To save the precious spawn thus inevitably lost two plans have been tried or suggested: (1) Collecting the egg lobsters from the canneries and fishermen and subsequently hatching and liberating the fry, and (2), placing the berried females thus obtained in suitable inclosures and allowing the young to hatch under more natural conditions. The former plan has been adopted and carried out on a rather large scale in Canada and less extensively in the United States. As a means of saving the eggs which might be otherwise totally lost, both methods are to be commended, but for the preservation of the fishery neither is adequate.

By use of the second method more eggs would doubtless be hatched and more vigorous larvæ produced, while, on the other hand, an unnatural concentration of the young at a few points near shore would lead to a greater destruction. The hatching and immediate liberation of the young, which is far less commendable, will be later discussed.

The most important things to consider first are (2) the legal length limit, and (4) the hatching and immediate liberation of the young, because they are fundamentally related, have been long on trial, and have entailed great expense. That they have had a fair trial and that they have signally failed all must admit.

THE GAUGE LAW.

No doubt there are many who are ready to affirm that the present laws would be good enough if enforced. Most people are aware that the gauge law has not been rigidly carried out, and that the illegal sale of short lobsters has become a trade of big proportions. I know very well that at many times of the year it is possible to buy short lobsters (said to come from Baltimore) in the markets of Cleveland and of other towns in the great Middle West. Nevertheless I can not share this idea. Both of these measures were bound to fail, and would have failed whether the short lobsters were destroyed or not.

To come back to our question, What is the matter with the lobster, or with our means of fostering it? We have committed a series of grave errors in dealing with this fishery, to the chief of which, the gauge law, the others have been contributory.

First, by legalizing the capture of the large adult animals, above 10½ inches in length, we have destroyed the chief egg-producers, upon which the race in this animal, as in every other, must depend. Second, as supporting or contributory causes, some of us now, like others in the past, have entertained false ideas upon the biology of this animal, especially (*a*) upon the value of the eggs or their rate of survival, that is, the ratio between the eggs and the adults which come from them, and (*b*) of the true significance to the fisheries of the breeding habits, especially in regard to the time and frequency of spawning and the fosterage or carriage of the eggs. Our practices have been neither logical nor consistent, for, while we have overestimated the amount of gold in the egg, we have killed the "goose" which lays it. We have thought the eggs so valuable that we have been to great trouble and expense in collecting and afterwards hatching them and committing the young to the mercy of the sea, while we have legalized the destruction of the great source of the eggs themselves—the large producing adults.

This fundamental error of destroying the adult lobster was first clearly pointed out in 1902 by Dr. George W. Field, chairman of the Commissioners on Fisheries and Game in Massachusetts, who in various reports since has ably advocated a sounder policy, based both on science and common sense, as will appear later in this chapter.

At first sight this question seems to be about as broad as long and suggests the problem of how to eat your bread and butter and save enough for another meal when the demands of hunger are strong. While we are dependent on the adult lobsters to yield a continuous supply of eggs, and let us say we will reserve them for that purpose, we also depend upon a continuous supply of the young to yield the adults; moreover, the young at 6 inches long are many thousand fold more useful to the fishery than the eggs.

In dealing with such questions comparisons are often made with the flocks or herds of domesticated animals, and are almost certain to be misleading. The shepherd knows his flock and its resources; every member of it is numbered and under his control, and he is able to select the young or the old for slaughter, as his interest or that of his flock may demand. Among wild animals the conditions are entirely changed, and especially in those that are aquatic like the lobster, which lives at the bottom of the sea and is seldom seen, except when caught and brought up in a trap. We can select or reject among the captured only and have no definite knowledge of the proportion of young to the adults, of the various sizes, or of their distribution at any given time. If the wild flock could be brought under our knowledge and control, the comparison sought would be of real value.

We might form a comparison, however, which would be parallel in every respect by assuming that the animals of a domestic herd became more valuable for breeding purposes with each added year of life. If instead of producing 1 young at each reproductive period, they were to give birth to 2 in the second year, 4 in the third, 8 in the fourth, and so on for a considerable time, would the ranchman sacrifice his old or his young breeders for the market?

In dealing with the problem we are reminded of the proceedings of a fisheries committee in Great Britain, quoted by Mr. Allen (*2*), and the answers of a stubborn witness on the proper legal size limit of crabs: "If they do not breed till they are much larger than 4¼ inches, do you not by killing all the crabs that are under the breeding size, stop the supply of crabs from those fish?" This fisherman thought not. "Then," said his questioner, "how is the supply to be kept up if you kill the crab before sufficient time is allowed for it to spawn once?" The witness was obdurate, and answered that they did not kill them all. "Then," said another member, "suppose all girls are killed when they are twelve years of age; there would be no young women or children. I think you understand that, and if young crabs under the age at which they can spawn be killed, it follows that there can be no crabs from them." "But crabs," replied the fisherman, "breed a deal different from what girls do; crabs when they spawn, spawn many thousand at a time."

While it is essential to recognize that the older the female lobster the more useful as an egg producer she becomes, we must also remember that nature kills far more of the young than of the adults. If man's almost unlimited power of destruction is allowed to supplement the destructive forces of nature, will the depleted stream of young be adequate to maintain a steady current of adults? We think that it would, since under Dr. Field's plan the number of breeding animals should tend to increase year by year.

Our lobster-fishery laws, which date in the main from 1873, are in principle like those which prevail elsewhere, and taken as a whole they illustrate the force of example and tradition, which were established long before the biology of this animal was even approximately understood. The past literature of this crustacean bristles everywhere with these false notions, which are more or less directly and mainly responsible for the enactment and maintenance of the present laws and practices of this fishery.

The legal length limits of 9 and 10½ inches, which sanction the destruction of the big egg-producers, but for these supporting causes would probably never have been retained, for these causes have led to a diversion of energy in various directions, such as the enactment of closed seasons and the practice of hatching and immediate liberation of the fry.

The reasoning which has led to the establishment of the gauge limit has been somewhat as follows: Lobsters come to breeding age when 9, 10, or 10½ inches long, and when they spawn they spawn many thousands at a time, which is true. Therefore, by placing the legal gauge at 9 or 10½ inches we allow this animal to breed at least once before it is sacrificed, which is also true in the main. Ten-inch lobsters lay on an average 10,000 eggs; the lobster, being a good mother to her unhatched progeny, and the best incubator known, will bring most of these eggs to term, and will emit to the sea her young by the tens of thousands. What more is needed to maintain this fishery? The answer is, Vastly more. This race needs eggs not by the tens of thousands merely, but by the tens of billions, and it must have them or perish. Moreover, it can get them only or mainly through the big producers, the destruction of which the present gauge laws have legalized. If the lobster is a good "incubator," the sea is a very poor nursery. We have put a false value upon the egg.

Before proceeding farther in this analysis, we shall review some of the most pertinent facts in the biology of the lobster, most of which have been fully discussed in earlier chapters. These facts concern chiefly (*a*) the period of maturity of adult lobsters; (*b*) the number of eggs borne by the females, or the size of the broods; (*c*) the frequency of spawning; (*d*) the treatment which these eggs receive, or the habits of spawning lobsters; (*e*) the habits of the fry or larvæ; and (*f*) possibly more important than all else, the death rate or the law of survival in the young.

(*a*) Lobsters do not mature at a uniform age or size, but females produce their first broods when from 7 to 11 inches long, approximately, the difference between these limits representing a period of from 4 to 5 years (age of female lobsters at these limits about 3 and 8 years, according to Hadley). Very rarely are eggs laid before the 8-inch stage is reached, and the majority are mature at 10 or 10½ inches, when some have reared more than one brood. Accordingly, by merely reducing the 10½-inch gauge to 9 or 8 inches we rob the animal of the very meager protection which it now enjoys.

(*b*) The number of eggs produced increases with surprising rapidity in proportion to the cube of the length or the total volume of the body, from the very beginning of sexual maturity. The approximate number of eggs at 8 inches is 5,000; at 10 inches, 10,000; at 12 inches, 20,000; at 14 inches, 40,000; at 16 inches, nearly 60,000; and at 18 inches, nearly 80,000. In the case of 532 10½-inch berried lobsters taken from the waters of Massachusetts, the smallest, average, and largest number of eggs borne were 5,000, 13,000, and 36,000. The smallest number probably represents a first brood, so that the average berried lobster at this size is probably carrying eggs for the second time. The maximum of production is reached at the 15 to 16 inch stage, when some individuals produce nearly 100,000 eggs at one time.

The average 10½-inch berried lobster is from 5 to 7 years old; and assuming that it has borne eggs once before, it has lived to produce 23,000 eggs. On the other hand, an egg-bearer 16 inches in length, which according to Hadley's estimate is nearly 18 years old, has had a succession of eight broods and has produced 210,000 eggs. The larger animal is thus worth nine times as much as the smaller; in other words, in the course of twelve years its value to the fishery has been increased 800 per cent.

Again, it should be noted that it is the class of small adults up to, but not including the 9 or 10½-inch animals, those which produce by the fives or tens of thousands, upon which we have relied to maintain the race, while it is the class of big lobsters, which produce the fifty and the hundred thousands, that has been nearly wiped out.

(*c*) There is a definite spawning period for the majority of adults, ranging on the coast of Massachusetts from July 15 to August 15, and averaging two weeks later in northern Maine. A relatively small per cent lay their eggs in fall and winter.

(*d*) It is a fact, though frequently denied, that the American lobster lays its eggs, as already stated, but once in two years (though rare exceptions to this rule may be looked for), and not annually, as was formerly supposed.

(*e*) The eggs are carried attached to the underside of the tail, and admirably guarded by parental instinct for nearly a year, or until they are hatched 10 or 11 months after deposition.

Ignorance of the fact that there is a definite spawning period, that the eggs are laid but once in 2 years, and that they are subsequently carried from 10 to 11 months, to hatch in June or July following the summer when laid, is responsible, in considerable measure, for erroneous ideas regarding the efficacy of closed seasons, laws protecting the berried lobster, and other matters of legislation, the effects of which have not yet worn away.

(*f*) The fry or young, when hatched, rise to the surface or toward it, and lead a free-swimming life for 3 weeks, hardly larger than a mosquito and infinitely more harmless, translucent, brilliant in reds and blues, and quite helpless in the presence of all but the minute animals upon which they prey. They perish quickly by the thousands before the storm and the countless fish and other enemies which they meet in their varied movements, and which do not disdain small fry.

At the third molt, or the fourth, counting that passed at the time of hatching, with what seems like a sudden leap and bound, they are transformed into the fourth or the lobsterling stage, which really looks like a little lobster. Either in this stage or in the fifth, which follows, they go to the bottom, hide under stones, burrow in the sand, and show an ability to protect themselves. The most critical period of infancy being now past, one lobster at this stage is worth many thousands in the first. Therefore, our efforts, to be of real avail, should not end with the hatching and immediate liberation of the fry; we should rear them to the bottom-seeking stage.

THE LIFE RATE OR LAW OF SURVIVAL.

What is the death rate or the rate of survival in the lobster? Upon the answer to this question hinges the gauge or legal-length law, as well as the expensive practice of hatching and turning loose the young, which has been pursued in this country and Canada for many years (since 1886 in the United States and since 1891 in Canada).

As was pointed out 10 years ago, too many fish culturists have been content to turn out so many thousands or millions of eggs of lobsters and fish, and confidently expect results, to the neglect of the most important question of the whole matter—the rate of survival in the young set free, or the number of adults which can be raised from them—the very end for which all the time, trouble, and money have been expended.

In the popular mind an egg is an egg, like that of the fowl which we eat for breakfast. An egg really represents opportunity or chance to survive, and its biological value to the race depends upon the law or rate of survival, which was definitely fixed in nature before the advent of man with his traps and hatching jars, and differs in every species of animal and plant known. When the gantlet of life is long and hazardous, especially in infancy, nature, as in the present case, multiplies the chances or multiplies the eggs. Many eggs always means death, under natural conditions, to all but a remnant of the host. The number of eggs alone serves as a rough gauge to determine the rate of survival.

At one end of the scale stand the birds and mammals, with few eggs and the highest life rate known, secured by guarding and parental instincts, with big yolks and rapid development in one case and the special conditions of fetal life in the other. At the other extreme we find a parasite like the tapeworm, where the conditions of early life are so unpromising—since it must run a long hazard of chances and be eaten by two distinct vertebrates—that its eggs are required by the hundreds of millions or even billions. The lobster needs more eggs than the trout, and of smaller size, but far less than the edible blue crab, which carries nearly five millions of eggs attached to its body. Each one of these is barely visible to the unaided eye and the young which issues from it must pass a long and dangerous larval period before reaching maturity.

What, then, is the life rate or rate of survival in the lobster? Probably not more than 2 in 30,000, and certainly not more than 2 in 10,000. This number would be exactly known, provided we knew the exact proportion of the sexes or the proportion of the total number of males to the total number of females and the average number of eggs laid by mature females during their entire life. The life rate accordingly would be expressed by the proportion $2 : x$, in which x represents the average number of eggs laid by mature females during the whole of life.

Since the sexes are about equal numerically, to maintain the species at an equilibrium it is only necessary for each pair of adults, or for each adult female to leave two children which attain adult age, whatever the actual length of life in either generation. If the adult progeny exceeds two, the race will increase; if less than two, it will diminish. Since under present conditions the race of this animal is falling off, the actual rate of survival for the individual having remained the same, the total number of survivals only has changed. In other words, there is at present a deficiency of eggs.

What is the average number of eggs for the entire life of this animal? We know the minimal and maximal limits of egg production in individuals (roughly, 3,000 and 100,000); we know the average number of eggs borne at the average age of maturity (at the 10-inch size, 10,000 eggs); but, as Allen (*3*) in discussing this question points out, we do not know the number of female lobsters destroyed at different ages. Many after laying their first eggs are killed before any young are allowed to hatch, and the number which survive to produce successive broods is a constantly diminishing one; but this is made good in part by the rapid increase in the number of eggs.

The average number of eggs borne by all the berried lobsters captured should give us an indication of the average number of eggs borne by all female lobsters during life—the number sought. In 4,645 egg lobsters from the Woods Hole region, Massachusetts, the average number of eggs was 32,000, which would correspond to a 13 or 13½ inch lobster which had produced three or more broods. Allen found the number of eggs borne by 96,098 lobsters caught in Newfoundland to be 2,247,908,000, which would give an average of 23,000 to each female. This number corresponds to an animal 12 or 12½ inches long, which, as he remarks, from the known average age at which female lobsters mature (10–10½ inches), would be carrying at least a second brood. Such a lobster must therefore have produced 13,000 eggs (the average product at 10½ inches) plus 23,000, or at least 36,000 in all. We are therefore right in concluding that the maximum rate of survival of 2 in 10,000, formerly given, was much too high, as it was known to be at the time, and that the proportion of 2 to 30,000 is much nearer the truth. Another estimate, by Meek (*200*), based upon the statistics of the fisheries of Northumberland, England, gives a life rate of 1 in 38,000.

If, then, it is true, as we are thoroughly convinced it is, that the normal rate of survival in the lobster is not greater than 2 in 30,000 or 1 in 15,000 (and it can not be greater than 2 in 10,000), the fact is big for the lobster fishery, and the sooner it is faced the better. It has a direct bearing upon our laws and fishery operations. It enables us to evaluate the egg and the egg lobster truly. It shows in a conclusive manner that the present gauge laws are indefensible, because they rob the fishery of the billions of eggs necessary to maintain it. It further shows that the method of hatching the eggs of this animal and immediately liberating its young is ineffective, because of the meager results which can come from it. On the other hand, it speaks loudly in favor of a law to protect the large egg producers, and of the newer plan of rearing the young to the bottom-seeking stage, as the only means by which pisciculture can hope to aid this fishery materially.

The importance of the law of survival to the operations of the fisheries, and especially in its bearing upon some of our present illogical laws, is the only excuse for dwelling upon it at this length. To illustrate further: With respect to period of maturity and value to the fishery, all lobsters in the sea may be divided into three classes—(1) the young and adolescents, mainly from egg or larva, to the 8-inch stage; (2) intermediate class of adolescents and adults, 8 or 9 to 10½ inches in length; and (3) large adults, mainly above 10½ inches long. The biological value of the individual increases with every stage from egg to adult of largest size, and therefore is greatest in class 3. The present laws sanction the destruction of class 3, but class 1, the beginning of the series, must, as we have seen, be mainly recruited from this class or from those animals which under present conditions are being wiped out. In other words, our policy shifts the duty of maintaining the race upon the small producers, which the law of survival plainly tells us it is unable to bear. There is no way of getting over this grave defect.

We speak of the "living chain" from egg to adult, but the metaphor is not a happy one. There is no "chain" relation in living nature, only a succession of individuals, of individual eggs, united in origin but discrete in each generation. The embryologist begins with the egg, but the fish culturist with the egg producer. Spare the egg producer, then, and nature will save the race. We can not wholly take the place of nature in dealing with the eggs, but we can defeat the ends of nature by killing the "bird" which lays them.

But, do you say, "We have the egg lobster law, and the protection of lobsters in spawn should remedy our difficulties?" In reply we have but to recall the fact that adults lay their eggs but once in two years, and consequently we should not expect to find more than one-half of this class with spawn attached to the body at any given time. This at once reduces the protection aimed at in the egg lobster law by one-half. The other half shrinks to small proportions when we consider that there is an overlap of four weeks in July between the climax of the periods of hatching and spawning, when the majority of all adult female lobsters are without eggs of any kind, and also when we further consider the ease with which a fisherman by a few strokes of the hand can make a berried lobster eggless.

When analyzed in the light of the law of survival, the showing of the lobster hatcheries is not very encouraging. The hatching and immediate liberation of the fry has been practiced for many years in Europe, where experiments were made in Norway as early as 1873, as well as in Canada and the United States. The whole number of fry hatched and liberated on the Atlantic coast for a period of ten years, according to official returns from the hatcheries of the United States, Canada, and Newfoundland, reached a grand total of 4,214,778,200. Detailed statistics are given in the following table.[a]

[a] H. F. Moore, of the United States Bureau of Fisheries, to whom we are indebted for collating these statistics, says that no definite annual records appear in the official reports of Newfoundland for 1896 and 1897. The number of fry for each of these years is stated to be an average of the output for the seven preceding years.

TABLE 15.

Fiscal year.	United States.	Canada.	Newfoundland.
1893	8, 818, 000	153, 600, 000	517, 353, 000
1894	78, 398, 000	160, 000, 000	463, 890, 000
1895	72, 253, 000	168, 200, 000	174, 840, 000
1896	97, 079, 000	100, 000, 000	435, 079, 200
1897	115, 606, 000	90, 000, 000	450, 000, 000
1898	95, 234, 000	85, 000, 000	171, 900, 000
1899	108, 463, 000	100, 000, 000	
1900	77, 166, 000	120, 000, 000	
1901	60, 879, 000	110, 000, 000	
1902	81, 020, 000	120, 000, 000	
Total for 10 years	794, 916, 000	1, 206, 800, 000	2, 213, 062, 200

In addition to the number of lobster fry planted by the United States Fish Commission in 1900, there were sent to Dr. H. C. Bumpus 3,767,000 for experimental use. In 1902 also, in addition to the plant recorded by the commission, 6,178,000 fry were used for the same purpose.

Applying the law of survival, with life rate of 2 in 30,000, which has been shown to be a fair allowance, this number of young would yield only 280,985, while there must have been captured on this coast in the same period nearly 1,000,000,000 lobsters. By applying the maximum rate of 2 in 10,000, which we are assured is far too large, the yield would be 842,955. To have held the fishery at an equilibrium by this means, there should have been hatched 5,000,000,000,000 young, or 1,250 times as many as were actually liberated.

To take another example, the total output of all the Canadian lobster hatcheries for the entire history of this fishery, 1880 to 1906, was as follows:

Bay View, Nova Scotia, 1891–1906	1, 889, 300, 000
Canso, Nova Scotia, 1905–6	79, 000, 000
Shemogue, New Brunswick, 1903–1906	291, 000, 000
Shippegan, New Brunswick, 1904–1906	220, 000, 000
Charlottetown and Dunk River, Prince Edward Island, 1880–1906	256, 085, 000
	2, 735, 385, 000

Again, allowing the too generous rate of 1 in 5,000, this product of the activity of 24 years would yield only 547,077 lobsters, or but little over the two-hundredth part of the numbers caught in certain years in Canada alone.

In cases of this kind it is as detrimental to overestimate the value of the egg as to undervalue it. The eggs are true gold, although the amount which each weighs is infinitesimal. Like drops of water and grains of sand, these eggs count for but little singly, but in mass the inanimate particles can make the oceans and the continents, while the living germs can fill them with teeming inhabitants.

We can not work on the colossal scale of nature in dealing with egg or larva, but we may frustrate nature by destroying the egg producers. Nature long ago provided for the cod and hundreds of other predaceous fishes; she took into account the tides, the storm, and the rock-ribbed coast also, by giving to this race billions of eggs each year; but no provision was made for millions of traps working night and day at the bottom of the sea to destroy the producers of these eggs.

THE PROPAGATION OF THE LOBSTER.

The method of rearing the young through their critical larval or pelagic period, until they finally go to the bottom in the fourth or fifth stages, promises material aid to this fishery. While opinions may differ upon most of the questions which have been hitherto discussed, here is a subject upon which all should be agreed, and we believe that the method can not be extended too far or adopted too widely. Accordingly we shall briefly review the history of lobster rearing.

The first successful attempts at the artificial breeding of fish in America were made upon the speckled trout by Dr. Theodatus Garlick and Prof. H. A. Ackley, of Cleveland, Ohio, in 1853, the eggs and sperm being forcibly removed from the bodies of the ripe animals, brought into contact, and young trout subsequently reared from the eggs thus artificially impregnated.

No such results have ever been obtained in the Crustacea, nor is such a procedure possible in an animal like the lobster, owing to the unyielding nature of its body, due to a hard external skeleton. In the case of this animal we can only remove the already naturally fertilized and developing eggs from the underside of the abdomen, to which they are attached by the female herself at the time of egg laying, and afterwards give them such favorable conditions that the processes of development will proceed in a normal course to the time of hatching, as in the case of the artificial incubation of the eggs of fowls.

Messrs. Guillon and Coste were apparently the first to rear lobsters in Europe in considerable numbers, and an account of their experiments, which were conducted at the laboratory of Concarneau on the coast of France, was published in 1865 by Moquin-Tandon and Soubeiron (*202*).

How sanguine were these pioneers of the success of their experiments is shown by the following extracts:

> The ease with which young lobsters are reproduced and developed in the basins of Concarneau is a sure token that upon our coasts suitable places should be readily found for establishing vivaria where one may obtain myriads of the young, but these should not be permitted to enter the sea until they are sufficiently advanced to resist most of the causes of destruction which constantly menace them. What we have seen since our first visit to Concarneau, namely, basins literally black with little lobsters hatched in a vivarium, and from what we know of the habits of a great number of fishes in coming in immense numbers to stock particular regions of the coast, we may hope that it will be possible to regenerate the fishery on parts of our shores. By means of reservoirs we should be able to create an abundant food supply.

It was also stated that at the island of Tudy, M. de Cresoles had designed aquaria for preserving, hatching, and feeding lobsters and the *Palinurus* or langouste, some of the compartments being shaded or otherwise adapted to the animals in different stages of growth.

The writers quoted above further add:

To surprise nature with the accomplishment, to see life develop down to the smallest details, to possess a world of the sea in miniature in a transparent house, where nothing could escape investigation, such are really the promises of the establishment at Concarneau. These promises, gentlemen, are to-day realized.

It is pleasant to read of this enthusiasm at the dawn of the period of marine laboratories, and so far as the lobster is concerned we can only regret that the difficult problems of its successful culture, which were then hardly appreciated, should have had to wait nearly 40 years for their solution.

According to Roché (*237*), Mr. S. H. Ditten, a pharmacist to the court at Christiania, proposed to collect the egg-bearing lobsters in large floating cars and keep them until the young hatched out and were set at liberty naturally.

In the years 1873 to 1875 experiments in the hatching and rearing of lobsters were again undertaken by several gentlemen at Stavanger, Norway (*227*), both independently and with the aid of the Kongeligt Selskab for Norges Vel. According to the reports of Professors Rasch and G. O. Sars they were eminently successful; many young lobsters were carried to the ambulatory or bottom-seeking stage, the necessity of which was duly emphasized, and incidentally important facts on the natural history of the lobster were brought to light. Again, whatever progress was made at the time, the work was not systematically continued.

In 1883 Saville Kent (*245*) contributed a paper on "The Artificial Culture of Lobsters," which later appeared in the proceedings of the International Fisheries Exhibition at London for that year. He stated that in 1877 1,000,000 lobsters, valued at £22,500, were imported from Norway into Great Britain; that the catch in both countries was falling off; and that the decadence of the fisheries was due to three main causes, as follows: (1) Overfishing of the inshore districts; (2) destruction of undersized lobsters, and (3) destruction of the spawn for culinary purposes; the destruction of the eggs being the chief cause, which should be combated by artificial propagation.

By feeding lobsters hatched in aquaria on minced fish he reared them to the 1-inch length, when they would go to the bottom and hide. As a result of his experience he made the following significant remarks:

The rearing of lobsters in thousands instead of in tens or units would, it is needless to assert, be but a matter of augmented apparatus, and what the results would be upon our depopulated lobster grounds if several thousands, or rather millions, of such young animals could be turned out upon them annually, those are best qualified to record a verdict who have already had practical experience in the cultivation of the Salmonidæ.

He would pay a bounty for the egg lobster in order to divert the supply of eggs, "at present only flowing to the saucepans of the cooks," into the hatcheries of the cultivator, advises the use of hatching jars, feeding upon minced fish and mussels, rearing to the ambulatory stage, and liberating on rocky ground.

Still later, in 1885, Captain Dannevig (*69*) also succeeded in hatching the eggs of the lobster and in rearing the young through the first three earliest stages, at Flödevig, Norway. He did not consider it of much service to hatch the eggs and set free the young immediately; and he rightly said that so great was the destruction in nature from storms and other causes that out of the 25,000 or 30,000 eggs which a lobster might produce not a single one might reach its full development.

This work gave the first impetus to lobster culture in this country, where the hatching of eggs was accomplished in the summer of the same year (1885) at the newly opened laboratory of the United States Fish Commission at Woods Hole, Mass., as reported by Doctor Rathbun (*229*).

In 1894 we urged the importance of finding a means of rearing the young through the free-swimming stages, and thereby reducing the terrible death rate which inevitably occurs under natural conditions. As we then remarked, "If we could save 100 instead of 2 out of every 10,000 hatched, every million young would produce 10,000 adults and every billion would yield 10,000,000 lobsters capable of reproduction" (*143*).

While results somewhat similar to those outlined above have been obtained in England and in other parts of Europe, signal success in providing the young with a proper food supply and in maintaining them in a healthy condition up to the lobsterling stage has only been obtained in recent years in this country through the admirable work of Messrs. Bumpus and Mead and their associates. These experiments were begun under the auspices of the U. S. Fish Commission, at Woods Hole, Mass., in 1900, and were continued at other points on the coast, and especially at Wickford, R. I., where, under the direction of Professor Mead and of the Commissioners of Inland Fisheries of Rhode Island, the most efficient apparatus yet devised for the culture of lobsters has been gradually perfected and installed. All who are interested in the problems of lobster rearing should consult Professor Mead's original papers. (See, especially, *198*.)

Given a water supply which has been found by experiment to offer favorable conditions for the growth of lobster larvæ, and a suitable food supply, such as minced clams, beef, or "scrambled" eggs, the apparatus mechanically aerates the water and at the same time holds both the lobsters and their food in suspension with little detriment to the larvæ themselves.

At an early stage in his work Professor Mead found that in no case was the number of lobsters reared to the fourth stage less than 16 per cent of the total number of fry placed in the brood chambers (scrim bags, or wooden boxes, as now in use). The ratio of survival may even exceed 50 per cent. In 1901, between 9,000 and 10,000 lobsterlings were thus reared at the Wickford station to the bottom-seeking stage; in 1908, between 300,000 and 400,000 fourth or fifth stage lobsters were reared and distributed on the coast.

The rate of survival of the young in the early ambulatory stage is not known, but it is probably not less than 1 in several hundred, or a fraction of 1 per cent.

Instead of striving to work on the vast scale of nature in dealing with the egg, this is an attempt to improve upon nature by lowering the death rate in the most critical period. Great care, however, is needed at every stage of the process, and especially at the last, since the young do not seek the bottom at a uniform time.

Had it been our attempt to destroy this animal, could we have acted more effectively than by destroying its great egg-producing class? When we attempt to rid this country of the English or house sparrow, will it help greatly to break its eggs and destroy its young ones, though so relatively few and with a far higher life rate than in the crustacean? Must we not eventually kill the producers of the eggs if we would be rid of the pest? This is the nature of the treatment which the lobster has received. If we would preserve this fishery, we must reverse our laws, as Doctor Field has ably pointed out, and follow the principles and practice of breeders of domestic animals everywhere—use the smaller and better animals for food, and keep the older, and in this case by far the most valuable, for propagation.

RECOMMENDATIONS.

In applying the principles already discussed the following suggestions are offered:

1. Adopt a double gauge or length limit, placing in a perpetual close season or protected class all below and all above these limits. Place the legal bar so as to embrace the average period of sexual maturity, and thus to include what we have called the intermediate class of adolescents, or smaller adults. These limits should be approximately 9 inches and 11 inches, inclusive, thus legalizing the destruction of lobsters from 9 to 11 inches long only when measured alive. In this way we protect the young as well as the larger adults, upon which we depend for a continuous supply of eggs. The precise terms of these limits are not so vital, provided we preserve the principle of protecting the larger adults.

2. Protect the "berried" lobster on principle, and pay a bounty for it, as is now done, whether the law is evaded or not, and use its eggs for constructive work, or for experimental purposes with such work in view.

3. Abolish the closed season if it still exists; let the fishing extend throughout the year.

4. Wherever possible, adopt the plan of rearing the young to the bottom-seeking stage before liberation, or cooperate with the United States Bureau of Fisheries or with sister states to this end.

5. License every lobster fisherman, and adopt a standard trap or pot which shall work automatically, so far as possible, in favor of the double gauge, the entrance rings being of such a diameter as to exclude all lobsters above the gauge, and the slats of the trap of such a distance apart as to permit the undersized animals to escape.

Many objections can be raised, but this plan is defensible on scientific grounds, while the older methods are not. The best thing which can be said of it is that it would eventually give us more eggs, and in an ever-increasing quantity—the greatest need of this fishery, both now and in the future. Under present conditions, the supply of eggs is yearly diminishing and at a tremendous rate.

The most striking objection to the proposed changes would be that if class 3, that of the big producers, has been nearly exterminated, and we proceed to wipe out class 2, the smaller adults, there will soon be no more lobsters; but this is not valid. No doubt if this change were made, the supply of smaller lobsters would be temporarily increased where the 10½-inch gauge law still prevails, as was the case in Massachusetts in 1907 when the 9-inch law went into effect; and this might be followed by a temporary stringency. No one can speak with positive assurance upon this subject, but the important point to bear in mind is that under such an arrangement we would have a perpetually protected class constantly growing and at work all the time.

Again, it may be asked, Will enough lobsters survive to enter the exempt class? We believe that there would, and that the answer to this question is to be found in the records of catches for every locality where lobsters are now trapped. Even in places where the average size is small, larger lobsters occasionally appear, and in sizes showing more than one year's growth. Why were not all such animals weeded out the previous year? Instead of waiting to be caught up in the end, these "escapes" would all enter the protected growing class, to enjoy a green old age of 50 years and possibly more, though we have no positive knowledge of the life span in this interesting race.

The trouble of a double gauge, such expense as would be needed in adjusting traps to admit and hold lobsters of the legal size, would have to be met, but it would be well worth while. In our opinion, the markets would not be seriously disturbed. Protect the big egg producers and nature will preserve the race.

Without doubt there are many who would consider any legal measure involving a double gauge impracticable because of the difficulty of carrying it out, for to be effective it must be uniformly adopted and enforced. If the present laws are to be maintained in principle, the following steps should be taken:

(1) Raise the legal gauge to 10½ inches wherever it now stands below this limit.

(2) License every lobster fisherman, and adopt a standard trap, with slats of sufficient distance apart to permit the undersized lobsters to escape.

(3) Destroy the present enormously destructive interstate commerce in short lobsters.

(4) Do not turn another larval lobster into the sea, but devote the energy expended in lobster hatcheries to rearing these young to the bottom-seeking stage after the methods now successfully practiced at Wickford, R. I.

BIBLIOGRAPHY OF THE LOBSTER—HOMARUS.

In the following bibliography we endeavor to give a record of the scientific literature of the lobsters (*Homarus gammarus* and *H. americanus*), embracing their anatomy, physiology, development, general habits, behavior, and habitat, as well as the lobster fisheries, and the preservation, artificial propagation, and economy of the species in general. In order to reflect the history of our knowledge of the subject, we have endeavored to include all papers, which were once, or are now, of any interest or value, from an early period to the present time. While keeping, in the main, within the limits of original research, we have given place to some minor works in which the knowledge of the day or period was reflected more or less clearly. The statistical records of the fisheries, however, are so widely scattered, and in some respects so unsatisfactory, that we have attempted to give only the most important references.

The voluminous literature of the related crayfishes, of the Norway lobster (*Nephrops norvegicus*), the spiny, thorny, or rock lobsters, "la langouste" of the French (*Palinurus*), and the Spanish lobsters (*Galathea*) is not generally included in this survey, and when referred to is noticed in the text. While we have endeavored to secure accuracy in giving titles, a few have been necessarily taken at second hand; further, we have not hesitated to add an occasional note, when in our opinion the use of this list to future students could thus be enhanced.

1. ALDROVANDI, ULYSSIS.

Philosophi et medici Bononiensis. De mollibus, crustaceis, testaceis et zoophytis. De animalibus exanguibus reliquis quattuor. Liber secundus, qui est de crustatis. De astaco, cap. iii. De leone, seu elephanto, cap. iv.

Crude figure of a lobster under name of *Astacus verus;* repeats Gesner's figure of "Chela Astaci marini ex Zoographo;" also figures of Olaus Magnus of *Astacus marinus* devouring a man, and of a marine rhinoceros eating an *Astacus* 12 feet long.

2. ALLEN, E. J.

Studies on the nervous systems of Crustacea. [The embryonic lobster] I, Quarterly Journal of Microscopical Science, vol. XXXVI (n. s.), p. 461–482; II–III, ibid., p. 483–498, 2 pl., 1894; IV, ibid., vol. XXXIX (n. s.), p. 33–50, 1 pl. 1897. London.

3. ———

Protection of crabs and lobsters. Journal Marine Biological Association of the United Kingdom, vol. IV (n. s.), p. 182–187. 1895–97. Plymouth.

4. ———

The reproduction of the lobster. Ibid., vol. IV (n. s.), p. 60–69. 1895–97. Plymouth.

5. ANDERTON, T.

The lobster (*Homarus vulgaris*). Report of the Marine Department of New Zealand for 1908–1909, p. 17–23, pl. I–IV. Wellington, 1909.

Cites cases of annual breeding in European lobster transplanted to New Zealand, and describes sexual union as following molting in the female.

6. APPELLÖF, A.

Mittheilungen aus der Lebensweise des Hummers. Mittheilungen des Deutschen Seefischerei-Vereins, bd. 15, p. 99. Berlin, 1899.

Apparently demonstrates by experiment that at Stavanger, Norway, the European lobster spawned but once in two years.

7. ———

Indberetning til Stavanger filial av Selskabet for de norske Fiskeriers Fremme om hummerundersokelser i 1892. (1) Aarsberetning for Selskabet for de norske Fiskeriers Fremme. Bergen, 1892.

8. ———

Researches on the development of the lobster, 1893–1901 (2–7). Ibid., 1893–1901.

9. ———

Researches on the development of the lobster, 1902. Ibid. Norsk Fiskeritidende, bd. 22, p. 114–132. Bergen, 1903.

10. ———

Researches on the development of the lobster in 1903. Ibid. Norsk Fiskeritidende, bd. 23, p. 112–119. Bergen, 1904.

11. ATWATER, W. O.

The chemical composition and nutritive values of food fishes and aquatic invertebrates. Report U. S. Commission of Fish and Fisheries for 1880, p. 679–868, pl. LXXVI–LXXXIX. Washington, 1892.

12. ATWOOD, N. E.

On the habits and geographical distribution of the common lobster. Proceedings Boston Society Natural History, vol. X, p. 11–12. Boston, 1866. See also Proceedings Essex Institute, vol. IV, p. clxxviii–clxxx. Salem, 1864–1865.

13. AUDUBON, JOHN JAMES.

Labrador Journal; from "Audubon and His Journals," by Maria R. Audubon, edited with the assistance of Elliott Coues, vol. II, p. 363.

Audubon refers to the lobster as "very scarce" in Labrador, and describes method of Indians in roasting them alive.

14. BANCEL, C., and HUSSON, C.

Sur la phosphorescence de la viande de homard. Comptes rendus de l'Académie des Sciences, t. 88, p. 191–192. Paris, 1879.

15. BARNES, EARNEST A.

Lobster culture in 1905. Thirty-sixth Annual Report of the Commissioners of Inland Fisheries of Rhode Island, for 1906, p. 111–119. Providence, 1906.

16. ———

Methods of protecting and propagating the lobster, with a brief outline of its natural history. Ibid., p. 120–152, 13 pl. Providence, 1906.

16 a. ———

Lobster culture at Wickford, R. I., in 1906. Thirty-seventh Annual Report of the Commissioners of Inland Fisheries of Rhode Island for 1907, p. 88–98. Providence, 1907.

17. BASTER, J.

Opuscula subseciva. De astacis, tom. II, lib. 1, tab. 1. Harlemi, 1762.

Figures egg-bearing lobster, and describes breeding habits of European lobster on testimony of "Norwegian friends."

18. BATE, C. SPENCE.

Fourth report on the fauna of South Devon. Report of British Association for the Advancement of Science for 1872. London, 1873.

Observations on development of lobster.

19. Bateson, William.
Materials for the study of variation treated with especial regard to discontinuity in the origin of species. i–xvi+1–598 p. London, 1894.
Best review of monstrosities in appendages of lobster, chiefly the big claws, with formulation of important principles. (Chap. xxi.)

20. Bell, Thomas.
A history of the British stalk-eyed Crustacea. i–lxvi+1–386 p. London, 1853.

21. Bernhardus, Martinus a Berniz.
Chela Astaci marini monstrosa. Miscellanea curiosa medico-physica Academiæ Naturæ Curiosorum, annus secundus, observatio C, p. 174, 1 pl. 1671.

22. Biedermann, W.
Über den Zustand des Kalkes im Crustaceenpanzer. Biologisches Centralblatt, bd. 21, p. 343–352 (3 fig.) Leipzig, 1901.

23. ———
Über die Structur des Chiten bei Inseckten und Crustaceen (*Astacus* und *Homarus*). Anatomischer Anzeiger, bd. 21, p. 485–490. Leipzig, 1902.

24. Boeck, Axel.
Om det norske Hummerfiske og dets Historie. Tidskrift for Fiskeri, 3die Aargangs. Kjöbenhavn, 1868–1869. Translated in Report U. S. Commission of Fish and Fisheries, 1873–1875, p. 223–258. Washington, 1876.

25. Bohn, G.
Théorie nouvelle du phototropisme. Comptes rendus de l'Académie des sciences, t. 139, p. 890–891. Paris, 1904.

26. ———
Impulsions motrices d'origine oculaire chez les Crustacés, 2e mémoire. Bulletin de l'Institut psychologie, t. 5, p. 412–456. Paris, 1905.

27. ———
Sur le phototropisme des larves de Homard. Comptes rendus de l'Académie des Sciences, t. 141, p. 963–966. Paris, 1905.

28. ———
Des tropismes et des états physiologiques. Comptes rendus de la Société de Biologie, t. 59, p. 515–516. Paris, 1905.

29. ———
Mouvements rotatoires chez les larves de Crustacés. Comptes rendus de la Société de Biologie, t. 59, p. 517–518. Paris, 1905.

30. ———
L'éclairement des yeux et les mouvements rotatoires. Comptes rendus de la Société de Biologie, t. 59, p. 564–566. Paris, 1905.

31. Botazzi, Fil.
Untersuchungen über das viscerale Nervensystem der decapoden Crustaceen. (II) Zeitschrift für Biologie, n. f., bd. xxv, p. 340–371. München und Berlin, 1902.
Refers chiefly to the spiny lobster, *Palinurus*.

32. Bouchard-Chantereaux.
Catalogue des Crustacés du observés dans le Boulonnais. 1833.

33. Brandes, G.
Zur Begattung der Dekapoden. Biologisches Centralblatt, bd. 17, p. 346–350. Leipzig, 1897.

34. Braun, Max.
Zur Kenntniss des Vorkommens der Speichel- und Kittdrüsen bei den Decapoden. Arbeiten aus dem zoologisch-zootomischen Institut in Würzburg, bd. III, p. 472–479, taf. 21. 1876–77.

35. BRIGHTWELL, T.
Description of the young of the common lobster, with observations relative to the questions of the occurrence and non-occurrence of transformations in crustaceous animals. Loudon's Magazine Natural History, 1st ser., vol. VIII, p. 482–486. London, 1835.
First notice of double monsters in larva of European lobster.

36. BROCCHI, P.
Recherches sur les organes génitaux mâles des Crustacés décapodes. Reprinted from Annales des Sciences Naturelles, 6e sér., p. 1–132, pl. 13–19. Paris, 1875.
Figures spermatophores and copulatory appendages of male lobster.

37. BROOK, GEORGE.
Notes on the reproduction of lost parts in the lobster (*Homarus vulgaris*). Proceedings Royal Physical Society of Edinburgh, p. 370–385, fig. 1–5. Edinburgh, 1887.

38. BRUES, C. T.
The internal factors of regeneration in Alpheus. Biological Bulletin, vol. VI, p. 319–320. 1904.

39. BUCKLAND, FRANK; WALPOLE, SPENCER, et al.
Reports on the crab and lobster fisheries of England and Wales, of Scotland and Ireland. p. i–xxii, I–XXVI, i–iv and 1–80, with appendices, 8 pl. London, 1877.

40. BUCKLAND, FRANK.
Reports on the fisheries of Norfolk, especially crabs, lobsters, herrings, and broads. Presented by Her Majesty's command. Ordered by the House of Commons to be printed. London, 1875.

41. BUMPUS, HERMON CAREY.
The embryology of the American lobster. Journal of Morphology, vol. V, p. 215–262, pl. XIV–XIX. 1891.
The first circumstantial account of the embryology of the lobster based upon modern methods. Describes structure and function of seminal receptacle.

42. ———
The American Lobster. Review, Science (n. s.), vol. IV, p. 536–537. New York, 1896.
Describes rayed sperm cells of lobster in "active movement."

43. ———
On the movements of certain lobsters liberated at Woods Hole during the summer of 1898. Bulletin U. S. Fish Commission, vol. XIX (1899), p. 225–230. Washington, 1901.

44. ———
The results attending the experiments in lobster culture made by the United States Commission of Fish and Fisheries. Science, n. s., vol. 14, p. 1013–1015. 1901.

45. CALMAN, W. T.
On a lobster with symmetrical claws. Proceedings Zoological Society of London, June 19, 1906, p. 633–634, 1 fig. London.

46. CANO, G.
Morfologia dell' apparecchio sessuale femminile, glandole del cemento e fecondazione nei crostacei decapodi. Mittheilungen der Zoologischen Station Neapel, bd. IX, p. 503–532, taf. 17. 1891.

47. CAREY, C. B.
Large lobster (*Homarus vulgaris*). Zoologist, 2d ser., vol. 8, p. 3654. London, 1873.
Record of lobster weighing 12 pounds.

48. CARRINGTON, JOHN T., and LOVETT, EDWARD.
Notes and observations on British stalk-eyed Crustacea. Zoologist, 3d ser., vol. VI, p. 9–15 (continued). London, 1882.

49. CAVOLINI.
Memoria sulla generazione dei pesci e dei granchi. Napoli, 1787. See also Über die Erzeugung der Fische und Krebse. Aus dem italienischen übersetzt von Zimmerman. Berlin, 1792.

50. CHADWICK, H. C.

Experiments in lobster rearing. Report for 1904 in the Lancashire Sea-Fisheries Laboratory at the University of Liverpool and the Sea-Fish Hatchery at Peel, p. 124–128, fig. 1–6. Liverpool, 1905. Also Proceedings Liverpool Biological Society, vol. XIX, p. 304–308, 6 fig. Liverpool, 1905.

Briefly describes and figures first five stages of *H. vulgaris*, but presents no new facts.

51. CLOUSTON, T. S.

The minute anatomy and physiology of the nervous system in the lobster (*Astacus marinus*). Edinburgh New Philosophical Journal, vol. 17, p. 17–51, 2 pl. Edinburgh, 1863.

52. COBB, JOHN N.

The lobster fishery of Maine. Bulletin U. S. Fish Commission, vol. XIX, p. 241–266, pl. 28–32. Washington, 1901.

52a. COLE, LEON J.

Description of an abnormal lobster cheliped. Biological Bulletin, vol. xviii, p. 252–268, fig. 1–9. Boston, 1910.

Describes crusher claw of lobster, bearing a double extra claw in secondary symmetry.

53. COLLINS, JOSEPH W.

Report upon a convention held at Boston in 1903 to secure better protection of the lobster. 1–52 p. Boston, 1904.

54. CORNISH, THOMAS.

Enormous lobster. Zoologist, 2d ser., vol. 2, p. 1018. London, 1867.

55. COSTE, M.

Voyage d'Exploration sur le littoral de la France et de l'Italie. Deuxième édit., suivie de nouveaux documents sur les pêches fluviales et marines. Publiée par ordre de S. M. l'Empereur sous les auspices de S. Exc. le Ministre de l'Agriculture, du Commerce et des Travaux publiques. 4to. Imprimerie Impériale, p. I–XXIV, 1–298, text fig. and 2 pl. Paris, 1861. Beginning on page 157 is a series of appendices under the general title "Documents relatifs aux Pêches Marines." Appendix VI (p. 201–208) is entitled "Rapport à S. E. le Ministre de la Marine sur la reproduction des Crustacés, au point de vue de la réglementation des Pêches."

Coste's original account of the reproduction and development of the European lobster, thus buried in an appendix to a public document, seems to have successfully escaped bibliographers up to the present. We are indebted to Dr. Richard Rathbun, who gave a summary of the paper in 1884 (see *226*, p. 803), for unearthing it at the present time. An indication of the accuracy of Coste's statements is given at page 190 of the present work.

56. ———

[Report of work of Gerbe.] Faits pour servir à l'histoire de la fécondation chez les crustacés. Comptes rendus d l'Académie des sciences, t. 56, p. 432. Paris, 1858.

57. ———

Étude sur les mœurs et sur la génération d'un certain nombre d'animaux marins. Ibid., t. 47, p. 45–50. Paris, 1858.

58. COUCH, JONATHAN.

Observations on some circumstances attending the process of exuviation in shrimps and lobsters. Magazine of Zoology and Botany, vol. 1, p. 170–173. 1837. For translation, see Bemerkungen über den Häutungsprocess der Krebse und Krabben. Archiv für Naturgeschichte von Wiegmann, jahrg. 4, bd. 1, p. 337–342. Berlin, 1838.

59. ———

On the process of exuviation and growth in crabs and lobsters and other British species of stalk-eyed crustacean animals. Eleventh Annual Report of the Royal Cornwall Polytechnic Society, 1843, p. 1–15. Falmouth, 1845.

60. COUCH, R. Q.
On the metamorphoses of the decapod crustaceans. Eleventh Annual Report of the Royal Cornwall Polytechnic Society, 1843, p. 28–43, pl. 1. Falmouth, 1843.

61. ———
On the metamorphosis of the crustaceans, including the Decapoda, Entomostraca, and Pycnogonidæ. The Twelfth Annual Report of the Royal Cornwall Polytechnic Society, p. 17–46. Falmouth, 1844.

62. COUES, ELLIOTT.
Notes on the natural history of Fort Macon, North Carolina, and vicinity (no. 2). Proceedings Academy Natural Sciences of Philadelphia, p. 120–148. Philadelphia, 1871.

63. CUNNINGHAM, J. T.
Contributions to the knowledge of the natural history of the lobster and crab. Journal of the Royal Institute of Cornwall, p. 1–4. Truro, 1897.

64. ———
Lobster rearing, Cornwall County Council Instruction Committee. Report of the Lecturer on Fishery Subjects for the years 1897–1898, 1899–1900. See also Nature, vol. LIX, p. 62. London, 1898.

65. ———
Experiments in hatching and rearing lobsters. Report Royal Cornwall Polytechnic Society, vol. 69, p. 25–30. Falmouth, 1901.
The author maintains that larvæ of the lobster do not habitually devour living prey, but feed on carrion, like the adults.

66. ———
Reports on the experiments in oyster and lobster culture in 1902, with summary of the work from 1897. Ibid., vol. 70, p. 27–39. Falmouth, 1903.

67. DAHLGREN, ULRIC, and KEPNER, WILLIAM A.
A text-book of the principles of animal histology. i–xiv+1–516 p. New York, 1908.
Gives histological analysis of the integument, the gill filament, and other organs of the lobster.

68. DALYELL, JOHN GRAHAM.
The powers of the Creator displayed in the Creation. London, 1827.

69. DANNEVIG, G. M.
Beretning over Virksomheden ved Udklæckningsanstalten for Saltvandsfisk. Arendal, 1885. Reprinted under title "Om Hummeravl in Dansk Fiskeritidende," von A. Feddersen, jahrg. 1886. See also Aarsberetning for Selskabet for de norske Fiskeriers Fremme. 1892.

70. DAVIES, H.
The lobster industry. 10 p. (pamphlet). Charlottetown, 1896.
Record of consultation with lobster packers of Prince Edward Island.

71. DEACON, J.
A huge lobster. The Zoologist, 2d ser., vol. 8, p. 3618. London, 1873.

72. DE KAY, JAS. E.
Zoology of New York, or the New York fauna. Part VI. Crustacea, p. 23–25, pl. XII, fig. 52–53. Albany, 1844.

73. DOGIEL, J.
Anatomie du cœur des Crustacés. Comptesrendus de l'Académie des sciences, t. 82, p. 1117. Paris, 1876.

74. ———
De la structure et des fonctions du cœur des crustacés. Archives de Physiologie normale et pathologique, 2e sér, t. 4, p. 400–408. Paris, 1877.

75. DEZSÖ, BÉLA.
Über das Herz des Flusskrebses und des Hummers. Zoologischer Anzeiger, vol. 1, p. 126–127. Leipzig, 1878.

76. DULK.
Chemische Untersuchungen eines Mageninhalts von Krebsen, die sich eben gehäutet haben. Archiv für Anatomie und Physiologie, jg. 1834, p. 523–527. Berlin, 1834.

77. ——
Chemische Untersuchungen der Krebssteine. Archiv für Anatomie und Physiologie, jg. 1835, p. 428–430. Berlin, 1835.

78. DUVAR, J. H.
Report on lobster fishery of Prince Edward Island. Annual Report of the Department of Fisheries, Dominion of Canada. Appendix no. 6, p. 240. 1884.

79. DUVERNOY.
Deuxième fragment sur les organes de génération de divers animaux. Des organes extérieurs de fécondation dans les crustacés décapodes. Comptes rendus de l'Académie des sciences, t. 31, p. 342–348. Paris, 1850.

80. EARLE, ALICE MORSE.
Home life in colonial days. New York, 1898. See p. 117 for early references to the abundance and great size of lobsters in the Massachusetts Bay Colony.

81. EDWARDS, H. MILNE.
Histoire naturelle des Crustacés. 3 vol., with atlas. Paris, 1834–1840.

82. ——
Observations sur la structure et les fonctions de quelques zoophytes mollusques et crustacés des côtes de la France. Annales des sciences naturelles, 2e sér., t. 18, p. 321–350, pl. 10–15. Paris, 1842.

83. EHRENBAUM, ERNST.
Der Helgolander Hummer, ein Gegenstand deutscher Fischerei. Wissenchaftliche Meeresuntersuchungen, herausgegeben von der Kommission zur Untersuchung der deutschen Meere in Kiel und der biologischen Anstalt auf Helgoland. Neue folge, bd. 1, p. 277–300. Kiel und Leipzig, 1894.

84. ——
Der Hummerfang von Helgoland auf der Ausstellung in Berlin 1896, nebst Mittheilungen über den Hummer. Der Hummer: Eine Zusammenstellung der Resultate neuerer Untersuchungen. Mittheilungen des Deutschen Seefischereivereins, nr. 9, p. 12. Berlin, 1896.

85. ——
Neuere Untersuchungen über den Hummer. Mittheilungen des Deutschen Seefischerei-Vereins, bd. 19, p. 146–159, taf. Berlin, 1903.

86. ——
Ueber den Hummer (Fang, Häutung, Wachstum, Vermehrung, künstliche Zucht). Fischerei-zeitung, bd. 6, p. 417–422, 433–436, 449–452, 465–477. Neudamm, 1903.

87. ——
Künstliche Zucht und Wachstum des Hummers. Mitteilungen des Deutschen Seefischerei-Vereins, no. 6, p. 1–23, fig. 1–4 mit Beilage. Berlin, 1907. For English translation see Thirty-eighth Annual Report of the Commissioners of Inland Fisheries of Rhode Island for 1908, p. 14–26. Providence, 1908.

88. EMMEL, VICTOR E.
The regeneration of the lost parts in the lobster (*Homarus americanus*). Preliminary report. Thirty-fifth Annual Report of the Commissioners of Inland Fisheries of Rhode Island, p. 81–117, pl. XXI, XXII. Providence, 1905.

89. ———
The relation of regeneration to the molting process in the lobster. Thirty-sixth Annual Report of the Commissioners of Inland Fisheries of Rhode Island, p. 258–313, 2 pl. with charts. Providence, 1906.

90. ———
Torsion and other transitional phenomena in the regeneration of the cheliped of the lobster (*Homarus americanus*). Journal of Experimental Zoology, vol. III, no. 4, p. 603–620, pl. 1–2. Baltimore, 1906.

91. ———
The regeneration of the "crusher-claws" following the amputation of the normal asymmetrical chelæ of the lobster (*Homarus americanus*). Archiv für Entwicklungsmechanik der Organismen, bd. XXII, hft. 4, p. 542–552, taf. XV. Leipzig, 1906.

92. ———
Regenerated and abnormal appendages in the lobster. Thirty-seventh Annual Report of the Commissioners of Inland Fisheries of Rhode Island, p. 99–152, pl. I–IX. Providence, 1907.

93. ———
Regeneration and the question of symmetry in the big claws of the lobster. Science, n. s., vol. XXVI, p. 83–87. New York, 1907.

94. ———
Note upon experiments on control of symmetry in big claws. Science, n. s., vol. XXVII, p. 779–780. New York, 1908.

95. ———
The problem of feeding in artificial lobster culture. Thirty-eighth Annual Report of the Commissioners of Inland Fisheries of Rhode Island for 1908, p. 98–114. Providence, 1908.

96. ———
The experimental control of asymmetry at different stages in the development of the lobster. Journal of Experimental Zoology, vol. V, p. 471–484. Philadelphia, 1908.

97. ———
A study of the differentiation of tissues in the regenerating crustacean limb. American Journal of Anatomy, vol. 10, p. 109–158, pl. I–VIII. Philadelphia, 1910.

98. ERDL, M. P.
Entwicklung des Hummereies von den ersten Veränderungen im Dotter an bis zur Reife des Embryo. i–x, 11–40 p., 4 taf. München, 1843.

The first independent work on the embryonic development of the lobster, with hand-colored copper-plate engravings of the egg-embryo, good for the period, but now of historical interest chiefly.

99. ———
Sur le développement de l'œuf du homard. (Abstract.) Comptes rendus de l'Académie des sciences, t. 17, p. 321–322. Paris, 1843.

99a. EWART, J. C., and FULTON, T. WEMYSS.
The Scottish lobster fishery. Annual Report of the Fishery Board for Scotland for 1887, p. 189–203. Edinburgh, 1888.

100. FARRE, A.
On the organ of hearing in Crustacea. Philosophical Transactions, p. 233–242, 2 pl. London, 1843. See also Das Gehörorgan des Hummers. Froriep's Notizen, bd. 28, p. 183–184. Weimar, 1843.

101. FABRE-DOMERGUE et BIÉTRIX.
Le mécanisme de l'émission des larves chez la femelle du Homard européen. Comptes rendus de l'Académie des sciences, vol. 136, p. 1408–1409. Paris, 1903.

102. FAXON, WALTER.
On some crustacean deformities. Bulletin Museum of Comparative Zoology at Harvard College, vol. VIII, no. 13, p. 257–274, pl. I–II. Cambridge, 1881.

103. ———
Selections from the embryological monographs, compiled by Alexander Agassiz, Walter Faxon, and E. L. Mark. I. Crustacea. Memoirs Museum of Comparative Zoology, vol. IX, no. 1, pl. 1–14, with descriptions. Drawings of lobster given in fig. 11–17, pl. XII. Cambridge, 1882.

104. FIELD, GEO. W.
A report upon the scientific basis of the lobster industry, the apparent causes of its decline, and suggestions for improving the lobster laws. Report of the Commissioners of Inland Fisheries and Game of Massachusetts for 1901, p. 111–130. Boston, 1902.

Advocates a policy of protecting the older lobsters, by prohibiting the capture and sale of adult animals over 10 inches long.

105. ———
The biological basis of legislation governing the lobster industry. Science, n. s., vol. XV, p. 612–616. New York, 1902.

106. ———
The lobster fisheries and the cause of their decline. From the 40th Annual Report of the Commissioners of Fisheries and Game of Massachusetts, p. 1–46, 2 pl. Boston, 1906.

107. ———
Lobsters and the lobster problem in Massachusetts. With discussion. Proceedings of the Fourth International Fishery Congress. Bulletin of the Bureau of Fisheries, vol. XXVIII, 1908, p. 211–217. Washington, 1910.

108. FOETTINGER, ALEXANDRE.
Recherches sur l'organisation de *Histriobdella homari*, P.-J. Van Beneden, rapportée aux Archiannélides. Archives de Biologie, t. V, p. 435–516, pl. XXV–XXIX. Paris, 1884. See also Van Beneden, Bulletin de l'Académie royale de Belgique, t. XX, p. 63, 1853; also later account, with plates, in memoirs of the same academy, t. XXXIV. Bruxelles, 1858.

109. FRAICHE, FÉLIX.
Guide pratique de l'ostréiculteur et procédés d'élevage et de multiplication des races marines comestibles. Translated by H. J. Rice, in Report of the Commissioner of Fish and Fisheries for 1880, p. 753–824. Washington, 1883.

110. FREDERICQ, LÉON.
Note sur le sang du Homard. Bulletin de l'Académie royale de Belgique, 2e sér., t. XLVII, p. 409–413. Bruxelles, 1879.

111. FREDERICQ, LÉON, and VANDEVELDE, G.
Vitesse de transmission de l'excitation motrice dans les nerfs du Homard. Archives de Zoologie expérimentale et générale, t. 8, p. 513–520. Paris, 1879–80.

112. ———
Physiologie des muscles et des nerfs du Homard. Bulletin de l'Académie royale de Belgique. 2e sér., t. 47, p. 771–779, with fig. Bruxelles, 1879. See also Archives de Biologie, t. 1, p. 1–24. Paris, 1880.

113. FULLARTON, J. H.
The European lobster; breeding and development. Fourteenth Annual Report of the Fishery Board for Scotland, p. 186–222, pl. VI–VIII. Glasgow, 1896.

114. GARMAN, S.
Report on the Lobster. Report of the Massachusetts Commission of Inland Fisheries and Game for 1891, p. 60–61. Boston, 1892.

115. ———
Lobster reproduction. Zoologischer Anzeiger, XVIII. jahrg., p. 38–40. Leipzig, 1895.

116. GEOFFROY LE JEUNE.
Observations sur les écrevisses de rivière [dated August 23, 1709]. Histoire de l'Académie royale des sciences, 1709, p. 309–314. Paris, 1711.
The first notice of gastroliths in the lobster.

117. GERBE, Z.
Appareils vasculaire et nerveux des larves des crustacés marins. Comptes rendus de l'Académie des sciences, Paris, t. LXII, p. 932–937. Paris, 1866.

118. GESNER, CONRAD.
Conradi Gesneri medici Tigurini Historiæ Animalium. De Astaco, Rondeletius. Lib. IIII, qui est de Piscium & Aquatilium animantium natura, p. 113–119. Tiguri, 1551–1558.
One of the first circumstantial accounts of the European lobster, with many curious ideas derived from an age of superstition, also many quaint figures. The same wood-cuts, made for the first edition of Gesner's work, were used by Conrad Forer in his German abridgment, which appeared forty-five years later (Franckfurt, 1598).

119. GILSON, G.
Étude comparée de la spermatogenèse chez les arthropodes. Recueil La Cellule, t. I and II, fasc. 1, 2, Crustacea, p. 140–310, pl. 8–14. Louvain, 1886.

120. GOODSIR, H. D. S.
A short account of the mode of reproduction of lost parts in the Crustacea. Annals and Magazine of Natural History, vol. XIII, p. 67. London, 1844. For abstract, see also: On the reproduction of lost parts in the Crustacea. Report British Association for the Advancement of Science, 1844, p. 68. London, 1845.

121. GORHAM, FREDERIC P.
Causes of death in artificially reared lobster fry. Report of the U. S. Commission of Fish and Fisheries for 1903, p. 175–194. Washington, 1905.

122. GROBBEN, C.
Beiträge zur Kenntniss der männlichen Geschlechtsorgane der Dekapoden, nebst vergleichenden Bemerkungen über die der übrigen Thoracostraken. Arbeiten aus dem Zoologischen Institut der Universität zu Wien und der Zoologischen Station in Triest, bd. 1, p. 1–94, taf. I–VI. Wien, 1878.

123. GURNEY, J. H.
Note on a pied lobster. The Zoologist, 2d ser., vol. 9, p. 4080. London, 1874.

124. HADLEY, PHILIP B.
Changes in form and color in successive stages of the American lobster (*Homarus americanus*), with drawings from life. Preliminary report. Thirty-fifth Annual Report of the Commissioners of Inland Fisheries of Rhode Island for 1905, p. 44–80, pl. VII–XVIII. Providence, 1905.

125. ———
Phototropism in the larval and early adolescent stages of *Homarus americanus*. Science, n. s., vol. XXII, p. 675–678. New York, 1905.

126. ———
Regarding the rate of growth of the American lobster. Thirty-sixth Annual Report of the Commissioners of Inland Fisheries of Rhode Island, p. 153–226, pl. XXVI–XXXVII and XL. Providence, 1906. See also Biological Bulletin, vol. X, p. 233–241. Boston, 1906.

127. HADLEY, PHILIP B.

Observations on some influences of light upon the larval and early adolescent stages of *Homarus americanus*. Preliminary report. Thirty-sixth Annual Report of the Commissioners of Inland Fisheries of Rhode Island for 1906, p. 237–257, 2 pl. Providence, 1906.

127 a. ———

Continued observations on some influences of light upon the larval and early adolescent stages of the American lobster. Thirty-seventh Annual Report of the Commissioners of Inland Fisheries of Rhode Island for 1907, p. 191–216. Providence, 1907.

128. ———

The relation of optical stimuli to rheotaxis in the American lobster, *Homarus americanus*. American Journal of Physiology, vol. XVII, p. 326–343. Boston, 1906.

129. ———

Galvanotaxis in larvæ of the American lobster (*Homarus americanus*). American Journal of Physiology, vol. XIX, p. 39–52. Boston, 1907.

130. ———

The reaction of blinded lobsters to light. American Journal of Physiology, vol. XXI, p. 180–199. Boston, 1908.

131. ———

The behavior of the larval and adolescent stages of the American lobster (*Homarus americanus*). Journal of Comparative Neurology and Psychology, vol. XVIII, no. 3, p. 199–302, fig. 1–22. Philadelphia, 1908.

In this paper the author brings together his previous researches, which are the most extended and valuable yet made upon the subject.

132. HARTING, PIETER.

Een slimmezeekreeft (*Homarus vulgaris*). Album der Natur, p. 24. Haarlem, 1878.

133. HELLER, CAMIL.

Die Crustaceen des südlichen Europa. Crustacea Podophthalmia, i–xii, 1–336 p., 10 taf. Wien, 1863.

134. HASSE, C. E.

Observationes de sceleto astaci fluviatilis et marini. Dissertatio, 1–38 p., 1 pl. Lipsiæ, 1833.

135. HENNSCHEN, F.

Zur Structur der Eizelle gewisser Crustaceen und Gastropoden. Anatomischer Anzeiger, bd. 24, p. 15–29, 14 fig. Jena, 1891.

Describes pseudochromosomes in Astacus and Homarus.

136. HERBST, J. F. W.

Versuch einer Naturgeschichte der Krabben und Krebse, nebst einer systematischen Beschreibung ihrer verschiedenen Arten. 3 bde. Berlin und Stralsund, 1790–1804.

For an account of the lobster *Cancer* (*Astacus*) *gammarus* and the lobster fishery in Europe in the eighteenth century, see vol. II, p. 42. A curious epitome of the lore and natural history of the higher Crustacea from the most ancient times.

137. HERDMAN, W. A.

Lobster hatching. Nature, vol. LXX, p. 296. London, 1904.

Note on hatching of lobsters at Biological Station, Port Erin, Isle of Man.

138. HERMANN, G.

Notes sur la structure et le développement des spermatozoïdes chez les Décapodes. Bulletin scientifique de la France et de la Belgique, t. XXII. Paris, 1890.

First record of amœboid movements in sperm cells, with description of hermaphrodite character of sexual organs.

139. HERRICK, FRANCIS H.
The development of the American lobster, *Homarus americanus*. Johns Hopkins University Circulars, vol. IX, no. 80, p. 67–68. Baltimore, 1890. See also Zoologischer Anzeiger, 1891, p. 133–137, and p. 145–149, fig. 1–6.

140. ———
Notes on the habits and larval stages of the American lobster. Johns Hopkins University Circulars, vol. X, p. 97–98. Baltimore, 1891.

141. ———
The reproductive organs and early stages of development of the American lobster. Johns Hopkins University Circulars, vol. X, p. 98–101. Baltimore, 1891.

142. ———
Cement glands and origin of egg-membranes in the lobster. Johns Hopkins University Circulars, vol. XII, p. 103. Baltimore, 1893.

143. ———
The habits and development of the lobster and their bearing upon its artificial propagation. Paper presented at the World's Fisheries Congress, Chicago, 1893. Bulletin U. S. Fish Commission for 1893, p. 75–86. Washington, 1894.

144. ———
The reproduction of the lobster. Zoologischer Anzeiger, p. 289–292. Leipzig, 1894. See also The Zoologist, vol. 18, p. 413–417. London, 1894.

145. ———
The lobster. Johnson's Universal Encyclopedia, vol. V, p. 317–318. New York, 1894.

146. ———
Notes on the biology of the lobster. Science, n. s., vol. I, p. 263–266. See also Science, n. s., vol. I, p. 382. New York, 1895.

147. ———
The reproduction of the lobster. Zoologischer Anzeiger, p. 226–228. Leipzig, 1895.

148. ———
Movements of the nucleolus through the action of gravity. Anatomischer Anzeiger, bd. X, p. 337–340, fig. 1–4. Jena, 1895.

149. ———
The American lobster: A study of its habits and development. Bulletin U. S. Fish Commission, vol. XV, 1895, p. 1–252, pl. A to J and 1–54. Washington, 1896.

150. ———
The protection of the lobster fishery. Bulletin U. S. Fish Commission vol. XVII, 1897, p. 217–224. Washington, 1896.

151. ———
The "great forceps" of the American lobster. Science, n. s., vol. XXI, p. 375–376. New York, 1905.

152. ———
The reproductive period of the lobster. Bulletin U. S. Fish Commission, vol. XXI, 1901, p. 161–166, fig. 1–5. Washington, 1902.

153. ———
Torsion of the crustacean limb. Biological Bulletin, vol. IX, p. 130–137, fig. 1–5. Boston, 1905. See also Science, n. s., vol. XXI, p. 376. New York, 1905.

154. ———
Effective protection for the lobster fishery. Science, n. s., vol. XXIII, no. 591, p. 650–655. New York, 1906.

155. HERRICK, FRANCIS H.
Symmetry in big claws of the lobster. Science, n. s., vol. XXV, p. 275–277. New York, 1907.

156. ———
The preservation and propagation of the lobster. Proceedings of the First New England Conference called by the Governors of the New England States, November 23–24, 1908, p. 41–60, with discussion by delegates. Published for the Governors. Boston, 1908. Reprinted in Report of the Commissioners on Fisheries and Game of Massachusetts for 1908. Public Document no. 25, p. 29–46. Boston, 1909.

This paper has been reproduced in part in chapter XII of the present work.

157. ———
Facts about the "lobster pearl." American Naturalist, vol. XLIV, p. 294–301, fig. 1–5. New York, 1910.

158. HOME, C.
Note on the phosphorescence of the lobster after death. Zoologist, 2d ser., vol. 4, p. 1725–1726. 1869.

159. HORNADAY, WILLIAM T.
A large lobster. Zoological Society Bulletin, no. 29, p. 425–426. New York, 1908.

Records the capture of a living male lobster January 23, 1908, at Cranberry Isles, Hancock County, Me., measuring 16 inches and weighing 14½ pounds.

160. HOVEY, E. O.
Measurements of two large lobsters recently added to the collections of the American Museum of Natural History. Proceedings American Association for the Advancement of Science for 1898, p. 365–366. New York, 1899.

For measurements given at reading of this communication, but not printed here, see *278*.

161. HUNT, BOOKMAN, and TIERNEY.
Einige allgemeine Eigenschaften des Herzmuskels vom amerikanischen Hummer (*Homarus americanus*). Centralblatt für Physiologie, bd. 11, p. 274–278, 7 fig. Leipzig, 1898.

162. HUXLEY, T. H.
On the classification and the distribution of the crayfishes. Proceedings of the Zoological Society, p. 752–788. London, 1878.

For branchial formula of Homarus, see p. 777.

163. HYATT, ALPHEUS.
Moulting of the lobster, *Homarus Americanus*. Proceedings Boston Society Natural History, vol. XXI, p. 83–90, 1880–1882. Boston, 1883.

164. ———
[Remarks on distortions of lobster's claws.] Proceedings Boston Society Natural History, vol. XXI, Report of general meeting, p. 278. Boston, 1883.

165. IRVINE and WOODHEAD.
Secretion of carbonate of lime by animals. Part II. Proceedings Royal Society Edinburgh, vol. 16, for 1888–1889, p. 324–354. Edinburgh, 1889.

166. JORDAN, H.
(2) Zur Frage nach der excretiven Function der Mitteldarmrüse (Leber) bei *Astacus fluviatilis*. Archiv für die gesammte Physiologie, 105. bd., p. 365–379. Bonn, 1904.

167. ———
(3) Zur physiologischen Morphologie der Verdauung bei zwei Evertebraten. Biologisches Centralblatt, bd. 24, p. 321–332, 5 fig. Erlangen, 1904.

168. JORDAN, H.

(4) Beiträge zur vergleichenden Physiologie der Verdauung. IV. Die Verdauung und der Verdauungsapparat des Flusskrebses (*Astacus fluviatilis*). Archiv für die gesammte Physiologie, bd. 101, p. 263–310, fig. 1–6 and taf. VII. Bonn, 1904.

Treats of physiology of digestion in lobster, and attributes both digesting and absorbing function to the midgut or liver.

169. KALM, PETER.

Travels into North America; containing its natural history and a circumstantial account of its plantations and agriculture in general, with the civil, ecclesiastical, and commercial state of the country, the manners of the inhabitants, and several curious and important remarks on various subjects. Translated by John Reinhold Forster. 3 vol. Warrington, 1770.

Curious reference to stocking of waters near New York with lobsters by wreck of "well boat," vol. 1, p. 240–241.

170. KÖLLIKER, ALBERT.

Beiträge zur Kenntniss der Geschlechtsverhältnisse und der Samenflüssigkeit wirbelloser Thiere, nebst einem Versuch über das Wesen und die Bedeutung der sogenannten Samenthiere. Berlin, 1841. See also Observations pour servir à l'histoire des organes sexuels et du liquide séminal des Crustacés et des Cirrhipèdes. Annales des Sciences Naturelles, 2e sér., t. 19, p. 335–350, pl. 9–13. Paris, 1841.

171. KOLTZOFF, NICHOLAS.

Untersuchungen über Spermien und Spermiogenese bei Decapoden. Vorläufige Mittheilung. Anatomischer Anzeiger, bd. 24, p. 83–95, Jena, 1903.

172. KOLTZOFF, N. K.

Studien über die Gestalt der Zelle. 1. Untersuchungen über die Spermien der Decapoden, als Einleitung in das Problem der Zellengestalt. Archiv für Mikroskopische Anatomie und Entwicklungsgeschichte, bd. 67, p. 364–572, taf. XXV–XXIX, fig. 1–36. Bonn, 1906.

The most complete account of the structure and movements of crustacean sperm cells, and their probable behavior in fertilization.

173. KRÖYER, HENRIK.

Monografisk Fremstilling af Slaegten Hippolyte's nordiske Arter. Med Bidrag til Dekapodernes Udviklingshistorie. Det Kongelige Danske Videnskabernes Selskabs naturvidenskabelige og mathematiske Afhandlinger. Niende deel, pl. I–VI, p. 133–144. Kjöbenhavn, 1842.

Erdl (*98*), writing in 1843, thus characterizes this work: "Gives figures and descriptions of form of the outer parts of the lobster-embryos, especially the palps and legs. The text is brief, and the figures leave much to be desired."

174. LABBÉ, ALPHONSE.

Sur la spermatogenèse des crustacés décapodes. Comptes rendus de l'Académie des Sciences, t. 137, p. 272–274. Paris, 1903.

175. ———.

La maturation des spermatoides et la constitution des spermatozoides chez les crustacés décapodes. Archives de Zoologie, sér. IV, vol. 2, p. 1–14, fig. 1–27. Paris, 1903.

176. ———.

Sur la formation des tétrades et les divisions maturative dans la testicule du homard. Comptes rendus de l'Académie des sciences, t. 138, p. 96–99. Paris, 1904.

177. LATASTE, FERNAND.

Fécondité de la femelle du homard Américain en fonction de sa taille. Actes de la Société des Science du Chili, t. VI, p. 106–109. Santiago, 1896.

178. LAVELLE.

Recherches d'anatomie microscopique sur le test des Crustacés décapodes. Annales des Sciences naturelles, 3e sér., Zoologie, t. 7, p. 352–377, pl. 7, fig. 10–12. Paris, 1847.

One of the earliest accounts of structure of shell in lobster, describing canals (of the tegumental glands) in relation to hairs, tubercles, and angles of the exoskeleton.

179. LEMOINE, VICTOR.
Recherches pour servir à l'histoire des systèmes nerveux, musculaire et glandulaire de l'écrevisse. Parts I–II. Annales des Sciences naturelles, 5e sér., Zoologie, t. IX., p. 99–280, pl. 6–11, Paris, 1868. Part III, Annales des Sciences naturelles, t. X, p. 5–54, Paris, 1868.

180. LLOYD, W. A.
Exuviation of lobsters. The Field, May 25, London, 1878. Extract in The Zoologist, 3d ser., vol. II, p. 225–226. London, 1878.

181. LOEB, LEO.
Untersuchungen über Blutgerinnung. 6. Mittheilung. Beiträge zur Chemie, Physiologie und Pathologie, bd. 6, p. 260–286. Braunschweig, 1905.
Study on coagulation of blood in *Homarus* and *Limulus*.

182. LOVETT, EDWARD.
Abnormal color of common lobster. The Zoologist. 3d ser., vol. VIII, p. 491. London, 1884.

183. ———
Notes and observations on British stalk-eyed Crustacea. The Zoologist, 3d ser., vol. IX, p. 100–104. London, 1885.

184. LUCAS, H.
Notice sur quelques monstruosités observées dans les crustacés appartenant aux genres *Carcinus*, *Lupa*, *Homarus* et *Astacus*. Annales de la Société entomologique de France, 2e sér., t. II, pl. I. Paris, 1844.

185. MACMUNN, C. A.
On the chromatology of the blood of some invertebrates. Quarterly Journal of Microscopical Science, n. s., vol. XXV, p. 469–499, pl. 33, 34. London, 1885.

186. ———
Contributions to animal chromatology. Ibid., vol. XXX, p. 51–96, pl. 6. London, 1890.

187. MACPHAIL, ANDREW.
Discoloration in canned lobsters. Supplement 2, Twenty-ninth Annual Report of the Department of Marine and Fisheries of Canada, p. 1–33. Ottawa, 1898.

188. MALARD, A. E.
Influence de la lumière sur la coloration des crustacés. Bulletin de la Société Philomathique de Paris, 8e sér., t. IV, pl. 24–30. Paris, 1892.

189. MARSHALL, C. F.
Some investigations on the physiology of the nervous system of the lobster. Studies from the biological laboratories of Owens College, vol. I, p. 313–323. Manchester, 1886.

190. MARTIN, M. J.
Sur un spécimen blanchâtre de homard. Bulletin de la Société Philomathique de Paris, 8e sér., t. IV, p. 17–19. Paris, 1892.

191. MATHER, FRED.
What we know of the lobster. Bulletin U. S. Fish Commission, vol. III, 1883, p. 281–286. Washington, 1904. See also Scientific American Supplement, Feb. 10, 1894, and The lobster, The Aquarium, vol. III, p. 89–92. Brooklyn, 1894.

192. MAYER, PAUL.
Zur Entwicklungsgeschichte der Dekapoden. Jenaische Zeitschrift-Naturwissenschaft, bd. XI, p. 188–269, taf. XIII–XV. Jena, 1877.

193. MEAD, A. D.
Habits and growth of young lobsters, and experiments in lobster culture. Thirty-first Annual Report of the Commissioners of Inland Fisheries of Rhode Island for 1901, p. 61–80, pl. I–IV. Providence, 1901.

194. MEAD, A. D.
Habits and growth of young lobsters and experiments in lobster culture. Thirty-second Annual Report of the Commissioners of Inland Fisheries of Rhode Island for 1902, p. 25–51, 7 pl. Providence, 1902.

195. MEAD, A. D. and WILLIAMS, L. W.
Habits and growth of the lobster, and experiments in lobster culture. Twenty-third Annual Report of the Commissioners of Inland Fisheries of Rhode Island for 1903, p. 57–86, 4 fig. Providence, 1903.

196. MEAD, A. D.
Experiments in lobster culture. Thirty-fourth Annual Report of the Commissioners of Inland Fisheries of Rhode Island for 1904, p. 74–82. 4 pl. Providence, 1904. See also the same, in Thirty-fifth Annual Report of same, p. 33–43, pl. I–VI.

197. ———
The problem of lobster culture. (Contributions from the Anatomical Laboratory of Brown University, vol. 4, no. 3.) Proceedings American Fisheries Society, p. 156–166, 3 fig. 1905.

198. ———
A method of lobster culture. Proceedings of the Fourth International Fishery Congress, Bulletin of the Bureau of Fisheries, vol. XXVIII, 1908, p. 219–240, pl. VII–XI. Washington, 1910. Reprinted in Thirty-ninth Annual Report of the Commissioners of Inland Fisheries of Rhode Island for 1909, p. 105–138, pl. 1–9. Providence, 1909.

199. ———
A new principle of aquiculture and transportation of live fishes. Proceedings of the Fourth International Congress, Bulletin of the Bureau of Fisheries, vol. XXVIII, 1908, p. 759–780, pl. XC–C. Washington, 1910. Reprinted in Thirty-ninth Annual Report of Commissioners of Inland Fisheries of Rhode Island for 1909, p. 79–100, fig. 1–23. Providence, 1909.

200. MEEK, A.
The crab and lobster fisheries of Northumberland. Report of the Northumberland sea Fisheries Committee on the Scientific Investigation for 1904, p. 21–67. Newcastle-upon-Tyne, 1904.

201. MIVART, ST. GEORGE.
The lobster. Popular Science Review, vol. 7, p. 345–353. London, 1868.

202. MOQUIN-TANDON et SOUBEIRAN, J. L.
Établissements de pisciculture de Concarneau et de Port-de-Bouc. Bulletin de la Société impériale zoologique d'Acclimatation, 2e sér., t. II, p. 533–545, with fig. Paris, 1865.

Gives measurements and weights from fourth to fourteenth molts of lobsters said to have been reared in the inclosures at this station.

203. MORGAN, T. H.
Notes on regeneration. Biological Bulletin, vol. VI, p. 159–172, Boston, 1904. (Note on regeneration of lobster's claws.)

204. NEWTON, EDWIN T.
The structure of the eye of the lobster. Quarterly Journal of Microscopical Science, n. s., vol. 13, p. 325–343, 2 pl. London, 1873.

205. NICHOLLS, F.
An account of the hermaphrodite lobster presented to the Royal Society on Thursday, May 7, by Mr. Fisher, of Newgate Market, examined and dissected pursuant to an order of the society. Philosophical Transactions Royal Society, vol. XXXVI, 1729–1730, p. 290–294, fig. 1–4. London, 1731. Abridgment, vol. VII, p. 421–423, pl. III, IV. 1734.

Described for first time the structure in the female lobster now known as the sperm-receptacle.

206. OWSJANNIKOW, M. P.
Sur la structure intime du système nerveux du Homard. Comptes rendus de l'Académie royale des sciences, t. 52, p. 378–381. Paris, 1861. See also Recherches sur la structure intime du système nerveux des crustacés et principalment du homard. Annales des Sciences Naturelles, sér. 4, t. xv, p. 129–141, pl. 6–7. Paris, 1861.

207. PACKARD, A. S.
The history of the lobster. Review of paper by S. I. Smith. American Naturalist, vol. VIII, p. 414–417. One pl. and fig. Salem, 1874.

208. ———
The molting of the lobster. American Naturalist, vol. XX, p. 173. Philadelphia, 1886.

209. ———
The Labrador coast. New York, 1891.

210. PARKER, G. H.
The histology and development of the eye in the lobster. Bulletin of the Museum of Comparative Zoology, vol. XX, p. 1–60, pl. I–IV. Cambridge, 1890.

211. PATTERSON, A. H.
Remarkable lobster claw. The Zoologist, vol. IX, p. 350–351, with fig. London, 1905.

212. PENNANT, THOS.
Article on lobsters, with letters by Travis. (See ref. no. *264*.) British Zoology, vol. IV, p. 8–19. London, 1777.

213. PHILIPPI, R. A.
Zoologische Bemerkungen. Wiegmann's Archiv für Naturgeschichte, bd. VI, p. 181–195, taf. III–IV. Berlin, 1840.

Treats especially of the metamorphosis of the lobster.

214. PLATEAU, FÉLIX.
Recherches physiologiques sur le cœur des Crustacés décapodes. Archives de Biologie, t. I, p. 595–696, pl. XXVI–XXVII. Gand, 1880.

Gives brief account of the anatomy of the heart of the lobster, and a detailed study of its movements by the graphic method.

215. POUCHET, G.
Note sur un muscle vibrant existant chez le homard. Comptes rendus de la Société de Biologie, sér. 6, t. 2, p. 358–360. Paris, 1876.

216. ———
Des changements de coloration sous l'influence des nerfs, Journal de l'Anatomie et de la Physiologie, p. 1–90, 113–165, 4 pl. Paris, 1876.

217. PRENTISS, C. W.
Otocyst of decapod Crustacea: Its structure, development and functions. Bulletin Museum of Comparative Zoology at Harvard College, vol. XXXVI, p. 167–252, pl. 1–10. Cambridge, 1901.

218. PRINCE, E. E.
Special report on the natural history of the lobster with special reference to the Canadian lobster industry. Supplement 1 to Twenty-ninth Annual Report of the Department of Marine and Fisheries of Canada, p. I–IV, 1–36. Ottawa, 1897.

219. ———
Report of the Canadian Lobster Commission for 1898. Supplement 1, to Thirty-first Annual Report of the Department of Marine and Fisheries, p. 1–42. Ottawa, 1899. See also Addendum.

220. PRZIBRAM, HANS.
Experimentelle Studien über Regeneration. (1st paper.) Archiv für Entwickelungsmechanik der Organismen, bd. 11, p. 321–345, taf. 11–14. Leipzig, 1901.

Describes reversal of asymmetry or compensatory hypertrophy in *Alpheus*, with records of experiments on regeneration in related Macrura.

221. PRZIBRAM, HANS.
Experimentelle Studien über Regeneration. (2d paper, Crustacea.) Ibid., bd. XIII, p. 507–527, taf. XXI–XXII. Leipzig, 1901–02.

222. ———
Beobachtungen über adriatische Hummer im Aquarium (und vorlaüfige Mittheilung uber Regenerations Versuche). Ibid., bd. 25, p. 76–82, fig. 1902.

223. ———
Die "Heterochelie" bei decapoden Crustaceen. (3d paper, Crustacea.) Ibid., bd. XIX, p. 181–247 taf. VIII–XIII. Leipzig, 1905.

224. RASCH.
Om Forsog med kunstig Udklaekning af Hummer. "Nordisk Tidsskrift for Fiskeri," ny raekke, 2 en aargang, p. 184–188. 1875. Translated in Report of the U. S. Fish Commission for 1873–74, and 1874–75, p. 267–269. Washington, 1876.

225. RATHBUN, MARY J.
Some changes in crustacean nomenclature. Proceedings of the Biological Society of Washington, vol. XVII, p. 169–172. Washington 1904.

226. RATHBUN, RICHARD.
The Fisheries and Fishery Industries of the United States. Prepared through the cooperation of the Commissioner of Fisheries and the Superintendent of the Tenth Census by George Brown Goode. Section I.—Natural history of the useful aquatic animals; part V, Crustaceans, Worms, Radiates, and Sponges, p. 759–850. With one volume of plates. Washington, 1884.

227. ———
In same work, section V, History and methods of the fisheries, vol. II, pt. XXI, The crab, lobster, crayfish, rock lobster, shrimp and prawn fisheries, p. 627–810. Washington, 1887.

228. ———
The transplanting of lobsters to the Pacific coast of the United States. Bulletin U. S. Fish Commission, vol. VIII, for 1888, p. 453–472. Washington, 1890.

229. ———
Notes on lobster culture. Bulletin U. S. Fish Commission, vol. VI, p. 17–32. Washington, 1886.

230. RATHKE, HEINRICH.
Zur Entwickelungsgeschichte der Dekapoden. Wiegmann's Archiv für Naturgeschichte, bd. VI, I, p. 241–249. Berlin, 1840. Translated by W. Francis in Annals and Magazine of Natural History, vol. VI, p. 263–269. London, 1841. Abstract of complete paper which follows (*231*).

Describes hatching of ripe embryos of *Astacus marinus* (*Homarus*) and other decapods.

231. ———
Beiträge zur vergleichenden Anatomie und Physiologie. Reisebemerkungen aus Skandinavien. Neueste Schriften der naturforschenden Gesellschaft in Danzig, bd. III, pt. II, Zur Entwickelungsgeschichte der Dekapoden, (I) *Astacus marinus*, p. 23–29, taf. II, fig. 11–21. 1842. Danzig.

Contains a description of the hatching of the lobster, with good figures.

232. ———
De Animalium Crustaceorum Generatione. Commentatio. 26 p. Regiomontii, 1844.

233. RAVERET-WATTEL.
L'aquiculture marine en Norvège. Revue des Sciences naturelles appliquées, t. 37, p. 147–156, 246–257. Paris, 1890.

An account of Captain Dannevig's experiments in rearing lobsters and fish.

234. RÉAUMUR.
Sur les diverses reproductions qui se font dans les écrevisses, les omars, les crabes, etc., et entre autres sur celles de leurs jambes et de leurs écailles. Mémoires de l'Académie royale des Sciences, p. 226–245, pl. 12. Paris, 1712.

235. Reed, Margaret.
The regeneration of the first leg of the crayfish. Archiv für Entwickelungsmechanik der Organismen, bd. 18, p. 307–316, 3 fig. Leipzig, 1904.

236. Ridegood, W. G.
Abnormal oviducts in the lobster. Annals of Natural History, vol. III, p. 1–7, fig. 1–2. London, 1909.

237. Roché, Georges.
La culture des mers en Europe. 328 p., illustrated. Bibliothèque Scientifique Internationale. Paris, 1898.
In chapter VI is given a general summary of the development and artificial propagation of the lobster and langouste, or *Palinurus*.

238. Ryder, J. A.
The metamorphosis of the American lobster, *Homarus americanus* H. Milne-Edwards. American Naturalist, vol. XX, p. 739–742. Philadelphia, 1886.

239. ———
Hatching, rearing, and transplanting lobsters. Science, vol. VII, p. 517–519. New York, 1886.

240. Sabatier, Armand.
De la spermatogenèse chez les crustacés décapodes. Travaux de l'Institut de Zoologie de Montpellier et de la Station Maritime de Cette. 394 p., 10 pl. Montpellier et Paris, 1893.

241. Salter, S. J. A.
On the molting of the common lobster (*Homarus vulgaris*) and the shore crab (*Carcinus mænas*). Journal Linnean Society, London, vol. 4, p. 30–35. London, 1860.

242. Sars, G. O.
Om Hummerens postembryonale Udvikling. Med 2 autographiske Plancher. Christiania Videnskabs-Selskabs Forhandlingar, p. 1–28. Christiania, 1874.
Gives first detailed and adequately illustrated account of metamorphosis of *Homarus gammarus*.

243. ———
Development of the European lobster. Abstract in American Journal Science and Arts, 3d ser., vol. 9, p. 231. New Haven, 1875.

244. ———
Reports made to the Department of the Interior of investigations of the salt-water fisheries of Norway during the years 1874–1877. (Indberetninger til Departmentet for det Indre fra Professor G. O. Sars om de afham i Aarene 1874–1877 anstillede Undersögelser vedkommende Saltvandsfiskerierne, Christiania, 1878.) Translated by Herman Jacobson, in Report U. S. Fish Commission for 1877, p. 663–705. Washington, 1879.

245. Saville-Kent.
The artificial culture of lobsters. International Fisheries Exhibition, London, 1883, The Fisheries Exhibition Literature, vol. VI, Conferences, part III, p. 327, 1 plate. London, 1884.

246. Say, Thomas.
An account of the Crustacea of the United States. Journal Academy Natural History, vol. 1, pt. 1, p. 155–169 (*Astacus*), 235–253, 316–319, 374–401, 423–459, with appendix, observations, and notes. Philadelphia, 1817.

247. Scherren, H.
Meristic variations in *Cancer pagurus* and *Astacus gammarus*. Proceedings Zoological Society, ser. 1903, vol. 2, p. 195–196. London, 1903.

248. Scott, Andrew.
On the spawning of the common lobster. Report for 1902 on the Lancashire and Sea-Fisheries Laboratory at University College, Liverpool, and the Sea Fishery Hatchery at Piel, p. 20–27. Liverpool, 1902.
Gives first detailed account of spawning of *Homarus gammarus*.

249. SEBA, ALBERTUS.
Locupletissimi rerum naturalium thesauri accurata descriptio et iconibus artificiosissimis expressio per universam physices historiam, t. III, tab. XVII, no. 3. Copper-plate figure of lobster, called *Astacus marinus americanus*. Amstelædami, 1758.

Seba was the first to describe and figure the American lobster as a distinct species.

250. SHERWOOD, GEORGE H.
Experiments in lobster rearing. Report of special commission for the investigation of the lobster and the soft-shell clam. Washington, 1903. Report of the U. S. Fish Commission for 1903, p. 149–174, pl. 1–3. Washington, 1905.

251. SMITH, A. C.
Notes on the lobster, *Homarus americcnus*. Bulletin U. S. Fish Commission, vol. V, p. 121–125. Washington, 1885.

252. SMITH, ALBERT W.
Composition of the shell and gastroliths of the lobster. In The American lobster (see *149*), Appendix II, p. 227–228, Bulletin U. S. Fish Commission, vol. XV, 1895. Washington, 1896.

253. SMITH, HUGH M.
In Report of the special commission for investigation of the lobster and the soft-shell clam. (Decline of the lobster fishery.) Report of the U. S. Fish Commissioner for 1903, p. 141–148. Washington, 1905.

253a. ———
A review of the history and results of the attempts to acclimatize fish and other water animals in the Pacific States. Bulletin U. S. Fish Commission, vol. XV, p. 379–472. (The American lobster, p. 459–463.) Washington, 1895.

254. SMITH, SIDNEY I.
The early stages of the American lobster (*Homarus americanus* Edwards). American Journal Science and Arts, vol. III, p. 401–406, pl. IX. New Haven, 1872. Abstract of fuller paper, no. *256*.

255. ———
The metamorphoses of the lobster and other Crustacea. Invertebrate Animals of Vineyard Sound, etc. (Verrill & Smith), in Report of the U. S. Fish Commissioner for 1871–1872, p. 522–537, fig. 4 and pl. IX. Washington, 1873.

256. ———
The early stages of the American lobster (*Homarus americanus* Edwards). Transactions Connecticut Academy of Arts and Sciences, vol. II, pt. 2, p. 351–381, pl. XIV–XVIII, fig. 1–4. New Haven, 1873.

The best account at date of the metamorphosis of the American lobster.

257. STAHR, H.
Neue Beiträge zur Morphologie der Hummerschere, mit physiologischen und phylogenetischen Bemerkungen. Jenaische Zeitschrift für Naturwissenschaft, bd. 33, p. 457–482, pl. XX–XXI. Jena, 1898.

First notice of periodic arrangement of teeth in toothed forceps of lobster.

258. ———
Über das Alter der beiden Chelæ von *Homarus vulgaris* und über die "similar claws" Herrick's. Zur Verständigung mit Herrn Przibram. Archiv für Entwickelungsmechanik der Organismen, bd. 12, p. 162–166. Leipzig, 1901.

259. STEBBING, THOMAS R. R.
A history of Crustacea: Recent Malacostraca. International Science Series, vol. LXIV. New York, 1893.

Advocates the use of the generic name of *Astacus* (Leach) instead of *Homarus* (Milne-Edwards), for the true lobsters.

260. STEBBING, THOMAS R. R.
The lobster in commerce and science. Natural Science, vol. IX, p. 38–42. London, 1896.
A review of no. *149*, with further arguments for the use of the name *Astacus* for the lobsters.

261. STEVENSON, CHARLES H.
The preservation of fishery products for food. Bulletin of U. S. Fish Commission, vol. XVIII, 1898, p. 337–564. Washington, 1899.
Describes methods of preserving lobsters, and of transporting them alive.

262. THOMPSON, J. V.
Letter to the editor of the Zoological Journal, dated "Cork, Dec. 16, 1830." Zoological Journal, vol. V, p. 383–384, pl. XV, fig. 13. London, 1831.
First announcement of discovery of metamorphosis in the European lobster (*Astacus marinus*).

263. THOMPSON, WILLIAM.
Description of a young lobster measuring only nine lines. The Zoologist, vol. IX, p. 3765. London, 1853.

264. TRAVIS.
Letter to Thomas Pennant dated "Scarborough, 25th October, 1768." Quoted in article on lobster by Thomas Pennant (see *212*), Pennant's British Zoology, vol. IV, p. 10–13. London, 1777.

265. TULLBERG, TYCHO.
Studien über den Bau und das Wachstum des Hummerpanzers und der Molluskenschalen. Kongliga Svenska Vetenskaps-Akademiens Handlingar, bd. 19, no. 3, p. 57, 12 taf. Stockholm, 1882.

266. VALENCIENNES, A.
Note sur la reproduction des Homards. Comptes rendus de l'Académie des Sciences, t. 46, p. 603–606. Paris, 1858.

267. VALENTIN, G.
Repertorium für Anatomie und Physiologie. Die Fortschritte der Physiologie im Jahre 1837, bd. III, p. 188. 1838.
Describes for the first time the rayed sperm cells of the lobster.

268. VAN BENEDEN, P. J.
Note sur une pince de homard monstrueuse. Bulletin de l'Académie Royale de Belgique, 2e sér., t. 17, p. 371–376, fig. Bruxelles, 1864.

269. ———
Bulletin de l'Académie Royale de Belgique, t. XXVIII, p. 444–456. Bruxelles, 1869. See also Development of Gregarinæ, Quarterly Journal Microscopical Science, vol. X, 1870, p. 290. London. For reference to other work, see Foettinger, *108*.
Describes gregarine parasitic in lobster.

270. VAN DER HOEVEN, J. E.
Handbook of zoology. Translated from 2nd Dutch edition by Rev. William Clark. 2 vol. Cambridge, Eng., 1856.
Discusses the function of the gastroliths, and was one of the first to protest against the theory that they served merely to provide a store of lime to be drawn upon for hardening the soft shell.

271. VERRILL, A. E.
Report upon the invertebrate animals of Vineyard Sound and the adjacent waters, with an account of physical characters of the region. Report U. S. Fish Commission for 1871–72, p. 295–778, pl. I–XXXVIII, with descriptions. Washington, 1873.

272. VITZOU, ALEXANDRE-NICOLAS.
Recherches sur la structure et la formation des tégumens chez les Crustacés décapodes. Archives de Zoologie expérimentale et générale, t. X, p. 451–576, pl. XXIII–XXVIII. Paris, 1882.

273. WAITE, F. C.
The structure and development of the antennal glands in *Homarus americanus* Milne-Edwards. Bulletin Museum Comparative Zoology, Harvard College, vol. XXXV, no. 7, p. 151–210, pl. 1–6. Cambridge, 1899.

274. ———
A large lobster. Science, n. s., vol. IV, p. 230–231. New York, 1896.

275. WALLENGREN, H.
Über das Vorkommen und die Verbreitung der sogenannten Intestinaldrüsen bei den Dekapoden. Zeitschrift für wissenschaftliche Zoologie, bd. 70, p. 321–346, 12 fig. Leipzig, 1901.

276. WELDON and FOWLER, G. H.
Notes on recent experiments relating to the growth and rearing of food-fish at the laboratory. I. The rearing of lobster larvæ. Journal Marine Biological Association of the United Kingdom, n. s., vol. I, no. 4, p. 367–375. London, 1890.

277. WHEILDON, WM. H.
The lobster (*Homarus americanus*): The extent of the fishery; the spawning season; food of the lobster; shedding of the shell; legislation on the fishery. Proceedings American Association for the Advancement of Science, vol. XXII, p. 133–141. 1875.

278. WHITFIELD, R. P.
Notice of two very large lobsters in the collection of the American Museum of Natural History. Bulletin American Museum Natural History, vol. XII, p. 191–194, pl. IX. New York, 1899.
The living weight of the animals described is given as 34 and 31 pounds.

279. WILLIAMS, LEONARD W.
The stomach of the lobster and the food of larval lobsters. Thirty-seventh Annual Report of the Commissioners of Inland Fisheries of Rhode Island, p. 153–180, pl. I–X. 1907.
Gives the first detailed and satisfactory account of the complex mechanism of the lobster's stomach.

280. WILLIAMSON, H. CHARLES.
Contributions to the life-history of the edible crab (*Cancer pagurus* Linn.). Eighteenth Annual Report of the Fishery Board for Scotland, pt. III, p. 77–142, pl. I–IV. Glasgow, 1900.
Argument for theory of annual spawning in the European lobster.

281. ———
Contributions to the life-histories of the edible crab (*Cancer pagurus*) and of other decapod Crustacea; Impregnation; Spawning; Casting; Distribution; Rate of Growth. Twenty-second Annual Report of the Fishery Board for Scotland, pt. III, p. 100–141, pl. I–V. Glasgow, 1904.
Advances a new theory to explain the attachment of the eggs to the swimmerets of decapods.

282. ———
A contribution to the life-history of the lobster (*Homarus vulgaris*). Twenty-third Annual Report of the Fishery Board for Scotland for 1904, pt. III, Scientific Investigations, p. 65–107, pl. I–IV. Glasgow, 1905.

283. WILSON, ANDREW.
The anatomy of the lobster. Science for All, vol. 2, p. 34–41, fig. 1–8 (appendages). London, 1879.

284. WILSON, E. B.
Notes on the reversal of asymmetry in the regeneration of the chelæ in *Alpheus heterochelis*. Biological Bulletin, vol. IV, p. 197–214. Boston, 1902–1903.

285. WOOD, R. K.
The lobster. Chiefly a translation from a work of M. Coste to the Minister of the French Marine, in "Land and Water," London. Extract in American Naturalist, vol. II, p. 494–496. Salem, 1869.

286. WOOD, W. M.
Transplanting lobsters to the Chesapeake: Experiments upon the temperature they can endure. Bulletin U. S. Fish Commission for 1885, vol. V, p. 31–32. Washington, 1885.

287. YOUNG, JOHN.

On the head of the lobster. Journal Anatomy and Physiology, vol. 14, p. 348–350, pl. XVIII. London, 1879.

288. YUNG, E.

De la structure intime et du système nerveux central des Crustacés décapodes. Thèse. Paris, 1879.

289. ZELENY, CHAS.

Compensatory regulation. Biological Bulletin, vol. II, p. 1–102, fig. 1–27. Boston, 1905.

290. ———

The regeneration of a double chela in the fiddler crab (*Gelasimus pugilator*) in place of a normal single one. Biological Bulletin, p. 152–155. Boston, 1905.

291. A MANUAL OF FISH-CULTURE, BASED ON THE METHODS OF THE UNITED STATES COMMISSION OF FISH AND FISHERIES.

Revised edition, i–x+1–340 p., numerous pl. Washington, 1900. The American lobster, p. 229–238.

Brief résumé of methods of collecting and hatching the eggs, and liberating the young fry.

292. ANNUAL REPORTS OF THE DEPARTMENT OF FISHERIES, WITH SUPPLEMENTAL FISHERIES STATEMENTS, DOMINION OF CANADA, FROM 1869 TO THE PRESENT. Ottawa.

Contain many special reports and notices of the lobster, the most important of which are here mentioned under separate titles.

293. A LARGE LOBSTER.

The Zoological Society Bulletin. New York, 1908.

Record of specimen measuring 23¾ inches in length, and weight 34 pounds.

294. A SUCCESSFUL EXPERIMENT IN LOBSTER REARING.

Nature, vol. LVI, p. 455. London, 1897.

295. CULTURE DU HOMARD EN AMÉRIQUE.

Bulletin Société d'Acclimatation, 2e sér., t. 10, p. 957–960. Paris, 1873.

296. FISHERIES STATEMENTS, 1880.

Supplement no. 2 to Eleventh Annual Report to Minister of Marine and Fisheries. Appendix no. 11, Report of J. H. Duvar, Inspector of Fisheries for the Province of Prince Edward Island, for 1880. Lobsters, p. 231. Ottawa, 1881.

297. FISHERIES STATEMENTS FOR THE YEAR 1882.

Supplement no. 2 to the Fifteenth Annual Report of the Department of Marine and Fisheries for the year 1882. Ottawa, 1883.

298. LES PÊCHERIES DE LA NORVÈGE.

Exposition Universelle de 1889 à Paris. Bergen, 1889.

Statistics on legal regulations of the lobster fishery, and upon the numbers and value of these animals annually exported, from 1883–87.

299. REPORT OF THE LOBSTER INDUSTRY IN CANADA FOR 1892.

Supplement to the Twenty-fifth Annual Report of the Department of Marine and Fisheries, p. 1–38. Ottawa, 1893.

300. REPORT OF THE CANADIAN LOBSTER COMMISSION.

Supplement no. 1 to the Thirty-first Annual Report of the Department of Marine and Fisheries, for 1898, p. 1–41, with map. Ottawa, 1899.

301. REVIEW OF THE REPORTS BY BUCKLAND AND SPENCER ON THE LOBSTER, CRAB, AND OYSTER FISHERIES OF GREAT BRITAIN.

Quarterly Review, vol. 144, art. VI, p. 249–262. 1877.

302. SAGACITY OF A LOBSTER.
Nature, vol. xv, p. 415. London, 1877.

303. THE CULTIVATION OF LOBSTERS.
Practical Magazine, vol. 2, p. 258–259. London, 1873.

ADDENDUM.

304. APPELLÖF, A.
Undersokelser over hummeren (*Homarus vulgaris*) med saerskilt hensyn til dens optraeden ved Norges kyster. Aarsberetning vedkommende Norges Fiskerier, 1 ste hefte. Bergen, 1909.

305. ———
Untersuchungen über den Hummer, mit besonderer Berücksichtigung seines Auftretens an den norwegischen Küsten. 1–80 p., pl. i–iii. Bergen, 1909.

This latest and in some respects the most detailed account of the habits and development of the European lobster was not received until the present work was in press.

306. ALEXANDROWICZ, JERZY STANILAW.
[The sympathetic nervous system of Crustacea.] Jenaische Zeitschrift für Naturwissenschaft, p. 395–444, 5 pl., 8 fig. Jena, 1909.

Attributes to the sympathetic system the control of peristalsis and the regulation of automatic movements.

307. DANNEVIG, G. M.
Hatching lobsters and cod in Norway. Bulletin U. S. Fish Commission for 1886, vol. VI, p. 13–14. Washington, 1887. See also on hatching of lobsters, Bulletin U. S. Fish Commission, vol. v, 1885, p. 280 and 446. Washington, 1885.

308. DITTEN, S. H.
De la protection et de la reproduction du Homard et des huitres. See L. Vaillant: Rapport du Jury international de l'Exposition universelle de 1878, groupe VIII, classe 84: Poissons, Crustacés et Mollusques.

309. FRIELE, M.
Notices sur les Pêcheries de la Norwège. Impression à part du catalogue spécial de la Norwège à l'exposition universelle de 1878 à Paris. Translated by J. Paul Wilson. Report U. S. Fish Commission, 1877, p. 707–740. Washington, 1879.

310. HADLEY, PHILIP B.
Additional notes upon the development of the lobster. Fortieth Annual Report of the Commissioners of Inland Fisheries of Rhode Island for 1910, p. 189–190, pl. 1–34, with descriptions of the appendages in stages 1 to 4. Providence, 1910.

311. (JACOBSON, HERMAN, translator.)
Transporting lobsters in Norway. Translation of "Forsendelse af Hummer," Norsk Fiskeritidende, Bergen, 1886. Bulletin U. S. Fish Commission, vol. VI, 1886, p. 319–320. Washington, 1887.

312. LUND, PETER WILHELM.
Zweifel an dem Dasein eines Circulationssystem bei den Crustaceen. Isis, p. 593–601, taf. II, fig. 2–4. Leipzig, 1825.

313. LUND, P. W., and SCHULTZ, A. W. F.
Fortgesetzte Untersuchungen über das System des Kreislaubes bei den Crustaceen. Isis, p. 1299. Leipzig, 1829.

314. MACMUNN, C. A.
On the gastric gland of Mollusca and Crustacea: its structure and functions. Philosophical Transactions of the Royal Society, vol. 193 B, p. 1–34; pl. 1–4. London, 1900.

315. MARCHAL, P.
Recherches anatomiques et physiologiques sur l'appareil excreteur des crustacés décapodes. Archives de Zoologie expérimentale et générale, sér. 2, t. 10, p. 57–275, pl. 1–9. Paris, 1892.

316. Meek, A.

The migration of lobsters. Report of the Northumberland Sea Fisheries Committee on the Scientific Investigation for 1902, p. 40. Newcastle-upon-Tyne, 1892.

317. Nickerson, W. S.

On *Stichocotyle nephropis* Cunningham, a parasite of the American lobster. Zoologische Jahrbücher; Abteilung fur Anatomie. 8 bd. p. 447–480, taf. 29–31. Jena, 1905.

Of 100 lobsters examined 2 were infested with this parasite, which was encysted at the rectal end of the intestine; 60–70 worms were found in one case, and but 1 in the other.

318. Owen, Richard.

Lectures on the comparative anatomy and physiology of the invertebrate animals, delivered at the Royal College of Surgeons. Hunterian lectures for 1852, 2d ed., p. i–viii, 1–690. London, 1855.

See also Descriptive and Illustrated Catalogue of the Museum of the College of Surgeons, Physiological series, vol. ii, p. 136, pl. fig. 1 and 2, and pl. 16, fig. 4, for figures and descriptions of Hunter's preparations of the arterial system of the lobster.

Figures heart of *Homarus gammarus*, and maintains that it carries mixed blood.

319. Perrin, Marshall L.

Transportation of lobsters to California. Report of the Commissioner, United States Commission of Fish and Fisheries for 1873–1875, p. 259–266. Washington, 1876.

320. Prince, E. E.

Report on fish culture in Canada for 1903, p. 1–48. Ottawa, 1904.

321. ———

Report on fish breeding in Canada for 1904, p. 1–48. Ottawa, 1905.

322. ———

Report on fish breeding operations in Canada for 1906, p. 1–62. Ottawa, 1906.

323. Rathbun, Richard.

Notes on the decrease of lobsters. Bulletin of the United States Fish Commission, vol. iv, 1884, p. 421–426. Washington, 1884.

324. Shearer, Cresswell.

On the anatomy of *Histriobdella homari*. Quarterly Journal of Microscopical Science, vol. 55, part 2, p. 287–360, pl. 17–20. London, 1910.

325. Smith, Hugh M.

The United States Bureau of Fisheries; its establishment, functions, organization, resources, operations, and achievements. Proceedings of the Fourth International Congress, Bulletin of the Bureau of Fisheries, vol. xxviii, 1908, p. 1365–1412. Washington, 1910.

326. Trybom, F.

Biologiska undersökningar 1901–1904. i. Hummerundersökningar vid Sveriges vestkust. Ur svenska hydrografisk-biologiska kommissionens skrifter. Häftet 2.

327. ———

Biologiska undersökningar 1901–1904. ii och iii. Tillägg till redogörelsen för hummerundersökningar vid Sveriges vestkust åren 1901–1903. Häftet 3.

328. Report of the Commissioners upon the lobster and oyster fisheries of Canada for 1887, p. 1–66, i–iv. Ottawa, 1888.

329. On the artificial propagation of the lobster; translated from Om Forsog med kunstig Udklaekning af Hummer, ny raekke, in Nordisk Tidsskrift for Fiskeri, ny raekke, Tidsskrift for Fiskeri, 2 en Aargang, p. 184–188. 1875. Report U. S. Fish Commission 1873–1875, p. 267–270. Washington, 1876.

HISTORY OF ECOLOGY

An Arno Press Collection

Abbe, Cleveland. **A First Report on the Relations Between Climates and Crops.** 1905

Adams, Charles C. **Guide to the Study of Animal Ecology.** 1913

American Plant Ecology, 1897-1917. 1977

Browne, Charles A[lbert]. **A Source Book of Agricultural Chemistry.** 1944

Buffon, [Georges-Louis Leclerc]. **Selections from Natural History, General and Particular, 1780-1785.** Two volumes. 1977

Chapman, Royal N. **Animal Ecology.** 1931

Clements, Frederic E[dward], John E. Weaver and Herbert C. Hanson. **Plant Competition.** 1929

Clements, Frederic Edward. **Research Methods in Ecology.** 1905

Conard, Henry S. **The Background of Plant Ecology.** 1951

Derham, W[illiam]. **Physico-Theology.** 1716

Drude, Oscar. **Handbuch der Pflanzengeographie.** 1890

Early Marine Ecology. 1977

Ecological Investigations of Stephen Alfred Forbes. 1977

Ecological Phytogeography in the Nineteenth Century. 1977

Ecological Studies on Insect Parasitism. 1977

Espinas, Alfred [Victor]. **Des Sociétés Animales.** 1878

Fernow, B[ernhard] E., M. W. Harrington, Cleveland Abbe and George E. Curtis. **Forest Influences.** 1893

Forbes, Edw[ard] and Robert Godwin-Austen. **The Natural History of the European Seas.** 1859

Forbush, Edward H[owe] and Charles H. Fernald. **The Gypsy Moth.** 1896

Forel, F[rançois] A[lphonse]. **La Faune Profonde Des Lacs Suisses.** 1884

Forel, F[rançois] A[lphonse]. **Handbuch der Seenkunde.** 1901

Henfrey, Arthur. **The Vegetation of Europe, Its Conditions and Causes.** 1852

Herrick, Francis Hobart. **Natural History of the American Lobster.** 1911

History of American Ecology. 1977

Howard, L[eland] O[ssian] and W[illiam] F. Fiske. **The Importation into the United States of the Parasites of the Gipsy Moth and the Brown-Tail Moth.** 1911

Humboldt, Al[exander von] and A[imé] Bonpland. **Essai sur la Géographie des Plantes.** 1807

Johnstone, James. **Conditions of Life in the Sea.** 1908

Judd, Sylvester D. **Birds of a Maryland Farm.** 1902

Kofoid, C[harles] A. **The Plankton of the Illinois River, 1894-1899.** 1903

Leeuwenhoek, Antony van. **The Select Works of Antony van Leeuwenhoek.** 1798-99/1807

Limnology in Wisconsin. 1977

Linnaeus, Carl. **Miscellaneous Tracts Relating to Natural History, Husbandry and Physick.** 1762

Linnaeus, Carl. **Select Dissertations from the Amoenitates Academicae.** 1781

Meyen, F[ranz] J[ulius] F. **Outlines of the Geography of Plants.** 1846

Mills, Harlow B. **A Century of Biological Research.** 1958

Müller, Hermann. **The Fertilisation of Flowers.** 1883

Murray, John. **Selections from *Report on the Scientific Results of the Voyage of H.M.S. Challenger During the Years 1872-76.*** 1895

Murray, John and Laurence Pullar. **Bathymetrical Survey of the Scottish Fresh-Water Lochs.** Volume one. 1910

Packard, A[lpheus] S. **The Cave Fauna of North America.** 1888

Pearl, Raymond. **The Biology of Population Growth.** 1925

Phytopathological Classics of the Eighteenth Century. 1977

Phytopathological Classics of the Nineteenth Century. 1977

Pound, Roscoe and Frederic E. Clements. **The Phytogeography of Nebraska.** 1900

Raunkiaer, Christen. **The Life Forms of Plants and Statistical Plant Geography.** 1934

Ray, John. **The Wisdom of God Manifested in the Works of the Creation.** 1717

Réaumur, René Antoine Ferchault de. **The Natural History of Ants.** 1926

Semper, Karl. **Animal Life As Affected by the Natural Conditions of Existence.** 1881

Shelford, Victor E. **Animal Communities in Temperate America.** 1937

Warming Eug[enius]. **Oecology of Plants.** 1909

Watson, Hewett Cottrell. **Selections from *Cybele Britannica.*** 1847/1859

Whetzel, Herbert Hice. **An Outline of the History of Phytopathology.** 1918

Whittaker, Robert H. **Classification of Natural Communities.** 1962